Food Associated Pathogens

Food Associated Pathogens

Editors

Wilhelm Tham

School of Hospitality
Culinary Arts and Meal Science
Örebro University
Grythyttan, Sweden

Marie-Louise Danielsson-Tham

School of Hospitality
Culinary Arts and Meal Science
Örebro University
Grythyttan, Sweden

CRC Press
Taylor & Francis Group
Boca Raton London New York

CRC Press is an imprint of the
Taylor & Francis Group, an **informa** business
A SCIENCE PUBLISHERS BOOK

CRC Press
Taylor & Francis Group
6000 Broken Sound Parkway NW, Suite 300
Boca Raton, FL 33487-2742

Printed in the United States of America on acid-free paper

International Standard Book Number: 978-1-4665-8498-3 (Hardback)

Library of Congress Cataloging-in-Publication Data

Food associated pathogens / editors, Wilhelm Tham, Marie-Louise Danielsson-Tham.
p. ; cm.
Includes bibliographical references and index.
ISBN 978-1-4665-8498-3 (hardcover : alk. paper)
I. Tham, Wilhelm, editor of compilation. II. Danielsson-Tham, Marie-Louise, 1947- editor of compilation.
[DNLM: 1. Food Microbiology. 2. Foodborne Diseases--microbiology. 3. Food Safety. QW 85]

RA601.5
664.001'579--dc23

2013028931

Visit the Taylor & Francis Web site at
http://www.taylorandfrancis.com

Science Publishers Web site at
http://www.scipub.net

CRC Press Web site at
http://www.crcpress.com

Preface

Food borne diseases associated with microbial pathogens account for a significant share of human illnesses. Changes in both, lifestyle and eating habits increase the chances of transmission of pathogenic microorganisms in foods. Examples, of this are the consumption of raw foods, which increases the risk of food borne infections and intoxications, and refrigerated ready-to-eat foods, as some food borne pathogens are capable of being active even at refrigeration temperatures. Extending shelf life of many food items is not always safe for consumers. In addition, personal hygiene standard of food workers has worsened, leading to enhanced food borne illnesses. Food safety education needs to be reintroduced, even at the primary level. The number of people susceptible to food borne disease has increased, especially among the population over 65 years of age. Thus, food microbiologists working with food-associated pathogens need to be familiar with the type of microorganisms associated with a food product in its natural state to be able to predict the general types of microorganisms expected in a particular food product. Food microbiologists must also be familiar with the incubation periods and clinical symptoms of different food borne diseases. The effective prevention of food borne diseases requires cooperation and open-mindedness among different authorities and professionals, e.g., physicians, medical officers, veterinarians, public health officers, food producers, national surveillance institutes, scientists, and government departments.

Future food safety should be an important factor in trade, and countries with the best food safety will gain market share. Change is the way of life, and new pathogens will emerge and old pathogens will reemerge in response to these changes; therefore, food microbiologists never will be short of work.

Food Associated Pathogens examines pathogenic bacteria, viruses, protozoan parasites, moulds, and mycotoxins in food, discusses food-associated antimicrobial resistance and lessons learned from actual food borne outbreaks, and explores the clinical aspects of food borne diseases. The book is intended for postgraduate students, physicians, veterinarians, scientists, technologists, and inspectors in public health-related fields who regularly contend with the issues related to food microbiology, food borne infections and intoxications, and food safety. The chapters in the book are entirely the work of the authors, who had freedom to write the text as they wanted, but were urged, to divide the chapter into four main parts: The Microorganism, The Patient, The Risk Food and its Control and Prevention. Pathogens are presented in an alphabetical order.

Grythyttan, Sweden
June 2013

Wilhelm Tham
Marie-Louise Danielsson-Tham

Contents

CHAPTER 1

Food-associated Pathogens—Insights and Reflections

Marie-Louise Danielsson-Tham[#]

Food Microbiology

To be a food microbiologist is special. You have to be a combination of a laborer and a detective and have an understanding of complex food systems and different food preparation systems, microbes, and, of course people. In addition, if something goes wrong, you must solve all problems as rapidly as possible.

The search for bacteria in foods is slightly different to that in clinical samples: sometimes, a food sample resembles a fecal sample more than what it actually is. As a food product is never sterile, the analyses are complicated by the high levels of indigenous microflora present in some foods. These microorganisms can interfere with the isolation and identification of a specific bacterium. Foods contain infinite arrays of ingredients, including proteins, carbohydrates, fats and numerous other components that can interfere with the detection of pathogens. The structure of foods also varies between liquid, semisolid, solid, or other forms, which can interfere with consistent extraction and isolation of bacteria and obtaining a uniform food homogenate for reproductive analysis.

Various treatments used in food processing such as heat, cold, freezing, drying, and the use of additives, e.g., preservatives, can cause sublethal injury to bacterial cells. The injured or stressed cells are extremely sensitive

[#] School of Hospitality, Culinary Arts and Meal Science, Örebro University, Box 1, S-712 02 Grythyttan, Sweden.
Email: marie-louise.danielsson-tham@oru.se

to ingredients used in selective microbiological media; consequently, standard microbiological methods can miss the stressed bacteria.

The tests for some specific pathogens (e.g., *Salmonella* and *Listeria monocytogenes*) need to be highly sensitive because of the zero tolerance for these organisms in all or certain foods. The psychotropic bacteria, e.g., *Yersinia enterocolitica* and *L. monocytogenes,* may be present at very low levels immediately after processing, but on storage and distribution, these bacteria can multiply to potentially hazardous levels in the food. In addition, low infective dose pathogens such as STEC and *Campylobacter* demand special isolation procedures.

Some food poisoning is caused by toxic compounds produced by bacteria actively growing in food and not through invasion of the human body by the organism, for example staphylococcal food poisoning. Therefore, a food could be free from staphylococci and yet be involved in an outbreak of food poisoning caused by it.

Therefore, a food microbiologist needs to be familiar with the type of microorganisms associated with a food product in its natural state. With this kind of knowledge, a food microbiologist can predict general type of microorganisms expected in a particular food product at a later stage in its history and organisms not usually present in a particular food product.

The food microbiologist must also be familiar with the incubation periods and clinical symptoms for different foodborne diseases. For example, the first reported gastrointestinal outbreak of listeriosis in Sweden was preceded by many specialists initially being skeptical about listeriosis as the diagnosis due to the short incubation period and atypical symptoms such as diarrhea and vomiting.

Food as Ecosystems

Foods are ecosystems and comprise the environment and the organisms that live or thrive in them. The microenvironment may be altered during processing, e.g., oxygen is driven out of the food during cooking and diffuses back into the food so slowly that most of the products remain anaerobic.

The characteristics of the food are intrinsic factors and include pH, water activity, oxidation-reduction potential (Eh), nutrient content, antimicrobial constituents, and biological structures, e.g., shells of eggs or skins on fruits. Through determination of the intrinsic factor for a food, the types of microorganisms likely to grow in the particular food can be generally predicted.

Extrinsic factors are external to the food and include the properties of the storage environment that affects both the foods and their microorganisms. The extrinsic factors with the largest influence on microbial growth are temperature of storage, gas composition, and relative humidity of the environment.

Estimating the Burden of Foodborne Illness

Foodborne diseases associated with microbial pathogens are an important cause of human illness. However, the true incidence of foodborne illness is unknown and depends on many circumstances. To begin with, someone must suspect the illness is caused by food and then the food and adequate samples from the patient must be analyzed. A number of events need to occur for a laboratory confirmed case: the ill person must seek medical care, a specimen must be submitted, the laboratory must test for the pathogen, and the case must be reported to health authorities. A notifiable disease is required by law to be reported to government authorities. However, the nature of notifiable foodborne diseases varies from country to country. Some countries report finding of certain pathogenic bacteria, e.g., salmonella, even though the patient does not present any symptoms, other countries only report findings if the patient is hospitalized. Furthermore, not all outbreaks are investigated for every possible etiology as it is too expensive and at the time of investigation, the agent may have decreased below the detection level in the food and patient. Often food and stool samples are only analyzed routinely, thus, rarer bacteria will be undetected. If only conventional culturing is used, bacterial food poisonings caused by enterotoxins produced directly into the food, such as by, e.g., staphylococci, may be missed. Viruses still cannot be cultivated, thus, *in vitro* detection of viruses in food is problematic and as the infective dose for food-mediated viruses associated with illness is low, this further complicates their detection.

The cause of many incidents of food poisoning is obscured as leftover remains of the meals are discarded before the environmental health officer is contacted. Outbreaks associated with hospital and nursing homes are more frequently reported than outbreaks in restaurants and private homes, as sometimes food plants and restaurants deny they have had a problem with food poisoning due to fear of negative publicity and financial loss. However, most culturing methods are developed for known pathogens, not new yet unidentified, unknown pathogens; therefore, it is important to be curious about peculiar growth on agar plates!

New Eating Habits

Changes in lifestyle, social attitudes, and eating habits increase the opportunity for transmission of pathogenic bacteria in foods. The consumption of raw protein foods, such as raw fish in the form of sushi, ceviche, raw seafood, or raw meats, such as steak tartar implies significant risks for foodborne infections and intoxications. Undercooked eggs (soft boiled, soft scrambled, fried "sunny-side-up") and the use of unpasteurized eggs in food products are major factors in the incidence of *Salmonella Enteritidis*. As raw milk often harbors pathogens such as *Campylobacter jejuni*, *Listeria monocytogenes* and Salmonella, milk and all dairy products should be pasteurized before consumption. In some countries, raw milk cheeses are stored at a minimum temperature of +2°C for at least 50 or 60 days to ensure the death of pathogenic microorganisms. However, subsequent research and documented outbreaks of foodborne illness caused by cheese have proved this storage is ineffective, particularly against *Salmonella* and *L. monocytogenes*.

Risk Groups

The number of people susceptible to foodborne disease has increased, especially among persons above 65 years. Susceptibility is increased in the rising number of people immunocompromised due to the HIV epidemic, with the increase in cancer and the use of immunosuppressive drugs, diabetes, alcoholism, malnutrition and antacids use. Pregnant women (and fetuses) are particularly susceptible to pathogens such as *L. monocytogenes* and *Toxoplasma gondi*.

New Food Processing and New Foods

Food production has changed with regard to new processes and new food products. For instance, the new generation of refrigerated foods, which are characterized by mild heat treatment, increased water activity, low salt content, low or no nitrite additive, a combination of several foods in the same product, new packaging methods, and long shelf life. *L. monocytogenes* and other foodborne pathogens are capable of growth even at refrigeration temperatures, in vacuum packages, and in controlled atmospheres. Thus, the extended shelf life of many food items can be a threat to the consumer's health.

Modified food processing has produced unusual opportunities for growth of bacteria. *Clostridium botulinum* has been identified in a variety of foods where it would never have been suspected previously. Botulism is reported to emanate from baked potatoes stored in the oven overnight, cloves of garlic immersed in oil stored at room temperature and from yoghurt with hazelnut cream.

Food reformulation is used as a strategy for producing foods for improved health; however, reformulation also changes the intrinsic physicochemical properties of the food, which may result in foods with shorter shelf lives and increased risk for the growth of pathogenic microorganisms. These foods often have higher water activity due to replacing NaCl with KCl or reducing the fat content.

Decreasing Budget—Poor Food Hygiene

During times with decreasing budgets and decreasing work force, there is no or too little space for preventive food hygiene. Inadequate cleaning procedures, lack of interest, complicated machinery and equipment increase the risk for foodborne infections and intoxications. In addition, personal hygiene has worsened, and improperly cleaned hands of food worker are increasingly responsible for foodborne illness. Cross-contamination from unclean cutting boards and knives and perishable food being left out of the refrigerator for too long are other important risk factors. The teaching of food safety needs to be reintroduced, even in primary education. Consumers, food handlers and food managers must be taught how to handle food products safely.

Analyzing Food Samples

In food microbiology, many selective culturing media are used. Some ingredients of the medium favor the growth of the targeted bacteria, other medium components inhibit the growth of competing microorganisms. However, if target organisms are injured, they may be killed by the selective component. Selective media are never entirely selective, in that, microorganisms other than the target microorganism will also grow. Thus, the identity of suspect pathogens isolated on selective media must be confirmed and the subtype determined through phenotyping and genotyping methods.

For some bacteria, e.g., Salmonella, there is zero tolerance in foods, and that means finding one Salmonella bacterium hiding among several million other bacteria. In such cases, the usual procedure is preenrichment in a nonselective medium, then enrichment in a selective medium: the pathogen can then be isolated on a selective agar. Injured bacterial cells and "viable but nonculturable" cells pose additional problems to food microbiologists, as these cells cannot form colonies and be detected, but can make people ill. Injured and "viable but non culturable" cells are also important from the aspect of food safety: if such bacterial cells are considered dead after, e.g., heat treatment, the effect of heating will be overestimated. Bacterial cells that escape detection during post-processing controls may repair during food storage and cause illness when the food is eaten.

Although a number of PCR-based methods have been developed to increase the speed of the isolation procedure, most foods contain inhibitory factors for PCR reaction. Traditional methods should never be replaced by alternative methods; therefore, molecular biological methods should be used in conjunction with conventional bacteriological methods. There will always be a need for viable bacterial cells obtained through standard cultural procedures and viable cells as bacterial strain collections will be important for the future.

Typing, Subtyping, and Characterization

The typing methods can be divided into phenotyping and genotyping methods.

Phenotypic typing comprises biotyping, serotyping, phage-typing and antimicrobial susceptibility testing. Genotyping methods characterize a nucleic acid target and include ribotyping, Multilocus Enzyme Electrophoresis (MEE), Random Amplification of Polymorphic DNA (RAPD), Pulsed Field Gel Electrophoresis (PFGE), Multiple-Locus Variable-Number Tandem Repeat Analysis (MLVA), and Multi-Locus Sequence Typing (MLST) or Multi-Virulence-Locus Sequence Typing (MVLST). In the investigation of foodborne diseases, the isolation of a bacterial species from patients and the isolation of the same species from a food product, direct suspicion towards a particular product. In an outbreak investigation, fast methods such as serotyping are useful as a first-step typing method for screening strains. However, a more discriminating method for identifying the epidemic strain is also needed; as it is preferable, the isolates from both the food and the patient are genetically identical. Typing methods can also be used when investigating routes of contamination in certain environments, for example, the spread of food borne pathogens, such as *L. monocytogenes*, in a food processing plant. Typing would clarify whether a variety of clonal types are just passing through the plant, or if the contamination is firmly established. In the case of a house flora consisting of a specific clonal type, the eradication regime may be different if new strains continuously enter the plant. Sometimes more than one clonal type of the pathogenic bacterium is present in the suspected food and/or the patient. Therefore, the isolation and typing of more than one colony from the primary culture medium is recommended.

Prevention

To address the problem of food safety, several countries have established national data collection and surveillance systems. National surveillance systems are useful for identifying trends in both the overall burden of foodborne illnesses and the changes in the frequency at which specific pathogens are identified. Surveillance systems are useful for detecting emerging pathogens and new routes or vehicles of transmission. This requires the data within the system to be correct and include samples analyzed from both patients and foods and the subtyping of a sufficient number of strains through discriminating methods. At times with decreasing budget, it could be tempting to reduce information collection and only analyze as many samples as necessary for the safety of the patients. However, this type of misdirected strict economy can have devastating consequences in the future as the improved comprehensive monitoring and surveillance systems, both intra- and inter-country and even for antibiotic drug resistance, need to be established.

Effective prevention to foodborne diseases is not possible through efforts just targeting one specific object. A complete picture of the overall situation is important, and the entire pathway should be represented from crop growing until consumption of the food product. This requires an interdisciplinary approach and collaboration between different disciplines, from the molecular biologist to the clinician. Inherent in this is cooperation and open-mindedness between different authorities and professionals, e.g., physicians, medical officers, veterinarians, public health officers, food producers, government ministers and departments.

Future food safety should be an important factor in trade and countries with the best food safety will gain new international markets. As change is the way of life, it is certain new pathogens will emerge and old pathogens will reemerge in response to change. Therefore, food microbiologists will never be short of work.

CHAPTER 2

Food Associated Antimicrobial Resistance

Annamari Heikinheimo[#]

Introduction

Throughout history, infectious diseases have been a major threat to human and animal health and a significant cause of mortality. Thus, it is not surprising that the discovery of antimicrobial drugs to treat bacterial infections has been termed as one of the greatest innovations in medicine. However, bacterial resistance to antimicrobial drugs has emerged rapidly and antimicrobial resistance has become a major public health problem globally (WHO 2011). In 2007, altogether 400,000 people in Europe suffered from infections caused by multidrug-resistant bacteria. More than 25,000 people died from these infections. The additional expenditure in terms of hospital costs and productivity losses exceeds €1.5 billion each year in Europe (ECDC and EMEA 2009). In the United States, antimicrobial-resistant infections cause $20 billion excess health care costs, $35 billion societal costs, and 8 million additional hospital days (Roberts et al. 2009).

Any kind of use of antimicrobial drugs in humans, animals or plants can promote the development of resistance. Antimicrobial resistance does not respect geographical or biological borders either. The global economy and the movement of foods, animals and humans have a massive and rapid impact over a wide area, enabling the spread of resistant bacteria.

[#] Department of Food and Environmental Hygiene, Faculty of Veterinary Medicine, P. O. Box 66, FI-00014 University of Helsinki, Helsinki, Finland.
Email: annamari.heikinheimo@helsinki.fi

The purpose of this article is to introduce antimicrobial resistance related to the food chain, and shed light on important food-borne pathogens with elevated resistance levels.

The History of Antimicrobial Drugs

The first commercially available antimicrobial drug was prontosil (sulfanilamide), a textile dyestuff which was found to be active against streptococci in experimental study in mice (Domagk 1935). At that time, other antibacterial agents were also discovered. In 1929, Alexander Fleming reported his observations on the antimicrobial action of a fungus from the genus *Penicillium* (Fleming 1929). Howard Florey and Ernst Chain succeeded in purifying the first penicillin further, procaine penicillin G. Both discoveries of prontosil and penicillin were rewarded with Nobel Prizes in Physiology and Medicine; Gerhard Domagk for prontosil in 1939, and Ernst Chain, Alexander Fleming and Howard Florey for penicillin in 1945.

A large number of new antimicrobial compounds continued to be discovered during the following decades. Many completely synthetic drugs became available, belonging to new and effective classes of antimicrobial drugs. Today, there are only few antibacterial agents under development. Due to the rapid antimicrobial resistance development in bacteria, there is need for new antibacterial agents with new mechanism of actions. There is a particular lack of effective agents to treat infections due to multidrug-resistant gram-negative bacteria (ECDC and EMEA 2009).

Antimicrobial Drugs and their Mechanisms

Antimicrobial drugs are essential for the health and welfare of both humans and animals. They are substances that have the capacity to selectively inhibit the growth and destroy microorganisms. Antimicrobial drugs may be substances of microbial origin, or synthetically manufactured. Originally the word antibiotic refers to substance of microbial origin, although the term is commonly used for all these compounds (Giguère 2006, Guardabassi and Courvalin 2006).

Antimicrobial drugs have different mechanisms by which they affect bacteria. Depending on the mechanism, the final result is either inactivation or actual death of the bacteria. Antimicrobial drugs are termed as bactericidal or bacteriostatic, depending on whether they inactivate or actually kill the bacteria. Bactericidal drugs are those that kill the target organisms, whereas bacteriostatic drugs inhibit or delay bactericidal growth and replication. Some antimicrobial drugs can have both bacteriostatic and bactericidal effects, depending on dosage, duration of exposure and the state of the invading bacteria. For example, some drugs exert concentration-dependent killing characteristics, since their rate of killing increases as the drug concentration increases (Giguère 2006).

Antimicrobial drugs are classified as broad-spectrum, intermediate-spectrum, or narrow-spectrum depending on the range of bacterial species susceptible to these agents. Narrow-spectrum antimicrobial drugs have a limited activity against different groups of bacteria and they are primarily useful against particular species of microorganisms. Intermediate-spectrum antimicrobial drugs are effective against group of bacteria. For example, some antimicrobial drugs are effective against gram-positive but not gram-negative bacteria. Usually, broad-spectrum antibiotics are active against both gram-positive and gram-negative bacteria, as well as against aerobes and anaerobes. Tetracyclines, phenicols, fluoroquinolones, third and fourth generation cephalosporins, and carbapenems are generally considered as broad-spectrum drugs (Giguère 2006, Guardabassi and Courvalin 2006).

Antimicrobial drugs act against bacteria by inhibiting the biochemical pathways that are involved in the biosynthesis of essential components of the bacterial cell. The main bacterial targets of the antimicrobial drugs are 1) inhibition of cell wall synthesis, 2) damage to cell membrane function, 3) inhibition of nucleic acid synthesis or function, 4) inhibition of protein synthesis, and 5) inhibition of other metabolic processes, such as folic acid synthesis (Guardabassi and Courvalin 2006).

Antimicrobial Resistance—Definitions and Mechanisms

In optimal conditions, the use of antimicrobial drugs would lead to the end of all infectious diseases. It is therefore understandable that antimicrobial drugs were referred as miracle drugs by the time they were discovered. However, bacterial resistance towards antimicrobial drugs has evolved rapidly both in human and animal isolates. The first case of multi-resistant bacteria transmitting from animals to humans was evidenced almost 50 years ago in Britain, when the multi-resistant strain of *Salmonella typhimurium* DT29 was transmitted from calves to humans (Anderson 1968). Today, antimicrobial resistance in the foodborne zoonootic bacteria is clearly linked to the use of antimicrobial drugs in food animals, and foodborne diseases caused by such resistant bacteria are well-documented in people (WHO 2011).

The resistance of a bacterial strain is a relative term, and may be termed as clinical, microbiological, genetical, or biochemical. The strain is termed clinically resistant when it survives antimicrobial therapy, whereas strain is defined microbiologically resistant if it grows in the presence of higher concentrations of the drug compared with phylogenetically related strains. It should be noted that natural levels of susceptibility to a drug may vary between bacterial species, and thus the antimicrobial resistance may be assessed only by comparing strains of the same species or genus (Guardabassi and Kruse 2008).

Bacteria have developed various mechanisms to become resistant, i.e., to neutralize the action of antimicrobial drugs. Bacteria are able to resist the effects of antimicrobial drugs through 1) preventing intracellular access, 2) immediately removing antimicrobial substances through efflux pumps, 3) modifying the antimicrobial agent through enzymatic breakdown, or 4) modifying the antimicrobial targets within the bacterial cell to render the substance ineffective. Successful development of resistance often results from a combination of two or more of these strategies (Guardabassi and Courvalin 2006).

Antimicrobial resistance can be either intrinsic or acquired. Intrinsic resistance is due to a structural or functional trait inherently associated with a bacterial species, a genus or even a larger group of bacteria. For example, gram-negative bacteria are intrinsically resistant to glycopeptides because their outer membrane is impermeable to such antimicrobial drugs. On the contrary, acquired resistance results from genetic changes in the bacterial genome, which can be a consequence of either random mutation in the genome or horizontal acquisition of foreign genes. Bacteria can acquire antimicrobial resistance genes by uptake of free DNA (transformation), via bacteriophages (transduction) or by cell-to-cell transfer (conjugation). Conjugation is the most important mechanism for the transfer of resistance genes due to its broad host-range and the frequent location of resistance genes on conjugative elements such as plasmids (Guardabassi and Kruse 2008).

Laboratory Methods for Measuring Antimicrobial Resistance

Antimicrobial susceptibility testing methods are *in vitro* procedures used to detect antimicrobial resistance in individual bacterial isolates. The purpose is to provide a reliable prediction as to whether an infection caused by a bacterial isolate will respond therapeutically to a particular antimicrobial drug treatment. This data may also be utilized at the population level as indicators of emergence and spread of resistance (Giguère 2006).

Determination of antimicrobial resistance can be done under laboratory conditions by determining the minimal inhibitory concentration (MIC) which is the lowest concentration of the drug that completely inhibits the growth of the bacterial isolate. Clinical or microbiological breakpoints are threshold values established for each pathogenic bacteria-antimicrobial drug combination. Based on these breakpoints, the strain is defined as 1) resistant, 2) intermediate or 3) susceptible. Guidelines and recommendations are updated regularly by certain organizations worldwide (Kahlmeter et al. 2003). There are several antimicrobial susceptibility testing methods available today and each one has its respective advantages and disadvantages (Turnidge and Paterson 2007). Some of these examples are 1) broth or agar microdilution methods, 2) disk-diffusion method, 3) e-test using stable gradient of different antibiotic concentrations on a plastic strip, 4) automated antimicrobial susceptibility testing methods, 5) mechanism-specific tests, such as chromogenic cephalosporin test and 6) genotypic methods such as PCR and DNA hybridization methods detecting antimicrobial resistance genes. Selection of the appropriate method will depend on the intended degree of accuracy, convenience, urgency, availability of resources, availability of technical expertise and cost. The two most commonly used methods are the disk-diffusion method and the broth microdilution method (Giguère 2006). From the clinical point of view, usually the phenotypic methods are sufficient to provide the information needed to perform the antimicrobial drug choice. However, there are several reasons when genotypic methods are reasonable. Phenotypic methods may be time-consuming and less suitable as rapid diagnostic tool. Therefore, direct detection of resistance genes in clinical specimens may allow early antimicrobial drug therapy. Genotypic methods are also useful in estimating the resistance risk when MIC of an isolate is near the breakpoints. In addition to molecular detection methods such as PCR and DNA hybridization, other molecular techniques may be used in unraveling genetic information of a bacterial isolate. For example, sequencing provides the nucleotide sequences of genes and elements and helps to determine the precise location of mutations. Single genes are no longer studied, the whole genome is seen as an entity (Aarts et al. 2006).

Antimicrobial Drug use in Food-Producing Animals and the Development of Antimicrobial Resistance

Antimicrobial drugs are administered to animals either to treat (therapy) or to prevent (prophylaxis) disease. In food animals, antimicrobial drugs may be also used as growth promoters to improve animal growth. When using antimicrobial drugs to treat infections in animals, they are predominantly used to treat respiratory and enteric infections in groups of intensively fed animals. Antimicrobial drugs are used especially for young animals, for example in broiler chickens or weaning pigs or calves. In individual animals, antimicrobial drugs are often used in dairy cows to treat mastitis. The global increase in intense fish farming has also increased the use of antimicrobial drugs in aquaculture (WHO 2011). Antimicrobial drugs used in animals generally belong to the same class of antimicrobials as antimicrobial drugs used on humans. Tetracyclines are one of the most commonly antimicrobial drugs used on animals, followed by macrolides, pleuromutilins, lincosamides, penicillins, sulfonamides, aminoglycosides, fluoroquinolones, cephalosporins, and phenicols (Schwarz and Chaslus-Dancla 2001). There are noticeable differences in market availability, regulation and total consumption of veterinary antimicrobial drugs in different countries.

When used as growth promoters, antimicrobials are used as feed supplement and continuously administered at sub-therapeutic doses. The rationality in using antimicrobials as growth promoters is that animals receiving antimicrobials in their feed has been shown to gain 4 to 5% more body weight than animals that do not receive antimicrobials (Witte 1998). The mechanism by which antimicrobial growth promoters exert their effects on feed efficiency and weight gain are still not fully understood (Guardabassi and Kruse 2008). In European Union, a total of 11 different compounds were approved as growth promoters until 1995 (Table 1). However, an association between usage of antimicrobials for

Table 1. Antimicrobials used for growth promotion in Europe and United States (Aarestrup and Jenser 2007).

Antimicrobial group	Antimicrobial growth promoter	United Sates	Europe	Related to antimicrobial drug used in humans
Polypeptides	Bacitracin	In use (swine, poultry)	Banned (1999)	Bacitracin
Flavofosfolipid	Flavomycin/ Bambermycin	In use (poultry)	Banned (2006)	None
Glycopeptides	Avoparcin	Not used	Banned (2006)	Vancomycin/ Teicoplanin
Ionophores	Monensin/ Salinomycin	Not used	Banned (2006)	None
Macrolides	Tylosin/ Spiramycin	In use (swine) Not used	Banned (1999)	Macrolides (erythromycin)
Oligosaccharides	Avilamycin	In use	Banned (2006)	Evernimicin[a]
Quinoxalines	Carbadox Olaquindox	In use (swine) Not used	Banned (1999)[b] Banned (1999)[b]	None None
Streptogramins	Virginiamycin	In use (poultry)	Banned (1999)	Quinupristin/ Dalfopristin/ Pristinamycin
Sulfonamides	Sulfathiazole	In use (swine)	Not used	Sulfonamides
Tetracyclines	Tetracyclines	In use (swine)	Not used	Tetracyclines
Penicillin	Penicillin	In use (swine)	Not used	Penicillin
Pleuromutilin	Tiamulin	In use (swine)	Prophylactic use	None

[a]Redrawn before released for human treatment due to side effects.
[b]Redrawn due to carcinogenic effects.

growth promotion and occurrence of antimicrobial resistance had been already established before 1969 (Swann 1969). It was understood that the use of antimicrobials as growth promoters causes antimicrobial resistance that is transferred through the food chain to humans. In 1969, the Swann committee recommended that antimicrobial drugs should not be used as growth promoters if they were used as therapeutic agents in human or animal medicine or associated with the development of cross-resistance to antimicrobial drugs used in humans. Thus, the Swann report was the first historical attempt to provide guidelines for use of antimicrobial drugs on animals, with particular focus on the use of growth promoters. During the following decades, many countries, including EU, banned or are in the process of phasing out the use of antimicrobials as growth promoters (WHO 2011). There is evidence that the withdrawal of antimicrobials as growth promoters has been an effective act since antimicrobial resistance levels in certain bacteria of animal origin has decreased after the ban (WHO 2002, Aarestrup and Jenser 2007).

Antimicrobial Resistance as a Food Safety Problem

Antimicrobial resistance was initially regarded only as a human medical problem in hospital-acquired infections, in critically ill and immune-compromised patients. Today, antimicrobial resistance is also spreading in the community. Many common bacterial infections are becoming difficult to treat and the general population is also considered to be at risk. The use and misuse of antimicrobial drugs in food animals has been considered as one of the contributing factors to the increase in resistance levels. Modern food production facilitates the emergence and spread of resistance through the intensive use of antimicrobial agents and international trade of both animals and food products. Food products of animal origin are often contaminated with bacteria and they are likely to transfer resistant bacteria and resistance genes in the food chain from animals to humans. Since global trade has increased and the geographical borders among countries and continents have become less distinct, antimicrobial resistance has become a global concern. As in other clinically important bacteria, antimicrobial resistance is also increasing in common food-borne pathogens such as *Salmonella* and *Campylobacter*. Resistance can spread also via transferable resistant genes in commensal bacteria. Extended spectrum betalactamase (ESBL) producers suchs as *E. coli*, *Klebsiella* and other *Enterobacteriaceae* have also became significant pathogens with food safety importance (WHO 2011, EFSA 2011).

Antimicrobial Resistance in *Campylobacter* species

Campylobacter is the most frequently diagnosed bacterial gastroenteritis in humans throughout the world. It is commonly acquired through ingestion of contaminated poultry meat. Most cases of *Campylobacter* infections are self-limiting and do not require antimicrobial therapy. However, when therapy is required, macrolides are considered as the drug of first choice in infections caused by *Campylobacter jejuni* or *Campylobacter coli*. Fluoroquinolones are also widely used when bacterial gastroenteritis is suspected (Engberg et al. 2006). Antimicrobial resistance, especially macrolide and fluoroquinolone resistance in *Campylobacter* spp. has increased and elevated resistance to these first line drugs which cause treatment failures. These treatment failures are often combined with longer duration of illness and more severe symptoms. Mortality from infections caused by *Campylobacter* is higher in patients with those infected with antimicrobial resistant strains (WHO 2011).

Campylobacter easily acquires resistance to antimicrobial drugs and thus *Campylobacter* isolated from poultry meat is often resistant to several antimicrobial drugs. Fluoroquinolones rapidly leads to resistance in *Campylobacter* and the resistance is principally due to single mutations in *gyr*A and occasionally in topoisomerase IV (*par*C). Resistance towards fluoroquinolones is frequent in many countries and human infections with fluoroquinolone resistant *Campylobacter* are a growing public health problem globally. Increased resistance to fluoroquinolones in *Campylobacters* isolated from human infections has been shown to be associated with the use of this antimicrobial drug in food animals. It is assumable that food animals may serve as reservoirs of antimicrobial resistant *Campylobacter,* since high levels of ciprofloxacin resistance has been reported in isolates from broiler chickens and their meat, pigs and cattle (Smith et al. 2010).

Antimicrobial Resistance in *Salmonella* species

The first case of multi-resistant bacteria transferring from food animals to humans was Salmonella Typhimurium DT29 transmitting from calves to humans (Anderson 1968). The increase of antimicrobial resistance in *Salmonella* spp. was already observed after the adoption of intensive farming methods and attempts to treat and control the disease with a range of antibiotics. This case underlined the fact that transferable drug resistance reaching humans from *Enterobacteriaceae* of animal origin may ultimately reach human pathogens. Today, *Salmonella* is a well-known zoonootic agent and one of the leading causes of bacterial foodborne illnesses worldwide

(WHO 2011). *Salmonella* spp. includes more than 2,500 serotypes and is carried by a variety of animals. The most common contaminated foods associated with salmonellosis include poultry, beef, pork, eggs, milk, seafood and fresh produce (Gomez et al. 1997). Generally, salmonellosis is self-limiting and the antimicrobial therapy is not needed. However, in some cases when the infection is severe or the patient is elderly, young, or immunocompromised, treatment with antimicrobials is needed.

Several multiple antimicrobial resistant clones of *Salmonella* Typhimurium have emerged globally. These isolates are typically resistant to a wide range of antimicrobial drugs. Resistance to quinolones in *Salmonella* from food animals has increased substantially in many countries over the last few years. Particularly, consumption of contaminated eggs and egg products are considered as sources of infection (Meakins et al. 2008). The resistance of *Salmonella* spp. towards cephalosporins is developing and spreading rapidly, and these resistant strains are referred to as Extended Spectrum Betalactamase (ESBL)- producers (Bradford 2001, Newell et al. 2010).

Extended Spectrum Betalactamase (ESBL)—producers

One of the greatest concerns is the rapid growth of extended-spectrum Betalactamase (ESBL) and carbapenemase resistance in *Enterobacteriaceae* family (Talbot 2008, Reynolds 2009, Oteo et al. 2010), often in *Klebsiella pneumoniae* or *Escherichia coli*. They cause different infections such as urinary tract infection, soft tissue infections, peritonitis, sepsis and others (Rolain et al. 2010). First they were found only in hospitals, but have now spread rapidly in the community also. These bacteria are resistant towards all third and fourth generation cephalosporins and some also to carbapenems (Pfeifer et al. 2010). Nowadays ESBL resistance can be found in many bacterial species, such as food-borne salmonella (Bradford 2001, Newell et al. 2010). The predominant ESBL families are plasmid mediated, and belong to enzyme types CTX-M, TEM, and SHV (Oteo et al. 2010). Transmission of these resistance genes is mainly driven by integrons, insertion sequences, transposons and plasmids, some of which are homologous in isolates from both food-production animals and humans. European Food Safety Authority considers that ESBLs are a major public health risk globally and that food animals serve as a source of infections (EFSA 2011). Especially bacteria in the poultry production pyramid have been considered of high importance in transferring the resistance genes vertically in the poultry production pyramid, following in local recirculation within flocks (EFSA 2011).

Future Aspects on Food-associated Antimicrobial Resistance

Antimicrobial resistance is a food safety challenge. Food products of animal origin are often contaminated with bacteria, and thus likely to constitute a route of transmitting resistant bacteria and resistance genes from food animals to humans. Modern agriculture is very intense and the industrialized production of food animals increase the need of antimicrobial drugs administered to animals. Food animal production uses large amounts of antimicrobials not only for therapeutic purposes but also to prevent disease and promote animal growth. In some countries, the quantity of antimicrobial drugs used in food production animals exceeds the amounts used in humans. This enables the development, spread and persistence of antimicrobial-resistant bacteria capable of causing infections to humans. The genes that encode the antimicrobial resistance can also be transferred from commensal bacteria to human pathogens. The transmission occurs generally via food but also through direct contact with animals or environmental spread. In the presence of antimicrobials, selection of resistance occurs, and the spread of selected resistant bacteria occurs easily. Such foods as fruits and vegetables contaminated by animal waste or contaminated water may also constitute a transmission route (WHO 2011).

The current data provides strong evidence that antimicrobial drug use is a powerful selector of antimicrobial resistance that is spreading from animals to humans via the food chain. Continued non-therapeutic use of antimicrobials in food animals will increase the pool of resistance genes. New, coordinated approaches are needed to solve the threat of antimicrobial resistance spreading via the food chain. For example, additional basic information is required, such as the knowledge of the principal reservoirs of resistance genes carried by humans and animals. Also, there is insufficient information about the conditions and factors that lead to the mobilization, selection and movement of these bacteria between animal and human populations. Public education, development of new antimicrobial drugs and controlled use of the present antimicrobial drugs are also needed in tackling antimicrobial resistance. The spread of antimicrobial resistance in the food chain is a global concern.

References

Aarestrup, F.M. and L.B. Jenser. 2007. Use of antimicrobials in food animal production. *In*: S. Simjee (ed.). Infectious diseases: Foodborne diseases. Humana Press Inc., Totowa, NJ, USA. pp. 405–417.

Aarts, H.J.M., B. Guerra and B. Malorny. 2006. Molecular methods for detection of Antibiotic resistance. *In*: F. Aarestrup (ed.). Antimicrobial Resistance in bacteria of animal origin. ASM Press, Washington D.C., USA. pp. 37–48.

Anderson, E.S. 1968. Drug resistance in *Salmonella* Typhimurium and its implications. Br. Med. J. 10: 333–339.

Bradford, P.A. 2001. Extended-spectrum betalactamases in the 21st century: Characterization, epidemiology and detection of this important resistance threat. Clin. Microbiol. Rev. 14: 933–951.

Domagk, G. 1935. Einbeitrag zur chemotherapie der bakterielle infectionen (On chemotherapy of bacterial infections). Deut. Med. Wochschr. 61: 250.

Engberg, J., M. Keelan, P. Gerner-Smidt and D.E. Taylor. 2006. Antimicrobial resistance in *Campylobacter*. *In*: F. Aarestrup (ed.). Antimicrobial Resistance in bacteria of animal origin. ASM Press, Washington D.C., USA. pp. 269–292.

EFSA, European Food Safety Authority. 2011. Scientific Opinion on the public health risks of bacterial strains producing extended-spectrum β-lactamases and/or AmpC β-lactamases in food and food-producing animals. EFSA Journal. 9: 2322.

European Centre for Disease Prevention and Control (ECDC) & European Medicines Agency (EMEA). 2009. Joint technical report: the bacterial challenge. Time to react (ECDC-EMEA, Stockholm).

Fleming, A. 1929. On the antibacterial action of cultures of a penicillium with special reference to their use in the isolation of *B. influenza*. Br. J. Experim. Pathol. 10: 226–236.

Giguère, S. 2006. Antimicrobial Drug Action and Interaction: An Introduction. *In:* S. Giguère, J.F. Prescott, J.D. Baggot, R.D. Walker and P.M. Dowling (eds.). Antimicrobial therapy in Veterinary Medicine Blackwell Publishing, Ames' Iowa, USA. pp. 3–26.

Gomez, T.M., Y. Motarjemi, S. Miyagawa, F.K. Käferstein and K. Stöhr. 1997. Foodborne salmonellosis. World Health Stat. Q. 50: 81–89. Review.

Guardabassi, L. and P. Courvalin. 2006. Modes of antimicrobial action and mechanisms of bacterial resistance. *In*: F. Aarestrup (ed.). Antimicrobial Resistance in Bacteria of Animal Origin. ASM Press, Washington D.C., USA. pp. 1–18.

Guardabassi, L. and H. Kruse. 2008. Principles of prudent and rational use of antimicrobials in animals. *In:* L. Guardabassi, L.B. Jensen and H. Kruse (eds.). Antimicrobial Use in Animals. Blackwell Publishing, Oxford, UK. pp. 1–10.

Kahlmeter, G., D.F. Brown, F.W. Goldstein, A.P. MacGowan, J.W. Mouton, A. Osterlund, A. Rodloff, M. Steinbakk, P. Urbaskova and A. Vatopoulos. 2003. European harmonization of MIC breakpoints for antimicrobial susceptibility testing of bacteria. J. Antimicrob. Chemother. 52: 145–148.

Meakins, S., I.S. Fisher, C. Berghold, P. Gerner-Smidt, H. Tschäpe, M. Cormican, I. Luzzi, F. Schneider, W. Wannett, J. Coia, A. Echeita, E.J. Threlfall and Enter-net participants. 2008. Antimicrobial drug resistance in human nontyphoidal Salmonella isolates in Europe 2000–2004: a report from the Enter-net International Surveillance Network. Microb. Drug. Resist. 14: 31–35.

Newell, D.G., M. Koopmans, L. Verhoef, E. Duizer, A. Aidara-Kane, H. Sprong, M. Opsteegh, M. Langelaar, J. Threfall, F. Scheutz, J. van der Giessen and H. Kruse. 2010. Food-borne diseases - the challenges of 20 years ago still persist while new ones continue to emerge. Int. J. Food Microbiol. 30: 139.

Oteo, J., M. Pérez-Vázquez and J. Campos. 2010. Extended-spectrum betalactamase producing *Escherichia coli:* Changing epidemiology and clinical impact. Curr. Opin. Infect. Dis. 23: 320–326.

Pfeifer, Y., A. Cullik and W. Witte. 2010. Resistance to cephalosporins and carbapenems in Gram-negative bacterial pathogens. Int. J. Med. Microbiol. 300: 371–379. Review.

Reynolds, R. 2009. Antimicrobial resistance in the UK and Ireland. J. Antimicrob. Chemother. 64: 19–23.

Roberts, R.R., B. Hota, I. Ahmad, R.D. Scott 2nd, S.D. Foster, F. Abbasi, S. Schabowski, L.M. Kampe, G.G. Ciavarella, M. Supino, J. Naples, R. Cordell, S.B. Levy and R.A. Weinstein. 2009. Hospital and societal costs of antimicrobial-resistant infections in a Chicago teaching hospital: implications for antibiotic stewardship. Clin. Infect. Dis. 49: 1175–1184.

Rolain, J.M., P. Parola and G. Cornaglia. 2010. New Delhi metallo-beta-lactamase (NDM-1): towards a new pandemia? Clin. Microbiol. Infect. 16: 1699–1701.

Schwarz, S. and E. Chaslus-Dancla. 2001. Use of antimicrobials in veterinary medicine and mechanisms of resistance. Vet. Res. 32: 201–225.

Smith, J., L. Fratamico and M. Pina. 2010. Fluoroquinolone resistance in *Campylobacter*. J. Food Prot. 73: 1141–1152.

Swann, M.M. 1969. Report of the Joint committee on the use of antibiotics in animal husbandry and veterinary medicine. HMSO, London.

Talbot, G.H. 2008. What is in the pipeline for Gram-negative pathogens? Expert Rev. Anti. Infect. Ther. 6: 39–49.

Turnidge, J. and D.L. Paterson. 2007. Setting and revising antibacterial susceptibility breakpoints. Clin. Microbiol. Rev. 20: 391–408.

WHO, World Health Organization. 2002. Impacts of antimicrobial growth promoter termination in Denmark. The WHO international review panel's evaluation of the termination of the use of antimicrobial growth promoters in Denmark. Geneva, World Health Organization. (http://whqlibdoc.who.int/hq/2003/WHO_CDS_CPE_ZFK_2003.1.pdf, accessed 29 February 2012).

WHO, World Health Organization. 2011. Tackling antibiotic resistance from a food safety perspective in Europe. Geneva, World Health Organization. (http://www.euro.who.int/__data/assets/pdf_file/0005/136454/e94889.pdf, accessed 29 February 2012).

Witte, W. 1998. Medical consequences of antibiotic use in agriculture. Science. 279: 996–7.

CHAPTER 3

Bacillus cereus Food Poisoning

Per Einar Granum[#]

Introduction

The *B. cereus* group today consists of seven different species. One of them is *Bacillus anthracis*. Several of the species can cause food poisoning and will in most cases not be distinguished in routine food laboratories, apart from *B. anthracis* which is usually not haemolytic and is sensitive to penicillin. It has also been suggested that all these species are so closely related that they should be considered as the same species. In this overview the *B. cereus* group (apart from *B. anthracis*) will be handled mostly as the same species.

B. cereus is a food poisoning bacterium of growing concern, although it does not usually cause the type of illness leading to news headlines. However, the spectrum of potential *B. cereus* toxicity ranges from strains used as probiotics for humans to highly toxic strains reported to be responsible for food-related fatalities (Stenfors Arnesen et al. 2008). As non-reportable diseases, the true number of cases and outbreaks is unknown, but large outbreaks have been reported from several countries. The fact that it is usually a relatively mild and short lasting disease (< 24 hours) is probably responsible for fact that it is under reported in official statistics.

[#] Department of Food Safety and Infection Biology, Norwegian School of Veterinary Science, PO Box 8146, Dep., N-0033 Oslo, Norway.
Email: PerEinar.Granum@nvh.no

The members of the *B. cereus* group consist of Gram-positive, spore-forming, motile, aerobic rods, also growing well anaerobically. It is a common soil saprophyte and can easily spread to many types of foods, especially of plant origin, but is also frequently isolated from meat, eggs and dairy products (through the udder of dairy animals) (Kramer and Gilbert 1989, Stenfors Arnesen et al. 2008). *B. cereus* and other members of the *B. cereus* group can cause two different types of food poisoning: the diarrhoeal type and the emetic type (Kramer and Gilbert 1989). The diarrhoeal type of food poisoning is most probably caused by enterotoxins, produced by *B. cereus* during vegetative growth in the small intestine (Stenfors Arnesen et al. 2008). At least three different enterotoxins are suspected to be involved, although other unknown factors could be in it as well. In contrast, the emetic toxin is preformed during the growth of emetic *B. cereus* strains in the food, and is therefore an intoxication and not an infection. For both types of food poisoning the food involved has usually been heat-treated, and surviving spores are the source of the food poisoning. *B. cereus* is not a competitive microorganism, but grows well after cooking and cooling (< 42–50°C). The heat treatment will cause spore germination, and in the absence of competing flora *B. cereus* grows well. The generation time can be as short as 12–13 minutes (Borge et al. 2001).

Characterization and Identification

The '*B. cereus* group', also known as *B. cereus sensu lato*, is an informal but widely used term describing a genetically highly homogeneous subdivision of the genus *Bacillus*, comprising seven recognized species: *B. cereus sensu stricto, B. anthracis, Bacillus thuringiensis, Bacillus mycoides, Bacillus pseudomycoides, Bacillus weihenstephanensis* and from 2012 also *B. cytotoxicus* that is a thermotolerant ecotype growing from 20 to 50°C (Guinebretière et al. 2012). *B. thuringiensis* is distinguished from *B. cereus* by the production of insecticidal δ-endotoxins during sporulation, and is commercially used for biological control of insects in crop protection (Aronson and Shai 2001). *B. anthracis* causes the fatal animal and human disease anthrax, and has in recent years become known for its use as a biological weapon (Mock and Fouet 2001). The species *B. mycoides* and *B. pseudomycoides* are phenotypically differentiated from *B. cereus* by rhizoidal colony shape and in fatty acid composition (Kramer and Gilbert 1989). During the last few decades, increasing numbers of psychrotolerant *B. cereus* strains were described, which led to the description of a new psychrotolerant species within the *B. cereus* group, named *B. weihenstephanensis*. This species is characterized by the ability to grow below 7°C but not at 43°C and specific signature sequences in 16S rRNA and cold-shock protein genes (Lechner et al. 1998). The newcomer *B. cytotoxicus* can be differentiated from the other species of the *B. cereus* group by maximum growth at 50°C and minimum growth at 20°C, by absence of starch hydrolysis, by absence of growth on synthetic media without tryptophan, and weak Voges-Proskauer reaction (Guinebretière et al. 2012).

B. cereus is a large (1.0–1.2 µm by 3.0–5.0 µm) Gram-positive rod-shaped bacterium which grows on common agar media to large colonies (3–8 mm diameter) with a rather flat, greyish and 'ground-glass' appearance, often with irregular borders. On blood agar, the colonies are surrounded by zones of β-haemolysis (Kramer and Gilbert 1989), the size of which is often large, but can vary between strains and is depending on culturing conditions. Most strains will form endospores within a few days on commonly used agar media. *B. cereus* spores are ellipsoidal, centrally or paracentrally placed, and do not distend the cell (Kramer and Gilbert 1989). Other commonly used features for identification are motility, haemolysis, carbohydrate fermentation (*B. cereus* does not ferment mannitol) and the very active lecithinase (phospholipase) production (Kramer and Gilbert 1989). Various plating media are used for the isolation, detection and enumeration of *B. cereus* from foods, including MYP (mannitol-egg yolk-phenol red-polymyxin-agar) and PEMBA (polymyxin-pyruvate-egg yolk-mannitol-bromthymol blue-agar). More recently, chromogenic media have been developed for several food pathogens, including *B. cereus* (for instance Cereus-Ident-Agar from heipha Dr. Müller GmbH, and Chromogenic *B. cereus* Agar from Oxoid Ltd.). These new media have been evaluated together with standard plating media (Fricker et al. 2007).

Virulence factors (Table 1)

Emetic toxin (cereulide)

The rapid onset of the emetic disease caused by *B. cereus*, generally from 0.5 to 6 hours after consumption of the meal, indicates that this is an intoxication by toxin pre-formed in the food. Cereulide, the emetic toxin, is a cyclic dodecadepsipeptide with molecular mass 1.2 kDa and the structure: $[\text{D-}O\text{-Leu-D-Ala-L-}O\text{-Val-L-Val}]_3$ (Agata et al. 1995). Cereulide is produced by a nonribosomal peptide synthetase, encoded by the 24-kb cereulide synthetase (*ces*) gene cluster (Ehling-Schulz et al. 2005), which is located on a megaplasmid related to pXO1 (Ehling-Schulz et al. 2006). Since cereulide is resistant towards acid conditions, proteolysis and heat, it will not be destroyed by gastric acid, the proteolytic enzymes of the intestinal tract, or by reheating foods that have been stored at room temperature after a first heating (Agata et al. 1995).

Table 1. Characteristics of the two types of illness caused by *Bacillus cereus* (Granum 2007).

Characteristics	**Diarrheal syndrome**	**Emetic syndrome**
Does causing illness	10^5–10^7 (total)	10^5–10^8 (cells per gram)
Toxin produced	In the small intestine of the host	Preformed in foods
Type of toxin Name	Protein; enterotoxins Nhe, Hbl and CytK	Cyclic peptide; emetic toxin cereulide
Incubation period	8–16 hours (occasionally > 24 hours)	0.5–5 hours
Duration of illness	12–24 hours (occasionally several days)	6–24 hours
Symptoms	Abdominal pain, watery diarrheal and occasionally nausea	Nausea, vomiting and malaise (sometimes followed by diarrheal, due to production of enterotoxin)
Foods most frequently implicated	Meat products, soups, vegetables, puddings, sauces, milk products	Fried and cooked rice, pasta, pastry and noodles

The mechanism of action in humans has not been determined definitely, but after release from the stomach into the duodenum, cereulide binds to the 5-HT_3 receptor, and stimulation of the vagus afferent causes vomiting in *Suncus murinus*, used as animal model (Agata et al. 1995). Several biological effects of cereulide have been described. The toxin acts as a cation ionopore, like valinomycin, and is therefore able to inhibit mitochondrial activity by inhibition of fatty acid oxidation (Mikkola et al. 1999). This effect of cereulide was the reason for the liver failure in four lethal cases of emetic food poisoning (Mahler et al. 1997, Dierick et al. 2005, Shiota et al. 2010, Naranjo et al. 2011). In all the published cases with fatal outcome paste or rice left at room temperature over a period of time has been the cause. In an experiment where mice were injected intraperitoneally with high doses of synthetic cereulide, massive degeneration of hepatocytes occurred. The serum values of hepatic enzymes were highest on days 2–3 after the inoculation of cereulide, and rapidly decreased thereafter. General recovery from the pathological changes, and regeneration of hepatocytes, were observed after 4 weeks (Yokoyama et al. 1999). Cereulide has also been shown to cause cellular damage and inhibit human natural killer cells of the immune system (Paananen et al. 2002).

Cereulide production commences at the end of logarithmic phase during vegetative growth of *B. cereus*, with the highest level of production at early stationary phase of growth. Cereulide synthesis takes place at temperatures ranging from approximately 12–37°C, although maximal production of emetic toxin appears to occur between 12–22°C (Häggblom et al. 2002). However, two isolates belonging to the psychrotolerant species *B. weihenstephanensis* were recently shown to produce cereulide at 8°C (Thorsen et al. 2006). Different foods have varying ability to sustain cereulide production. In infant formulas, levels from 0.02–2 µg cereulide ml^{-1} food were reached after 24 hours incubation at room temperature. Most cereulide producing strains are unable to degrade starch, and none of the emetic strains have the *hbl* enterotoxin gene. All the emetic strains belong to a single evolutionary lineage of closely related strains, and closer related to *B. anthracis* than the other species of the *B. cereus* group (Ehling-Schulz et al. 2006).

Enterotoxins

It is now apparent through cloning and sequencing studies that *B. cereus* produces at least three different proteins (or protein complexes) referred to as enterotoxins (Stenfors Arnesen et al. 2008), that may cause food borne illness (Table 1). Two of the enterotoxins are multicomponent and structurally related, whereas the third (cytotoxin K) is a single protein of 34 kDa. The three-component haemolysin (Hbl; consisting of three proteins: B, L_1 and L_2) with enterotoxin activity was the first to be fully characterized (Schoeni and Wong 2005). This toxin also has dermonecrotic and vascular permeability activities, and causes fluid accumulation in ligated rabbit ileal loops. Evidence has shown that all three components of Hbl are necessary for maximal enterotoxin activity (Schoeni and Wong 2005). Much more research is needed to make the picture of the mode of action for Hbl clear, but a 1:1:1 ratio of the three components seems to give the highest biological activity (Schoeni and Wong 2005). Substantial heterogeneity has been observed in the components of Hbl, and individual strains produced various combinations of single or multiple types of each component (Schoeni and Wong 2005). This is probably due to multiple genes of *hbl* with sequence variation, but this has to be established genetically. About 50% of *B. cereus* strains harbour the *hbl* genes (Stenfors Arnesen et al. 2008).

A non-haemolytic three-component enterotoxin (Nhe) was characterized later (Stenfors Arnesen et al. 2008). The three components of this toxin were different from the components of Hbl, although there are similarities. The three components of Nhe enterotoxin were first purified from a *B. cereus* strain isolated after a large food-associated outbreak in Norway in 1995. All *B. cereus* strains harbour the *nhe* gene and more than 90% produce detectable amounts of this enterotoxin complex, although not all at 37°C (but at lower temperatures). The enterotoxin is most active when the ratio between NheA, NheB and NheC is close to 10:10:1. Recently it has been shown both NheB and NheC bind to the surface of epithelial cells prior to NheA and the following model has been proposed: The first step in the mode of action of

Nhe is associated with binding of NheC and NheB to the cell surface and probably accompanied by conformational changes. Probably they bind together, and NheC might be a chain length regulator. These events allow subsequent binding of NheA to NheB leading to cell lysis (Lindbäck et al. 2010, Didier et al. 2012).

The last of the characterized *B. cereus* toxins involved in food poisoning is cytotoxin K (CytK). This toxin belongs to a family of β–barrel toxins and is similar to the β-toxin of *Clostridium perfringens* and was the cause of symptoms in a severe outbreak of *B. cereus* food borne illness in a nursing home in France in 1998 (Lund et al. 2000), and caused by *B. cytotoxicus*. In this outbreak several people developed bloody diarrhoea and three patients died. This was an outbreak of *B. cereus* necrotic enteritis, although it was not nearly as severe as the necrotic enteritis caused by *C. perfringens* type C (Lund et al. 2000). In this outbreak the normal rice water diarrhoea was not observed. The outbreak strain (together with only a few other strains so far) produced a more virulent type of CytK than most other strains (*cytK* is present in about 35–40% of strains). This type of CytK (from *B. cytotoxicus*) is 89% identical in amino acid sequence to the more commonly version of CytK (Stenfors Arnesen et al. 2008), isolated form most of the *B. cereus* strains. The most virulent version of CytK is about 5 times more toxic on epithelial cells than the most common CytK, although the activity on erythrocytes seems to be very similar (Stenfors Arnesen et al. 2008). It has been show that, like for the other members of this β-barrel toxins, CytK is inserted into membranes as a heptamer, and makes a pore of 7Å in diameter (Stenfors Arnesen et al. 2008).

Psychotropic Strains

The majority of the cold adapted strains from the *B. cereus* group are *B. weihenstephanensis*, although a few *B. cereus* (mesophilic) have also been isolated (Stenfors Arnesen et al. 2008). *B. weihenstephanensis* is the fastest growing in milk and milk procucts stored at temperatures below 8°C, and although they produce enterotoxins they are hardly able to grow and produce toxins at 37°C. This means that such products are usually safe even with high numbers of *B. weihenstephanensis* present in the food. Analysis for *B. cereus* in food products will not differentiate between the different species within the *B. cereus* group, apart for *B. anthracis* (non haemolytic and not motile).

Heat Resistance of Spores

Spores of *B. cereus* are usually quite heat stable, and can survive cooking for several minutes, although there are huge differences in stability among strains and among spores produced under different conditions, with variation in D values at 100°C from 1–8 minutes (Kramer and Gilbert 1989). The most stable spores are those that are newly made at high temperatures.

Characteristics of Disease

There are two types of *B. cereus* foodborne illness. The first type, which is caused by the emetic toxin, results in vomiting, whereas the second type, which is caused by enterotoxins, results in diarrheal. In a small number of cases both types of symptoms occur due to ingestion of *B. cereus* cells in addition to the emetic toxin that is preformed in the food. The characteristics of the diseases are given in Table 1.

Some recent reports of *B. cereus* foodborne illness have been of increased concern, including three outbreaks with fatal outcome. An outbreak in Norway associated with eating stew containing approximately 10^4–10^5 *B. cereus* per serving affected 17 people of which 3 were hospitalized, one for 3 weeks. The onset for these 3 patients was late, greater than 24 h, and only spores were found in the ingested food. In another case of *B. cereus* foodborne illness, the emetic toxin, cereulide, was responsible for the death of a 17-year-old Swiss boy, due to fulminant liver failure (Mahler et al. 1997). A large amount of *B. cereus* emetic toxin was detected in residue in the pan used to reheat the implicated food (pasta) and in the boy's liver and bile. Similarly, a seven year old girl died in Belgium only 13 hours after ingesting a pasta salad that had been stored over several days in a fridge where the temperature was 14°C (Dierick et al. 2005). A report on a serious outbreak of the emetic type is from Japan (Shiota et al. 2010), where three family members began vomiting 30 minutes after consuming reheated fried rice. After 6 hours, a 1-year-old boy died of acute encephalopathy. A 2-year-old girl recovered rapidly only after plasma exchange and subsequent hemodialysis. Their mother recovered after fluid therapy. The latest report on fatal outcome is from 2008, (Naranjo et al. 2011) and describes a 20 year old man that had eaten pasta stored at room temperature for several days. He also died from liver failure after starting out vomiting.

The most recently discovered *B. cereus* enterotoxin cytotoxin K (CytK) is similar to the β-toxin of *Clostridium perfringens* (and other related toxins) and was the cause of a severe outbreak of *B. cereus* foodborne illness in France in 1998 (Lund et al. 2000). Several people developed bloody diarrhea during this outbreak and three died. This was the first recorded outbreak of *B. cereus* necrotic enteritis, although it is not nearly as severe as *C. perfringens* type C foodborne illness (Lund et al. 2000). The *B. cereus* strain isolated from this outbreak is different from the other members of the *B. cereus* group and is growing at higher temperatures (thermo tolerant), and the new species, *B. cytotoxicus*, has been recognized for such strains.

Usually no treatment is needed after *B. cereus* food poisoning, but in some few cases liquid and electrolytes has to be replaced in hospitals. In severe cases of the emetic type of disease hemodialysis can possibly save patients if the correct diagnosis is done rapidly.

Foodborne Outbreaks

The number of outbreaks of *B. cereus* food borne illness is highly underestimated. The reasons for this are mainly due to the normally short duration of both types of illnesses, and that complete recovery is rapid after the symptoms are fading out. The dominating type of disease caused by *B. cereus* differs from county to country. In Japan, the emetic type is reported about 10 times more frequently than the diarrheal type (Shinagawa et al. 1995), whereas in Europe and North America the diarrheal type is most frequently reported (Kramer and Gilbert 1989). This is likely attributed to eating habits. Some patients experience both types of *B. cereus* food borne illness concurrently (Kramer and Gilbert 1989). About 5% of *B. cereus* strains can produce both types of toxins (Kramer and Gilbert 1989).

Because surveillance of food borne illnesses differs greatly among countries, it is not possible to directly compare the incidence of outbreaks reported by different countries. The percentage of outbreaks and cases attributed to *B. cereus* in Japan, North America and Europe varies from approximately 1 to 47% for outbreaks and from approximately 0.5 to 33% for cases (reports from different periods between 1960 and 2005) (Kramer and Gilbert 1989, Schmidt 2000). The greatest number of reported *B. cereus* outbreaks and cases are from Iceland, Netherlands and Norway. The reason for the high percentage is the relatively few outbreaks of salmonellosis and campylobacter enteritis in Norway and Iceland, which are the two most frequent reported causes of food borne illness in the most of Europe and the US. *B. cereus* in the Netherlands, between 1993 and 1998, was responsible for 12% of outbreaks for which the causative agent was identified. However, the actual incidence of *B. cereus* was only 2.0% of the total number of cases because most cases of food borne illness were of unknown etiology (Schmidt 2000). Examples of foods involved in food borne outbreaks are shown in Table 2. Although most of the cases are mild at least five of them have resulted in mortality: one case caused by the diarrhoeal type, by the species now recognised as *B. cytotoxicus*, and four by ingestion of the emetic toxin (Switzerland 1995, Belgium 2003, 2008, Japan 2008) (Lund et al. 2000, Mahler et al. 1997, Dierick et al. 2005, Shiota et al. 2010, Naranjo et al. 2011).

Table 2. Examples of the variety of foods involved in *Bacillus cereus* food poisoning (Granum 2007).

Type of food	Country	Number of people involved	Type of syndrome*
Barbecued chicken	Many countries	-	E, D
Cooked noodles	Spain	13	D
Cream cake	Norway	5	D
Fish soup	Norway	20	D
Hibachi steak	USA	11	E, D
Lobster pâté	UK	-	D
Meat loaf	USA	-	D
Meat with rice	Denmark	> 200	D
Milk	Many countries	-	E, D
Milkshake	USA	36	?
Pea soup	The Netherlands	-	D
Sausages	Ireland, China	-	D
School lunch	Japan	1877	E
Scrambled egg	Norway	12	D
Several rice dishes	Many countries	-	E, D
Stew	Norway	152	D
Turkey	UK, USA	-	D
Vanilla sauce	Norway (many countries)	> 200	D
Vegetable sprouts	USA	3	E, D
Wheat flour dessert	Bulgaria	-	D

*E: emetic syndrome; D: diarrhoeal syndrome

Reservoirs

Bacillus cereus is widespread in nature, and frequently isolated from soil and growing plants (Kramer and Gilbert 1989). From its natural environment it is easily spread to foods, especially those of plant origin. It is frequent in raw material and ingredients used in the food industry such as vegetables, starch, spices (30% of samples with 10^2–10^5 cfu/g). Through cross-contamination it may then spread to other foods, such as meat products (Kramer and Gilbert 1989). High numbers of *B. cereus* has also been described in faeces from cows, so the possibility to be spread directly to meet is present. The problems in milk and milk products appear through spreading from soil and grass to the udder of the cows and into raw milk. *B. cereus* spores survive milk pasteurisation and, after germination, the cells are free from competition from other vegetative cells (Andersson et al. 1995). In addition to rice, pasta and spices, dairy products are among the most common food vehicles for

B. cereus. The majority (if not all) of *B. weihenstephanensis* is unable to cause food poisoning (Stenfors Arnesen et al. 2007) and will outcompete mesophilic *B. cereus* in foods stored at temperatures below 8–10ºC.

The closely related *B. thuringiensis* is producing enterotoxins (Ray 1991, Rivera et al. 2000), and can cause foodborne illness when administered to human volunteers (Ray 1991). This may develop into a serious problem, as spraying of this organism to protect crops against insect infestations has become a common practice in several countries. *B. thuringiensis* reportedly caused an outbreak of foodborne illness (Jackson et al. 1995). However, because the procedures normally used for identification of *B. cereus* would not differentiate between the two species, outbreaks caused by *B. thuringiensis* may have been unrecognized. To assure safe spraying of *B. thuringiensis*, the organism should not produce enterotoxins. Although all members of the *B. cereus* group harbour the genes for at least one of the enterotoxins (Nhe), some strains are not producing detectable amounts of the toxin(s).

Problems for the Food Industry

There are many signs pointing in the direction that food borne diseases caused by spore formers are reduced over the last few years, mainly due to better understanding of how to treat foods containing spores in the serving industry (canteens and restaurants). There is still a challenge for the food industry that is producing higher quantities and steadily new recipes of more lightly heat treated food products every year. If hygiene is not fully evaluated prior to production new problems may occur. These problems cannot be addressed here as it would need at least another chapter to be highlighted.

Prevention and Control of *Bacillus* Food Poisoning

Prevention and control of *B. cereus* is relatively easy, apart from in the dairy industry, where it is causing major problems. There are also a growing numbers of precooked long life products on the market that is difficult to produce completely free from *Bacillus* spp. spores. Rapid cooling and proper reheating (> 60–70°C) of cooked food is essential if the food is not consumed immediately. Long term storage must be at temperatures at least below 8°C. Low pH foods (pH < 4.8) can be considered safe from growth of the food poisoning *B. cereus*. *B. cereus* spores are commonly isolated from spices, cereals and dried foods.

The emetic syndrome of *B. cereus* food poisoning is often connected with consumption of rice or pasta. The predominance of the episodes involved Chinese restaurants are linked with the common practice of saving portions of boiled rice from bulk cooking. The boiled rice is then stored, usually at room temperature, overnight and *B. cereus* is then able to multiply. The same problem may occur when foods as pasta and pizza are stored for long periods of time at room temperature.

Detection of the *Bacillus Cereus* Toxins

No commercial kit for detection of the emetic toxin (cereulide) is yet available. With the known structure of cereulide (Agata et al. 1995), it is likely that a kit would be available in the near future. However, since the genes encoding the NRPS (non ribosomal peptide synthase) responsible for the cereulide production is known (Ehling-Schulz et al. 2005) PCR can be used for detection of such genes. Screening for cereulide production by *B. cereus* strains can also easily be done by using a sperm motility assay (Haggblom et al. 2002), and several methods based on LC-MS have been published.

There are two available commercial immunoassays for detection of the *B. cereus* enterotoxins. The assay from Oxoid measures the presence of the HblC component, whereas the Tecra kit mainly detects the NheA component (Stenfors Arnesen et al. 2008). However, if one or both of the commercial kits reacts positively with proteins from *B. cereus* supernatant fluids, it is likely that the strain is enterotoxin positive. If culture supernatant fluids are cytotoxic (on epithelial cells in culture) the strains can be regarded as enterotoxin positive. At present there is no commercial method available for detecting CytK.

Concluding Remarks

B. cereus is a normal inhabitant of soil but it is also well adapted to grow in the gut of warm blooded animals. It is frequently isolated from a variety of foods, including spices, vegetables, dairy products and meat. It can cause an emetic or a diarrheal type of food-associated illness which is becoming increasingly important in developed countries. The diarrheal type of illness is most prevalent in the western hemisphere, whereas the emetic type is most prevalent in Japan and other Asian countries. Desserts, meat dishes, dairy products are most frequently associated with diarrheal illness, whereas rice and pasta are the most common vehicles of emetic illness.

The *B. cereus* emetic toxin has been isolated and characterized and also synthesized. Three types of *B. cereus* enterotoxins involved in outbreaks of foodborne illness have been identified. Two of these enterotoxins possess three components and are related, whereas the third is a one-component protein (CytK). Deaths have been caused by the emetic toxin and by a strain producing only CytK. Yet unknown factors might be involved in the diarrheal disease.

B. cereus spores are adhesive to many surfaces and will survive normal cleaning and disinfection (but not hypochlorite, some peroxides and UVC) procedures. *B. cereus* foodborne illness is likely to be highly underreported because of its relatively mild symptoms with short duration. Some strains of the *B. cereus* group are able to grow at refrigeration temperature and increased consumer interest for precooked, chilled food products with long shelf lives may lead to products well suited for *B. cereus* survival and growth.

References

Agata, N., M. Ohta, M. Mori and M. Isobe. 1995. A novel dodecadepsipeptide, cereulide, is an emetic toxin of *Bacillus cereus*. FEMS Microbiol. Lett. 129: 17–20.

Andersson, A., U. Rönner and P.E. Granum. 1995. What problems does the food industry have with the sporeforming pathogens *Bacillus cereus* and *Clostridium perfringens*? Int. J. Food Microbiol. 28: 145–156.

Aronson, A.I. and Y. Shai. 2001. Why *Bacillus thuringiensis* insecticidal toxins are so effective: unique features of their mode of action. FEMS Microbiol. Lett. 195: 1–8.

Borge, G.A., M. Skeie, T. Langsrud and P.E. Granum. 2001. Growth and toxin profiles of *Bacillus cereus* isolated from different food sources. Int. J. Food Microbiol. 69: 237–246.

Didier, A., R. Dietrich, S. Gruber, S. Bock, M. Moravek, T. Nakamura, T. Lindbäck, P.E. Granum and E. Märtlbauer. 2012. Monoclonal antibodies neutralize *Bacillus cereus* Nhe enterotoxin by inhibiting ordered binding of its three exoprotein components. Neutralization of *Bacillus cereus* Nhe enterotoxin. Infect. Immun. 80: 832–838.

Dierick, K., E. Van Coillie, I. Swiecicka, G. Meyfroidt, H. Devlieger, A. Meulemans, G. Hoedemaekers, L. Fourie, M. Heyndrickx and J. Mahillon. 2005. Fatal family outbreak of *Bacillus cereus*-associated food poisoning. J. Clin. Microbiol. 43: 4277–4279.

Ehling-Schulz, M., N. Vukov, A. Schulz, R. Shaheen, M. Andersson, E. Märtlbauer and S. Scherer. 2005. Identification and partial characterization of the nonribosomal peptide synthetase gene responsible for cereulide production in emetic *Bacillus cereus*. Appl. Environ. Microbiol. 71: 105–113.

Ehling-Schulz, M., M. Fricker, H. Grallert, P. Rieck, M. Wagner and S. Scherer. 2006. Cereulide synthetase gene cluster from emetic *Bacillus cereus*: structure and location on a mega virulence plasmid related to *Bacillus anthracis* toxin plasmid pXO1. BMC Microbiol. 6: 20.

Fricker, M., R. Reissbrodt and M. Ehling-Schulz. 2007. Evaluation of standard and new chromogenic selective plating media for isolation and identification of *Bacillus cereus*. Int. J. Food Microbiol. 121: 27–34.

Granum, P.E. 2007. *Bacillus cereus*. *In*: M. Doyle and L. Beuchat (eds.). Food Microbiology. Fundamentals and Frontiers 3. edition ASM Press. Washington DC, USA. pp. 445–455.

Guinebretière, H.M., S. Auger, N. Galleron, M. Contzen, B. De Sarrau, M.L. De Buyser, G. Lamberet, A. Fagerlund, P.E. Granum, D. Lereclus, P. De Vos, C. Nguyen-The and A. Sorokin. 2012. *Bacillus cytotoxicus* sp. nov. is a new 1 thermotolerant species of the *Bacillus cereus* Group occasionally associated with food poisoning. Int. J. Syst. Evolut. Microbiol. (2013) 63: 31–40.

Häggblom, M.M., C. Apetroaie, M.A. Andersson and M.S. Salkinoja-Salonen. 2002. Quantitative analysis of cereulide, the emetic toxin of *Bacillus cereus*, produced under various conditions. Appl. Environ. Microbiol. 68: 2479–2483.

Jackson, S.G., R.B. Goodbrand, R. Ahmed and S. Kasatiya. 1995. *Bacillus cereus* and *Bacillus thuringiensis* isolated in a gastroenteritis outbreak investigation. Lett. Appl. Microbiol. 21: 103–105.

Kramer, J.M. and R.J. Gilbert. 1989. *Bacillus cereus* and other *Bacillus* species. *In*: M.P. Doyle [ed.]. F oodborne Bacterial Pathogens. Marcel Dekker, New York, USA. pp. 21–70.

Lechner, S., R. Mayr, K.P. Francis, B.M. Pruss, T. Kaplan, E. Wiessner-Gunkel, G.S. Stewart and S. Scherer. 1998. *Bacillus weihenstephanensis* sp. nov. is a new psychrotolerant species of the *Bacillus cereus* group. Int. J. Syst. Bacteriol. 48: 1373–1382.

Lindbäck, T., S.P. Hardy, R. Dietrich, M. Sødring, A. Didier, M. Moravek, A .Fagerlund, S. Bock, C. Nielsen, M. Casteel, P.E. Granum and E. Märtlbauer. 2010. Cytotoxicity of *Bacillus cereus* Nhe enterotoxin requires specific binding order of the three components. Infect. Immun. 78: 3813–3821.

Lund, T., M.L. De Buyser and P.E. Granum. 2000. A new cytotoxin from *Bacillus cereus* that may cause necrotic enteritis. Mol. Microbiol. 38: 254–261.

Mahler, H., A. Pasi, J.M. Kramer, P. Schulte, A.C. Scoging, W. Bar and S. Krahenbuhl. 1997. Fulminant liver failure in association with the emetic toxin of *Bacillus cereus*. N. Engl. J. Med. 336: 1142–1148.

Mikkola, R., N.E. Saris, P.A. Grigoriev, M.A. Andersson and M.S. Salkinoja-Salonen. 1999. Ionophoretic properties and mitochondrial effects of cereulide: the emetic toxin of *B. cereus*. Eur. J. Biochem. 263: 112–117.

Mock, M. and A. Fouet. 2001. Anthrax. Annu. Rev. Microbiol. 55: 647–671.

Naranjo, M., S. Denayer, N. Botteldoorn, L. Delbrassinne, J. Veys, J. Waegenaere, N. Sirtaine, R.B. Driesen, K.R. Sipido, J. Mahillon and K. Dierick. 2011. Sudden Death of a Young Adult Associated with *Bacillus cereus* Food Poisoning J. Clin. Microbiol. 49: 4379–4381.

Paananen, A., R. Mikkola, T. Sareneva, S. Matikainen, M. Hess, M. Andersson, I. Julkunen, M.S. Salkinoja-Salonen and T. Timonen. 2002. Inhibition of human natural killer cell activity by cereulide, an emetic toxin from *Bacillus cereus*. Clin. Exp. Immunol. 129: 420–428.

Ray, D.E. 1991. Pesticides derived from plants and other organisms. *In:* W.J. Hayes and E.R. Jr., Laws (eds.). Handbook of Pesticide Toxology. Academic Press, Inc., New York, USA. pp. 585–636.

Rivera, A.M.G., P.E. Granum and F.G. Priest. 2000. Common occurrence of enterotoxin genes and enterotoxicity in *Bacillus thuringiensis*. FEMS Microbiol. Lett. 190: 151–155.

Schmidt, K. 2000. WHO surveillance programme for control of foodborne infections and intoxications in Europe. Seventh Report - FAO/WHO Collaborating Centre for Research and Training in Food Hygiene and Zoonoses, Berlin, Germany.

Shinagawa, K., H. Konuma, H. Sekita and S. Sugii. 1995. Emesis of rhesus monkeys induced by intragastric administration with the HEp-2 vacuolation factor (cereulide) produced by *Bacillus cereus*. FEMS Microbiol. Lett. 130: 87–90.

Schoeni, J.L. and A.C. Wong. 2005. *Bacillus cereus* food poisoning and its toxins. J. Food Prot. 68: 636–648.

Shiota, M., K. Saitou, H. Mizumoto, M. Matsusaka, N. Agata, M. Nakayama, M. Kage, S. Tatsumi, A .Okamoto, S. Yamaguchi, M. Ohta and D. Hata. 2010. Rapid detoxification of cereulide in *Bacillus cereus* food poisoning. Pediatrics. 125: 951–955.

Stenfors Arnesen, L.P. and A. Fagerlund and P.E. Granum. 2008. From soil to gut: *Bacillus cereus* and its food poisoning toxins. FEMS Microbiol. Rev. 32: 579–606.

Stenfors Arnesen, L.P., K. O'sullivan and P.E. Granum. 2007. Food poisoning potential of *Bacillus cereus* strains from Norwegian dairies. Int. J. Food Microbiol. 116: 292–296.

Thorsen, L., B.M. Hansen, K.F. Nielsen, N.B. Hendriksen, R.K. Phipps and B.B. Budde. 2006. Characterization of emetic *Bacillus weihenstephanensis*, a new cereulide-producing bacterium. Appl. Environ. Microbiol. 72: 5118–5121.

Yokoyama, K., M. Ito, N. Agata, M. Isobe, K. Shibayama, T. Horii and M. Ohta. 1999. Pathological effect of synthetic cereulide, an emetic toxin of *Bacillus cereus*, is reversible in mice. FEMS Immunol. Med. Microbiol. 24: 115–120.

CHAPTER 4

Brucellosis

Jacques Godfroid[#]

The Genus *Brucella*

Introduction

Brucellosis remains the world's most common bacterial zoonosis, with over half a million new cases annually and prevalence rates in some countries exceeding ten cases per 100,000 (Pappas et al. 2006). Human brucellosis is known to be highly endemic in the Mediterranean basin, Middle East, Western Asia, Africa, and South America. Although animal brucellosis has been brought under control in several industrialized countries, human brucellosis occurs sporadically in individuals who acquire the infection abroad or from ingestion of illegally imported unsafe animal products and in occupationally exposed groups (Al Dahouk et al. 2005, Al Dahouk et al. 2007).

In the Roman Empire, milk of small ruminants was used to make cheese, one of the primary ingredients in Roman cuisine. It was therefore hypothesised that milk and milk products were important sources of an infectious food-borne disease that was later described as Maltese fever. Recently, bone lesions typical of brucellosis in adult skeletal remains of people killed during the first volcanic surge of Mount Vesuvius in 79 AD

[#] Section for Arctic Veterinary Medicine, Department of Food Safety and Infection Biology, Norwegian School of Veterinary Science, Stakkevollveien 23, 9010 Tromsø, Norway.
Email: jacques.godfroid@nvh.no

Christ, have been described (Capasso 2002). The presence of cocco-like forms that were morphologically consistent with *Brucella* spp. was demonstrated by scanning electron microscopy analysis of a buried carbonized cheese (Capasso 2002). In 1887 Sir David Bruce isolated the organism (*Micrococcus melitensis*) responsible for the Maltese fever from a British soldier who died of the disease in Malta. This bacterium was later renamed *Brucella melitensis* in his honour. In 1905, Zammit demonstrated, again in Malta, the zoonotic nature of *B. melitensis* by isolating it from goat's milk (Wyatt 2005).

In April 2003, the first report of community-acquired human infections with marine mammal-associated *Brucella* spp. was published. The authors described the identification of these strains in two patients with neurobrucellosis and intracerebral granulomas (Sohn et al. 2003).

These facts illustrate that throughout history, *Brucella* spp. has always been of zoonotic importance. Although some human-to-human transmission of *Brucella* spp. have been described either by sexual contact or breast feeding, human brucellosis is almost always associated with an animal (domestic or wild) reservoir. New *Brucella* strains or species emerge and existing *Brucella* spp. adapt to changing social, cultural, and agricultural environment and thus interactions between *Brucella* spp., animals and humans have increased, providing multiple opportunities for a "species jump". As a result, sources of infection(s) and routes of contamination are multiple. Hence, the global human and animal brucellosis picture is, in essence, always changing and will remain incomplete (Pappas 2010).

Brucella species and biovars

The genus *Brucella* consisted until 2007 of 6 species, *Brucella melitensis, Brucella abortus, Brucella suis, Brucella ovis, Brucella neotomae* and *Brucella ovis,* known as the six classical species with their corresponding biovars. All these *Brucella* species are genetically highly related (Osterman and Moriyon 2006). Since 2007, *Brucella ceti* and *Brucella pinnipedialis* (infecting preferentially cetaceans and pinnipeds, respectively) have been recognized as new *Brucella* species (Foster et al. 2007). In 2008, another new *Brucella* species, i.e., *Brucella microti* was first isolated in the common vole (*Microtus arvalis*) (Scholz et al. 2008) and lastly, *Brucella inopinata* was recently isolated from a breast implant infection in an elderly woman with clinical signs of brucellosis (Scholz et al. 2010). This species is the only one that has not been isolated from any animal reservoir. To date the genus *Brucella* consists thus of 10 species. The preferential hosts and the pathogenicity for humans of the 10 recognized *Brucella* species are depicted in Table 1.

Table 1. *Brucella* species, preferred host and pathogenicity for humans.

***Brucella* Species**	**Biovars**	**Preferential host(s)**	**Pathogenicity for humans**
B. melitensis	1–3	Sheep, Goat	High
B. abortus	1–6, 9	Cattle	High
B. suis	1, 3	Pig	High
	2	Wild boar, Hare	No*
	4	Reindeer, Caribou	High
	5	Rodents	No
B. neotomae	-	Desert wood rat	No
B. ovis	-	Ram	No
B. canis	-	Dog	Moderate
B. ceti	-	Cetaceans	Unknown**
B. pinnipedialis	-	Pinnipeds	Unknown**
B. microti	-	Soil, Vole, Fox	Unknown
B. inopinata	-	Unkown	High

*: One case of *B. suis* biovar 2 infection in an immuno-compromised hunter has been described in France

**: One human laboratory contamination has been described in the UK. Two naturally acquired cases have been described, although the source of infection could not be traced back to marine mammals

Prospective *Brucella* species have also been isolated from three native rat species in Australia, but not yet been included in the genus (Tiller et al. 2010) as well as in association with two cases of stillbirth in non-human primates (Schlabritz-Loutsevitch et al. 2009). Lastly, a potentially novel *Brucella* spp. was recently isolated from wild-caught African bullfrogs (*Pyxicephalus edulis*) originating from Tanzania (Eisenberg et al. 2012).

Currently, only three *nomen* species of the genus *Brucella* have an essential impact on public health, i.e., in order of their significance *B. melitensis* infecting preferentially sheep and goats, *B. abortus*, infecting preferentially cattle and *B. suis* infecting preferentially pigs. Differences in the pathology induced by different *Brucella* species in humans are difficult to determine, since few studies have compared the clinical manifestations of sufficient cases of each species (Franco et al. 2007).

Epidemiology of brucellosis

Distribution in humans and animals

The global epidemiology of the disease in humans has seen the emergence of Central Asia, along with the Middle East, as the primary worldwide foci. However, by 2010 the disease travelled over almost all the Balkans which are now leading the map of *Brucella* endemicity in Europe (Pappas et al. 2006). Increasing recognition of the disease in sub-Saharan Africa is promising, although public health policies in these settings would be extremely difficult to implement (McDermott and Arimi 2002, Marcotty et al. 2009).

The information provided in this section is related to the epidemiological situation in Europe and has been collected in "The community summary report on trends and sources of zoonoses and zoonotic agents and food-borne outbreaks in the European Union (EU) in 2008" published by the European Food Safety Agency in 2010 (European Food Safety Agency 2010b). For further information related to the worldwide distribution of brucellosis in humans and animals, the reader is directed to recent reviews (Godfroid et al. 2005, Pappas et al. 2006, Franco et al. 2007, Lucero et al. 2008).

Distribution in animals in Europe

In Europe, for more than four decades, control and eradication programs of brucellosis in cattle and small ruminants have been implemented by national veterinary services. Classically after a first phase in which the infection is controlled by compulsory vaccination, vaccination is gradually restricted and eventually prohibited whereas a "test and slaughter" policy is implemented in order to eradicate the infection. More than a decade is usually needed to complete the brucellosis eradication program by a "test-and-slaughter" policy and the key for success is a sufficient financial compensation scheme for farmers for their culled livestock.

In the EU, such national programs are co-financed by the EU and the Member States (MSs). This policy has been successfully implemented for bovine as well as ovine and caprine brucellosis in Northern MS (with the notable exception of bovine brucellosis on the island of Ireland), whereas eradication programs, particularly ovine and caprine brucellosis eradication programs, are not yet completed in Spain, Cyprus, Greece and Bulgaria as well as regions of France and Italy (European Food Safety Agency 2010b).

The status regarding freedom of bovine brucellosis (OBF) and the occurrence of the disease in MSs and non-MSs in Europe in 2008 is as follows: Austria, Belgium, the Czech Republic, Denmark, Finland, France, Germany, Luxembourg, the Netherlands, Slovakia, Slovenia, Sweden, Norway and Switzerland, were officially free of brucellosis in cattle (OBF).

In addition, some new officially free areas were recognised in Italy (one province, Brindisi, and one region, Tuscany) and there are now eight regions and 13 provinces OBF in Italy. In the United Kingdom, Great Britain is OBF. In Portugal, four islands of the Azores are OBF.

In 2008, the 15 non-OBF MSs reported a total population of 2,587,376 bovine herds, of which 0.12% were found infected. Five non-OBF MSs: Estonia, Hungary, Latvia, Lithuania and Romania, reported no positive cattle herds whereas positive herds were detected in ten non-OBF MSs: Bulgaria, Cyprus, Greece, Ireland, Italy, Malta, Poland, Portugal, Spain, UK (Northeren Ireland) (European Food Safety Agency 2010b).

In 2008, 16 MSs (Austria, Belgium, the Czech Republic, Denmark, Finland, Germany, Hungary, Ireland, Luxembourg, the Netherlands, Poland, Romania, Slovakia, Slovenia, Sweden, the United Kingdom) and Norway and Switzerland, were officially *B. melitensis*-Free (ObmF). ObmF regions have been declared in France (64 departments), Italy (nine regions and seven provinces), Portugal (all the Azores Islands) and Spain (the two provinces of the Canary Islands). In 2008, 11 non-ObmF MSs reported a total population of 466,407 herds of sheep and goat, of which 1.1% were found infected with or positive for *B. melitensis*. This was a substantial decrease compared to 2007 (1.7%). In 2008, four non-ObmF MSs (France, Estonia, Latvia and Lithuania), reported no infected herds. *B. melitensis* has never been detected in Latvia and has not been detected in Estonia since the 1960s and France since 2004. Bulgaria, Cyprus, Greece, Romania, Portugal and Spain reported infected flocks. A decreasing trend can be observed for the proportion of existing positive herds in the non-ObmF MSs from 2.0% in 2005 to 1.1% in 2008 (European Food Safety Agency 2010b).

Cases of cross-infections with *B. melitensis* were observed in cattle herds mixed with sheep and goat flocks in southern Europe (Verger et al. 1989) and are regularly reported in the middle east where it has become an emerging veterinary and public health problem (Samaha et al. 2008). In April 2012, a dairy herd has been found infected with *B. melitensis* in Haute Savoie, France and resulted in one confirmed human contamination (http://www.promedmail.org/direct.php?id=20120414.1101283).

Although no eradication programs have been implemented in Europe, brucellosis in pigs occurs very seldom in Europe for more than 5 decades and occurs mainly in pigs remaining outdoor and in contact with *Brucella suis* biovar 2 infected wildlboars. In this species, an endemic infection with *Brucella suis* biovar 2 exists throughout Europe (European Food Safety Agency 2010a).

Distribution in humans in Europe

In 2008, 26 MSs provided information on brucellosis in humans; eight MSs (Cyprus, Estonia, Finland, Hungary, Latvia, Lithuania, Luxembourg and Malta) reported no human cases. In all, 709 cases of human brucellosis were reported in the EU in 2008, of which 619 (87.3%) were reported as confirmed cases. As expected, MSs with the status as officially free of brucellosis in cattle (OBF), as well as in sheep and goats (ObmF), reported low numbers of cases, whereas the non-OBF/non-ObmF MSs: Greece, Italy, Portugal and Spain, accounted for 85.1% of all confirmed cases reported in 2008. Greece accounted for 48.8% of the total confirmed cases, including one food-borne outbreak with 111 of the cases. In the EU, the notification rate of brucellosis in 2008 was 0.1 cases per 100,000. The number of confirmed cases was slightly higher in the 25 to 44 year old age group (201 cases) compared to the 45 to 64 year old age group (208 cases) exhibiting a seasonal pattern, with more cases occurring in late spring (May) followed by a smaller peak of cases occurring in mid-summer (July) (European Food Safety Agency 2010b).

Seventeen MSs providing information on confirmed human cases reported whether the cases were imported or domestically-acquired. In total 63.0% were domestically-acquired. Slovakia and Spain reported all cases to be acquired domestically and France, Germany and Sweden reported most of their cases as imported. Austria, the Czech Republic, the Netherlands and Poland reported all cases as imported. However, more than one in four cases of confirmed brucellosis was of unknown geographical origin (i.e., imported or domestic). More, in the majority of the human confirmed cases, the *Brucella* species was unknown. Indeed, *Brucella melitensis* and *B. abortus* were isolated in 11.3% of the brucellosis cases while *B. suis* accounted for only 0.2% of the confirmed cases in the EU. In 91.8% (562) of the confirmed cases, the route of transmission was unknown. The suspected routes of transmission were contact with farm animals (31 cases), consumption of cheese (15 cases), milk (two cases), dairy products (one case) and sheep meat (one case) (European Food Safety Agency 2010b).

It is important to note that *B. suis* biovar 2, which is highly prevalent in wildboar populations throughout Europe, does not seem to be a human pathogen. Indeed, only one infection in an immuno compromised hunter has been reported (European Food Safety Agency 2010a).

Characterization of strains by phenotyping and genotyping methods

Classically, detection and identification of *Brucella* spp. has been based on cultural and phenotypic analysis (biotyping). Although undoubtedly providing valuable information, biotyping was, and remains, a highly specialised and time-consuming approach requiring experienced staff and well-optimised non-commercial reagents ideally used under secured biological containment (Alton et al. 1975, Alton et al. 1988).

Initial PCR methods were based on the 16S rRNA and bcsp31 genes (Baily et al. 1992, Herman and Deridder 1992). The IS*711* element became the preferred target for general identification purposes due to its restricted occurrence in *Brucella* spp. and the presence of multiple copies (Halling et al. 1993, Ouahrani et al. 1993).

Until a few years ago, the most popular PCR assay developed for differentiating *Brucella* spp. at species level was the so-called AMOS PCR (Bricker and Halling 1994). However, *B. canis, B. neotomae*, some biovars of *B. abortus* and *B. suis* and the *Brucella* species isolated from marine mammals described later on, cannot be detected by AMOS PCR. AMOS-PCR provided the basis for other multiplex PCR assays, such as the Bruce-ladder-PCR, able to successfully discriminate isolates of all six classical species and the marine mammal brucellae (Lopez-Goni et al. 2008, Mayer-Scholl et al. 2010).

A major recent genome-driven advance has been the identification and exploitation of tandem DNA repeats as typing tools. These repeats have been exploited in many bacteria to develop a new generation of VNTR (Variable Number of Tandem Repeat) based typing approaches but are likely to prove particularly valuable in *Brucella* spp. which previously lacked any epidemiological tool with adequate resolution to facilitate reliable epidemiological trace-back (Bricker et al. 2003, Whatmore et al. 2006, Le Fleche et al. 2006). VNTR has recently proven highly efficient in confirming laboratory or other professionally acquired infections (Marianelli et al. 2008, Valdezate et al. 2010) in distinguishing relapse from reinfection (Kattar et al. 2008, Scholz et al. 2009), in characterizing outbreaks (Valdezate et al. 2007), in identifying associations of different genotypes with different pathogenic profiles (Nockler et al. 2009), in comparing *B. abortus* strains isolated in cattle and humans in Korea (Her et al. 2010), and in assessing the stability of veterinary vaccine preparations (Garcia-Yoldi et al. 2007).

Virulence/virulence factors

In contrast to other pathogenic bacteria, no classical virulence factors, such as exotoxins, cytolysins, capsules, fimbria, flagella, plasmids, lysogenic phages, resistant forms, antigenic variation, endotoxic lipopolysaccharide (LPS) or apoptotic inducers have been described in *Brucella* organisms (Gorvel and Moreno 2002). Instead, the true virulence elements of *Brucella* are those molecular determinants that allow them to invade cells, resist intracellular killing and reach their replicating niche in professional and non-professional phagocytes. Upon cell contact *Brucella* spp. are internalized via receptor molecules associated to lipid rafts. Once inside cells, *Brucella* spp. localize in early phagosomes, where they avoid fusion with late endosomes and lysosomes. Then, *Brucella* spp. are redirecting their trafficking to autophagosomes and finally they reach the endoplasmic reticulum, where they replicate extensively without modifying basic cellular functions or inducing obvious damage to the cell. Some of the bacterial molecular determinants involved in the internalization and early events in the infected cell are controlled by the BvrS/BvrR two component regulatory system, whereas the intracellular trafficking are controlled by the VirB type IV secretion system.

Thus, the emerging global picture is an extremely efficient adaptation to shield itself from the immune recognition and to manipulate key aspects of host cell physiology (apoptosis, vacuolar trafficking) (Gorvel and Moreno 2002, Letesson et al. 2002, Gorvel 2008). It also becomes more and more evident, though still poorly studied, that one of the *in vivo* adaptation keystones is the ability to finely tune the metabolism according to the various nutrients encountered during the infectious cycle (Lamontagne et al. 2010).

The Patient

Clinical picture

Human brucellosis is seldom a fatal disease. It has a wide spectrum of clinical manifestations. The clinical features of brucellosis depend on the stage of the disease and the organs and systems involved. Typically acute brucellosis cases present 2–3 weeks after infection with chills, fever, fatigue, sweating, weight loss, and arthralgia or arthritis. Subacute cases show a protean clinical presentation, although with less severe symptoms compared to the acute form. Patients with the chronic form of the disease (i.e., clinical symptoms persisting for 12 months or more from the time of diagnosis) usually present complaints of malaise, nervousness, emotional lability, depression, or generalized musculoskeletal pain. In case of focal complications, imaging studies provide useful information as demonstrated in cases of endocarditis and neurobrucellosis. Life-threatening focal complications are *Brucella* endocarditis and neurobrucellosis but the overall case fatality rate is low (less than 1%). Relapses (i.e., recurrence of typical symptoms occurring at some time after the completion of treatment) usually occurs within 6 months after therapy is discontinued. The most common osteoarticular finding in children is monoarticular arthritis (usually of the knees and hips), whereas in adults, sacroiliitis is most frequent (Franco et al. 2007, Buzgan et al. 2010, Al Dahouk and Nockler 2011).

Because of the protean clinical manifestations of brucellosis, it is essential to gather relevant epidemiological information, particularly whether there has been ingestion of contaminated dairy products or contact with infected animals. Such detailed informations are of paramount importance especially in urban and non-endemic areas, and in cases of imported brucellosis, in which travellers acquire the disease abroad and will show symptoms when coming back in non-endemic regions. The diagnosis of a patient with brucellosis should prompt the clinician to consider contacts that might have been exposed to the same risk (e.g., ingestion of contaminated dairy products or contact with infected animals) (Al Dahouk and Nockler 2011).

Laboratory diagnosis

Isolation and identification of Brucella *spp. from clinical samples*

Definitive diagnosis of human brucellosis requires the isolation of the etiologic agent from blood, bone marrow or other tissues and body fluids. The bacterial isolation rates are variable depending on the stage of disease, previous use of antibiotics, the clinical specimen and, last but not least, the culture methods (Al Dahouk and Nockler 2011). Since the isolation of *Brucella* spp. from clinical samples is hampered by its slow growth, culturing the fastidious bacterium might take several days or even weeks before visible, punctate, nonpigmented and nonhemolytic *Brucella* colonies may appear. Colonies of smooth brucellae are raised, convex, circular, translucent and 0.5–1 mm in diameter. The oxidase- and ureasepositive *Brucella* spp. are very small (0.5–0.7 µm in diameter and 0.6–1.5 µm in length), faintly stained Gram-negative coccobacilli. The reader is directed to the textbook *Techniques for the Brucellosis Laboratory* for a thorough description of bacteriological methods (Alton et al. 1988).

Biotyping of *Brucella* spp. is not necessary to arrive at a decision on therapeutic measures. In contrast, the rapid identification of the genus *Brucella* is crucial in order to initiate antibiotic treatment early in the course of disease, thus preventing chronic courses and focal complications (Al Dahouk and Nockler 2011). Conversely, biotyping and fingerprinting of *Brucella* spp. is critical on order to trace back the source of infection (Godfroid et al. 2011).

Serologogical diagnosis

The major *Brucella* antigens that are useful for diagnosing human brucellosis are the smooth (S) lipopolysaccharide (LPS) of the outer cell membrane. The LPS is the immunodominant antigen as also the molecule carrying the epitopes that may cross-react with other Gram-negative bacteria including *Yersinia enterocolitica* O:9, *Escherichia coli* O:157, *Francisella tularensis*, *Salmonella urbana* O:30, *Vibrio cholerae*, and others (Alton et al. 1988).

Antibody profile in the course of disease

Theoretically, in acute brucellosis, the first anti-*Brucella* immunoglobulin isotype is IgM, in the first week post-infection (PI). Subsequently (in the second week PI), there is a switch to anti-*Brucella* IgG isotype in patients who have not received treatment. Both isotypes peak 4 weeks PI. The initial IgM response may not be seen in patients with a slow insidious onset of disease, in those seen late in the course of the disease or in those with relapses. The serological titres of agglutinins (IgM, IgA and IgG), should decline after successful treatment; if they do not, it is necessary to evaluate the patient for the possibility of a relapse or chronic focal disease. IgG and IgA titres increase in relapses (Corbel 2006, Al Dahouk and Nockler 2011).

Recommended tests are the standard agglutination test (SAT), the Rose Bengal test (RBT), the Coombs antiglobulin test (CT), and ELISAs, detecting anti-*Brucella* IgM and IgG. The results of a combination of tests can be used to assess the stage of evolution of the disease at the time of diagnosis. In this respect the ELISA, with a conjugate of the appropriate IgM or IgG specificity and S-LPS, could replace established tests but requires further standardization (Corbel 2006).

The SAT, RBT and Lateral Flow Assay (LFA)

The SAT is generally accepted as the reference method in the serological diagnosis of human brucellosis. SAT titers ≥ 1:160 are generally considered indicative of active brucellosis if accompanied by a suggestive clinical course in patients and the identification of a potential exposure (Corbel 2006). An alternative is the RBT. Its intrinsic value is excellent in patients without previous exposure to *Brucella* spp., but poor in patients who were previously exposed to the etiologic agent or were formerly infected (Al Dahouk and Nockler 2011). The lateral flow assay is a promising rapid point-of-care test. It uses a drop of blood obtained by fingerprick and is easy to interpret. However, further studies are needed to confirm the clinical usefulness of such tests (Franco et al. 2007).

Coombs' test and Brucellacapt®

The Coombs' test (CT) is most commonly used as an extension of SAT to detect incomplete, blocking or nonagglutinating antibodies. A major drawback of the SAT and CT, is that they are labor-intensive and time-consuming. Brucellacapt (Vircell, Santa Fé, Granada, Spain) is an inmunocapture agglutination technique, which detects all antibodies against *Brucella* spp. is a valuable alternative to the CT (Al Dahouk and Nockler 2011).

Enzyme Linked Immunosorbent Assays

Commercially available ELISAs can reliably be used in the diagnosis of human brucellosis, especially in chronic brucellosis. It has been suggested that IgG and IgM antibodies have to be determined by ELISA to reliably diagnose human brucellosis and classify the stage of disease (Al Dahouk and Nockler 2011).

Risk groups

Laboratory exposure to Brucella *spp.*

Microbiology laboratory workers are at increased risk of brucellosis through unsuspected exposure to cultures from clinical specimens. Transmission occurs mainly via inhalation of organisms aerosolized through laboratory procedures such as making bacterial suspensions and pipetting. *Brucella* spp. are among the most frequently reported laboratory-acquired infections, and if these organisms are suspected they should be handled under biosafety level 3 conditions (Singh 2009).

Because of the high risk of laboratory infections when handling cultures and viable brucellae, serological and molecular methods play a key role in the routine diagnosis of brucellosis by clinical microbiology laboratories (Al Dahouk and Nockler 2011).

Exposure to Brucella *spp. in endemic countries*

Most of the brucellosis cases occur due to occupational exposure. High-risk occupations for the disease are clinical and research microbiology, butchery, farming, and veterinary intervention during delivery of the foetus. It is worth reiterating that although the number of reported human brucellosis cases is low in the EU (with an incidence of 0,1/100,000), 85.1% of all confirmed cases reported in 2008 originated from non-ObmF MSs: Greece, Italy, Portugal and Spain (European Food Safety Agency 2010b).

Exposure to Brucella *spp. in nonendemic countries*

In a study performed in 2002–2003 in Germany (officially ObmF MS free since 2000), the majority of the 31 bacteriologically confirmed brucellosis patients (30 *B. melitensis* isolates and 1 *B. suis* isolate) were infected in endemic countries while visiting friends and relatives during their summer holidays, while one case was suspected to be laboratory acquired. Brucellosis was transmitted mainly by the consumption of contaminated unpasteurized milk or cheese from goats and sheep (Al Dahouk et al. 2005). In 2010, brucellosis was diagnosed in the UK in a 14-year old boy showing pyrexia of unknown origin. The contamination was acquired by eating an imported *B. melitensis* contaminated cheese from Syria (Brough et al. 2011). In 2011, a retrospective study conducted in Denmark, showed that 14 isolates clustered with *B. melitensis* and all except 1 displayed typical "East

Mediterranean" profiles when fingerprinted by VNTR. The majority of patients originated from Turkey or the Middle East and had been on a family visit to these endemic areas before falling ill (Aftab et al. 2011). VNTR based genotyping of 66 *Brucella* strains isolated from German travellers and Turkish immigrants living in Germany revealed epidemiological concordance with 20 sheep isolates originating from Eastern Anatolia, Turkey (Gwida et al. 2012).

Treatment

The World Health Organization (WHO) issued recommendations for the treatment of human brucellosis in 1986 (Corbel 2006): doxycycline, 100 mg twice daily for six weeks combined with either rifampicin, 600–900 mg daily for six weeks, or streptomycin, 1 g daily for 2–3 weeks. These regimens are still valid today.

However, they allow for a small, albeit significant percentage of therapeutic failures, most commonly in the form of relapses, ranging from 5 to 15% of uncomplicated cases. This is the reason why a common specialist statement has been reached on the treatment of brucellosis made during a consensus meeting hold in November 2006 in Ionnina, Greece. The authors suggest that the optimal treatment of uncomplicated brucellosis should be based on a six-week regimen of doxycycline combined either with streptomycin for 2–3 weeks, or rifampicin for six weeks. Gentamicin may be considered an acceptable alternative to streptomycin, while all other regimens/combinations should be considered second-line (Ariza et al. 2007). So far, there is no evidence to suggest that drug resistance has an important part in treatment failure and relapse (Turkmani et al. 2008).

Notification/surveillance of agent

Brucellosis is a notified disease in all EU MSs, but Denmark. It is also notifiable in Norway, Iceland and Switzerland (European Food Safety Agency 2010b).

Foodborne Brucellosis

As a foodborne infection, brucellosis is rare in Europe and is generally contracted by eating unpasteurised dairy products or undercooked offal, the later one not being recorded in Europe.

It is striking that only six MSs provided information on *Brucella* spp. in milk, cheese and dairy products in the last EU report on "Trends and Sources of Zoonoses and Zoonotic Agents and Food-borne Outbreaks in the European Union in 2008" published in 2010 (European Food Safety

Agency 2010b). The majority of samples were of raw cow's milk and none of these samples were found positive. *Brucella* spp. was only reported to be isolated in Italy from raw milk from unspecified animals other than cattle and small ruminants (presumably water buffalo) and from two samples of dairy products.

Control and prevention

Although *Brucella* spp. can be transmitted directly from its animal reservoir to humans, indirect transmission remains the highest overall risk and occurs mainly through the consumption of unpasteurized milk or dairy products (Godfroid et al. 2005). The best prevention against human brucellosis is thus the control and possibly the eradication of brucellosis in small ruminants, cattle and pigs. Such programs have been implemented in Europe since 1964 (EU Directive 64/432). However, there are still a certain number of EU MSs that did not reach the OBF and ObmF status (see above) and thus domestic contamination in the EU remains responsible for more than 60% of human cases. The other human cases are very often linked to the consumption of milk or milk products produced in countries where *B. melitensis* is endemic in small ruminants (European Food Safety Agency 2010b).

Food industry—problems

In most Western countries commercially available foods are considered safe and do not contain brucellae due to the identification and removal of infected animals during the brucellosis eradication programs, the brucellosis free status achieved in the majority of MSs and the high hygienic standards (Falenski et al. 2011).

Pasteurizing milk at 72–75°C for 15–30 seconds kills brucellae and thus there should be no risk of acquiring brucellosis from cheese produced from pasteurized milk. Conversely, there is a significant risk of acquiring brucellosis by consuming raw milk or milk products made from unpasteurized sheep and goat milk in non-ObmF countries (Colmenero et al. 2011).

Surprisingly, the information related to the survival of *Brucella* spp. in cheese produced from raw milk is very little. Recently, a study conducted in Mexico concluded that whey fractions of ripened goat cheese, produced with raw milk, may contain *B. melitensis* in numbers capable of inducing disease in humans (i.e., 10^6 CFU/ml) (Mendez-Gonzalez et al. 2011).

A study conducted in Germany in 2011, showed that under normal storage conditions *Brucella* spp. survived in UHT milk for 87 days, for 60 days in water and less than a week in yogurt. When milk was inoculated with low bacterial numbers, *Brucella* spp. multiplied by five log units within three weeks. The authors concluded that contamination of these food products with *Brucella* spp. (both natural and deliberate) can pose a serious public health threat. Considering this scenario, suitable methods for the rapid detection and typing of this pathogen from the relevant food matrices should be developed (Falenski et al. 2011).

Concluding Remarks

Eradication of ovine and caprine brucellosis due to *B. melitensis* seems to be very difficult to achieve in all MSs in the EU and is likely beyond reach in the developing world. Unsuccessful programs may have dramatic consequences both for public health and the food producing sector. Moreover, *B. melitensis* infection in cattle has emerged as a serious public health problem in some southern European countries and Israel as a result of the consumption of unpasteurized cow milk since *B. melitensis* is capable of colonizing the bovine udder (Banai 2002). Therefore, the implementation of sanitary measures such as the pasteurization of milk and cheese as well as vaccination of the livestock reservoir in endemic countries are to be advocated in order to contribute significantly to lower incidence of human cases.

References

Aftab, H., R. Dargis, J.J. Christensen, P. Le Fleche and M. Kemp. 2011. Imported brucellosis in Denmark: Molecular identification and multiple-locus variable number tandem repeat analysis (MLVA) genotyping of the bacteria. Scand. J. Inf. Dis. 43: 536–538.

Al Dahouk, S., P. Le Fleche, K. Nockler, I. Jacques, M. Grayon, H.C. Scholz, H. Tomaso, G. Vergnaud and H. Neubauer. 2007. Evaluation of Brucella MLVA typing for human brucellosis. J. Microbiol. Meth. 69: 137–145.

Al Dahouk, S., K. Nockler, A. Hensel, H. Tomaso, H.C. Scholz, R.M. Hagen and H. Neubauer. 2005. Human brucellosis in a nonendemic country: a report from Germany, 2002 and 2003. Euro. J. Clin. Microbiol. Infect. Dis. 24: 450–456.

Al Dahouk, S. and K. Nockler. 2011. Implications of laboratory diagnosis on brucellosis therapy. Expert Rev. Anti-Infect. Ther. 9: 833–845.

Alton, G.G., L.M. Jones, R.D. Angus and J.M. Verger. 1988. Techniques for the brucellosis laboratory. Institut National de la Recherche Agronomique, Paris, France.

Alton, G.G., L.M. Jones and D.E. Pietz. 1975. Laboratory techniques in brucellosis. World Health Organization, Geneva.

Ariza, J., M. Bosilkovski, A. Cascio, J.D. Colmenero, M.J. Corbel, M.E. Falagas, Z.A. Memish, M.R.H. Roushan, E. Rubinstein, N.V. Sipsas, J. Solera, E.J. Young and G. Pappas. 2007. Perspectives for the Treatment of Brucellosis in the 21st Century: The Ioannina Recommendations. PLoS Med. 4, e317.

Baily, G.G., J.B. Krahn, B.S. Drasar and N.G. Stoker. 1992. Detection of Brucella-Melitensis and Brucella-Abortus by Dna Amplification. J. Trop. Med. Hyg. 95: 271–275.

Banai, M. 2002. Control of small ruminant brucellosis by use of Brucella melitensis Rev. 1 vaccine: laboratory aspects and field observations. Vet. Microbiol. 90: 497–519.

Bricker, B.J., D.R. Ewalt and S.M. Halling. 2003. *Brucella* 'HOOF-Prints': strain typing by multi-locus analysis of variable number tandem repeats (VNTRs). Bmc Microbiology 3.

Bricker, B.J. and S.M. Halling. 1994. Differentiation of *Brucella-Abortus*-Bv-1, *Brucella-Abortus*-Bv-2, and *Brucella-Abortus*-Bv-4, *Brucella-Melitensis, Brucella-Ovis*, and *Brucella-Suis*-Bv-1 by PCR. J. Clin. Microbiol. 32: 2660–2666.

Brough, H.A., A.W. Solomon, R.A. Wall, F. Isaza and G. Pasvol. 2011. Brucellosis acquired by eating imported cheese. J. Paediatr. Child Health. 47: 840–841.

Buzgan, T., M.K. Karahocagil, H. Irmak, A.I. Baran, H. Karsen, O. Evirgen and H. Akdeniz. 2010. Clinical manifestations and complications in 1028 cases of brucellosis: a retrospective evaluation and review of the literature. Int. J. Infect. Dis. 14: E469–E478.

Capasso, L. 2002. Bacteria in two-millennia-old cheese, and related epizoonoses in Roman populations. J. Infect. 45: 122–127.

Colmenero, J.D., E. Clavijo, P. Morata, M.J. Bravo and M.I. Queipo-Ortuno. 2011. Quantitative real-time polymerase chain reaction improves conventional microbiological diagnosis in an outbreak of brucellosis due to ingestion of unpasteurized goat cheese. Diagn. Microbiol. Infect. Dis. 71: 294–296.

Corbel, M.J. 2006. Brucellosis in humans and animals. M.J. Corbel, S.S. Elberg and O. Cosivi. Geneva, Switzerland, World Health Organization. Ref Type: Report.

Eisenberg, T., H.P. Hamann, U. Kaim, K. Schlez, H. Seeger, N. Schauerte, F. Melzer, H. Tomaso, H.C. Scholz, M.S. Koylass, A.M. Whatmore and M. Zschock. 2012. Isolation of Potentially Novel *Brucella* spp. from Frogs. Appl. Environ. Microbiol. 78: 3753–3755.

European Food Safety Agency. 2010a. Scientific Opinion of the Panel on Animal Health and Welfare (AHAW) on a request from the Commission on porcine brucellosis (*Brucella suis)*. EFSA Journal 1144: 1–112.

European Food Safety Agency. 2010b. The community summary report on trends and sources of zoonoses and zoonotic agents and food-borne outbreaks in the European Union in 2008. EFSA Journal 8, 1496.

Falenski, A., A. Mayer-Scholl, M. Filter, C. Gollner, B. Appel and K. Nockler. 2011. Survival of *Brucella* spp. in mineral water, milk and yogurt. Int. J. Food Microbiol. 145: 326–330.

Foster, G., B.S. Osterman, J. Godfroid, I. Jacques and A. Cloeckaert. 2007. *Brucella ceti* sp. nov and *Brucella pinnipedialis* sp. nov. for *Brucella* strains with cetaceans and seals as their preferred hosts. Int. J. Syst. Evol. Microbiol. 57: 2688–2693.

Franco, M.P., M. Mulder, R.H. Gilman and H.L. Smits. 2007. Human brucellosis. The Lancet Infect. Dis. 7: 775–786.

Garcia-Yoldi, D., P. Le Fleche, M.J. De Miguel, P.M. Munoz, J.M. Blasco, Z. Cvetnic, C.M. Marin, G. Vergnaud and I. Lopez-Goni. 2007. Comparison of multiple-locus variable-number tandem-repeat analysis with other PCR-based methods for typing *Brucella suis* isolates. J. Clin. Microbiol. 45: 4070–4072.

Godfroid, J., A. Cloeckaert, J.P. Liautard, S. Kohler, D. Fretin, K. Walravens, B. Garin-Bastuji and J.J. Letesson. 2005. From the discovery of the Malta fever's agent to the discovery of a marine mammal reservoir, brucellosis has continuously been a re-emerging zoonosis. Vet. Res. 36: 313–326.

Godfroid, J., H.C. Scholz, T. Barbier, C. Nicolas, P. Wattiau, D. Fretin, A.M. Whatmore, A. Cloeckaert, J.M. Blasco, I. Moriyon, C. Saegerman, J.B. Muma, S. Al Dahouk, H. Neubauer and J.J. Letesson. 2011. Brucellosis at the animal/ecosystem/human interface at the beginning of the 21st century. Prevent. Vet. Med. 102: 118–131.

Gorvel, J.P. 2008. *Brucella*: a Mr "Hide" converted into Dr Jekyll. Microb. Infect. 10: 1010–1013.

Gorvel, J.P. and E. Moreno. 2002. *Brucella* intracellular life: from invasion to intracellular replication. Vet. Microbiol. 90: 281–297.

Gwida, M., H. Neubauer, Z. Ilhan, G. Schmoock, F. Melzer, K. Nockler, P. Janczyk, H. Tomaso, U. Rosler and S. Al Dahouk. 2012. Cross-border molecular tracing of brucellosis in Europe. Comp. Immunol. Microbiol. Infect. Dis. 35: 181–185.

Halling, S.M., F.M. Tatum and B.J. Bricker. 1993. Sequence and characterization of an insertion-sequence, IS711, from *Brucella-Ovis*. Gene. 133: 123–127.

Her, M., D.H. Cho, S.I. Kang, Y.S. Cho, I.Y. Hwang, Y.C. Bae, H. Yoon, Y.R. Heo, S.C. Jung and H. Yoo. 2010. The development of a selective medium for the *Brucella abortus* strains and its comparison with the currently recommended and used medium. Diag. Microbiol. Infect. Dis. 67: 15–21.

Herman, L. and H. Deridder. 1992. Identification of *Brucella* spp. by using the polymerase chain-reaction. Appl. Environ. Microbiol. 58: 2099–2101.

Kattar, M.M., R.F. Jaafar, G.F. Araj, P. Le Fleche, G.M. Matar, R.A. Rached, S. Khalife and G. Vergnaud. 2008. Evaluation of a multilocus variable-number tandem-repeat analysis scheme for typing human *Brucella* isolates in a region of brucellosis endemicity. J. Clin. Microbiol. 46: 3935–3940.

Lamontagne, J., M. Beland, A. Forest, A. Cote-Martin, N. Nassif, F. Tomaki, I. Moriyon, E. Moreno and E. Paramithiotis. 2010. Proteomics-based confirmation of protein expression and correction of annotation errors in the *Brucella abortus* genome. Bmc Genom. 11.

Le Fleche, P., I. Jacques, M. Grayon, S. Al Dahouk, P. Bouchon, F. Denoeud, K. Nockler, H. Neubauer, L.A. Guilloteau and G. Vergnaud. 2006. Evaluation and selection of tandem repeat loci for a *Brucella* MLVA typing assay. Bmc Microbiol. 6.

Letesson, J.J., P. Lestrate, R.M. Delrue, I. Danese, F. Bellefontaine, D. Fretin, B. Taminiau, A. Tibor, A. Dricot, C. Deschamps, V. Haine, S. Leonard, T. Laurent, P. Mertens, J. Vandenhaute and X. De Bolle. 2002. Fun stories about Brucella: the "furtive nasty bug". Vet. Microbiol. 90: 317–328.

Lopez-Goni, I., D. Garcia-Yoldi, C.M. Marin, M.J. De Miguel, P.M. Munoz, J.M. Blasco, I. Jacques, M. Grayon, A. Cloeckaert, A.C. Ferreira, R. Cardoso, M.I.C. De Sa, K. Walravens, D. Albert and B. Garin-Bastuji. 2008. Evaluation of a multiplex PCR assay (Bruce-ladder) for molecular typing of all *Brucella* species, including the vaccine strains. J. Clin. Microbiol. 46: 3484–3487.

Lucero, N.E., S.M. Ayala, G.I. Escobar and N.R. Jacob. 2008. Brucella isolated in humans and animals in Latin America from 1968 to 2006. Epidemiol. Infect. 136: 496–503.

Marcotty, T., F. Matthys, J. Godfroid, L. Rigouts, G. Ameni, N.G. van Pittius, R. Kazwala, J. Muma, P. van Helden, K. Walravens, L.M. de Klerk, C. Geoghegan, D. Mbotha, M. Otte, K. Amenu, N. bu Samra, C. Botha, M. Ekron, A. Jenkins, F. Jori, N. Kriek, C. McCrindle, A. Michel, D. Morar, F. Roger, E. Thys and P. van den Bossche. 2009. Zoonotic tuberculosis and brucellosis in Africa: neglected zoonoses or minor public-health issues? The outcomes of a multi-disciplinary workshop. Annal. Trop. Med. Parasitol. 103: 401–411.

Marianelli, C., A. Petrucca, P. Pasquali, F. Ciuchini, S. Papadopoulou and P. Cipriani. 2008. Use of MLVA-16 typing to trace the source of a laboratory-acquired Brucella infection. J. Hosp. Infect. 68: 274–276.

Mayer-Scholl, A., A. Draeger, C. Gollner, H.C. Scholz and K. Nockler. 2010. Advancement of a multiplex PCR for the differentiation of all currently described *Brucella* species. J. Microbiol. Meth. 80: 112–114.

McDermott, J.J. and S.M. Arimi. 2002. Brucellosis in sub-Saharan Africa: epidemiology, control and impact. Vet. Microbiol. 90: 111–134.

Mendez-Gonzalez, K., R. Hernandez-Castro, E.M. Carrillo-Casas, J.F. Monroy, A. Lopez-Merino and F. Suarez-Guemes. 2011. *Brucella melitensis* survival during manufacture of ripened goat cheese at two temperatures. Foodborne Pathog. Dis. 8: 1257–1261.

Nockler, K., R. Maves, D. Cepeda, A. Draeger, A. Mayer-Scholl, J. Chacaltana, M. Castaneda, B. Espinosa, R. Castillo, E. Hall, S. Al Dahouk, R.H. Gilman, F. Cabeza and H.L. Smits. 2009. Molecular epidemiology of Brucella genotypes in patients at a major hospital in central Peru. J. Clin. Microbiol. 47: 3147–3155.

Osterman, B. and I. Moriyon. 2006. International Committee on Systematics of Prokaryotes; Subcommittee on the taxonomy of Brucella: Minutes of the meeting, 17 September 2003, Pamplona, Spain. Int. J. Syst. Evol. Microbiol. 56: 1173–1175.

Ouahrani, S., S. Michaux, J.S. Widada, G. Bourg, R. Tournebize, M. Ramuz and J.P. Liautard. 1993. Identification and sequence-analysis of IS6501, an insertion-sequence in *Brucella* spp: relationship between genomic structure and the number of IS6501 copies. J. Gener. Microbiol. 139: 3265–3273.

Pappas, G. 2010. The changing *Brucella* ecology: novel reservoirs, new threats. International J. Antimicrobial Agents. 36: S8–S11.

Pappas, G., P. Papadimitriou, N. Akritidis, L. Christou and E.V. Tsianos. 2006. The new global map of human brucellosis. Lancet Infect. Dis. 6: 91–99.

Samaha, H., M. Al-Rowaily, R.M. Khoudair and H.M. Ashour. 2008. Multicenter study of brucellosis in Egypt. Emerg. Infect. Dis. 14: 1916–1918.

Schlabritz-Loutsevitch, N.E., A.M. Whatmore, C.R. Quance, M.S. Koylass, L.B. Cummins, E.J. Dick, C.L. Snider, D. Cappelli, J.L. Ebersole, P.W. Nathanielsz and G.B. Hubbard. 2009. A novel Brucella isolate in association with two cases of stillbirth in non-human primates—first report. J. Med. Primatol. 38: 70–73.

Scholz, H.C., Z. Hubalek, I. Sedlacek, G. Vergnaud, H. Tomaso, S. Al Dahouk, F. Melzer, P. Kampfer, H. Neubauer, A. Cloeckaert, M. Maquart, M.S. Zygmunt, A.M. Whatmore, E. Falsen, P. Bahn, C. Gollner, M. Pfeffer, B. Huber, H.J. Busse and K. Nockler. 2008. *Brucella microti* sp. nov., isolated from the common vole *Microtus arvalis*. Int. J. Syst. Evol. Microbiol. 58: 375–382.

Scholz, H.C., K. Nockler, C. Gollner, P. Bahn, G. Vergnaud, H. Tomaso, S. Al Dahouk, P. Kampfer, A. Cloeckaert, M. Maquart, M.S. Zygmunt, A.M. Whatmore, M. Pfeffer, B. Huber, H.J. Busse and B.K. De. 2010. Brucella inopinata sp nov., isolated from a breast implant infection. Int. J. Syst. Evol. Microbiol. 60: 801–808.

Singh, K. 2009. Laboratory-Acquired Infections. Clin. Infecti. Dis. 49: 142–147.

Sohn, A.H., W.S. Probert, C.A. Glaser, N. Gupta, A.W. Bollen, J.D. Wong, E.M. Grace and W.C. McDonald. 2003. Human neurobrucellosis with intracerebral granuloma caused by a marine mammal *Brucella* spp. Emerg. Infect. Dis. 9: 485–488.

Tiller, R.V., J.E. Gee, M.A. Frace, T.K. Taylor, J.C. Setubal, A.R. Hoffmaster and B.K. De. 2010. Characterization of Novel *Brucella* Strains Originating from Wild Native Rodent Species in North Queensland, Australia. Appl. Environ. Microbiol. 76: 5837–5845.

Turkmani, A., A. Psaroulaki, A. Christidou, D. Chochlakis, D. Tabaa and Y. Tselentis. 2008. *In vitro*-selected resistance to fluoroquinolones in two Brucella strains associated with mutational changes in gyrA. Int. J. Antimicrobial Agents. 32: 227–232.

Valdezate, S., I. Cervera, P. Hernandez, A. Navarro and J.A.S. Nieto. 2007. Characterisation of human outbreaks of brucellosis and sporadic cases by the use of hyper-variable octameric oligonucleotide fingerprint (HOOF) variable number tandem repeats. Clin. Microbiol. Infect. 13: 887–892.

Valdezate, S., A. Navarro, P. Villalon, G. Carrasco and J.A. Saez-Nieto. 2010. Epidemiological and phylogenetic analysis of spanish human *Brucella melitensis* strains by multiple-locus variable-number tandem-repeat typing, hypervariable octameric oligonucleotide fingerprinting, and rpoB typing. J. Clin. Microbiol. 48: 2734–2740.

Verger, J.M., B. Garin-Bastuji, M. Grayon and A.M. Mahe. 1989. Brucella-Melitensis Infection in Cattle in France. Annal. Recherch. Vet. 20: 93–102.

Whatmore, A.M., S.J. Shankster, L.L. Perrett, T.J. Murphy, S.D. Brew, R.E. Thirlwall, S.J. Cutler and A.P. MacMillan. 2006. Identification and characterization of variable-number tandem-repeat markers for typing of *Brucella* spp. J. Clin. Microbiol. 44: 1982–1993.

Wyatt, H.V. 2005. How themistocles zammit found Malta fever (brucellosis) to be transmitted by the milk of goats. JRSM. 98: 451–454.

CHAPTER 5

Campylobacter

Eva Olsson Engvall[#]

Introduction

Campylobacter is a leading cause of bacterial gastroenteritis in humans in developed countries worldwide. Since 2004, campylobacteriosis has been the most commonly reported zoonosis in EU with between 175,000 and 200,500 laboratory confirmed cases per year (Anon 2011a). *Campylobacter* of the so-called thermophilic or thermotolerant species, which are the main causes of human campylobacteriosis, are found in the gastrointestinal tract of a great variety of animals, both wild and domestic mammals and avian species. The animals usually carry the organisms without showing any signs of disease (Altekruse et al. 1994). The transmission of *Campylobacter* to humans could be via direct contact with colonized animals but the most common route is foodborne, i.e., by consumption of contaminated food or water. The thermophilic *Campylobacter* spp. will be the focus of this chapter.

The Microorganism

General description

Members of the genus *Campylobacter* are Gram-negative, curved or spiral shaped rods, 0.2 to 0.8 µm wide and 0.5 to 5 µm long. They are motile, with a polar flagellum at one or both ends of the cell (Fig. 1). *Campylobacter*

[#] EU Reference Laboratory for Campylobacter, National Veterinary Institute, SE- 751 89 Uppsala, Sweden.
Email: Eva.olsson@sva.se

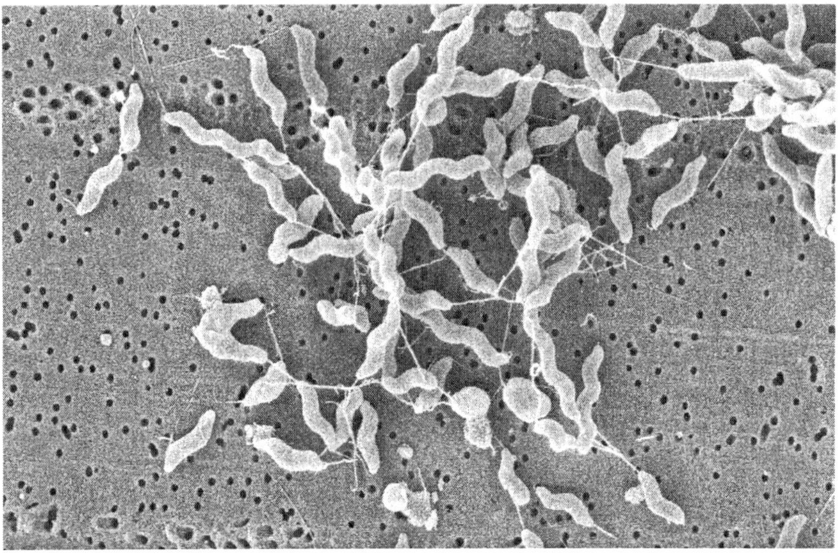

Figure 1. Electron microscopy image of *Campylobacter jejuni* with polar flagella.
Photo: Tapio Nikkilä and Leif Ljung

grow under microaerobic conditions (a few species are anaerobic) and have a respiratory and chemoorganotrophic type of metabolism. Taxonomically, genus *Campylobacter* belong to phylum Proteobacteria, class Epsilonproteobacteria. More than 20 species of *Campylobacter* have been described (http://www.bacterio.cict.fr). *Campylobacter* (*C.*) *jejuni*, *C. coli*, *C. lari* and *C. upsaliensis*, are referred to as thermophilic or thermotolerant species, since they exhibit optimal growth at a temperature of 42–43°C (Doyle and Roman 1981). *Campylobacter* organisms have reduced ability to multiply outside the animal (human) host, they have fastidious growth requirements and are sensitive to environmental stress (Park 2002). The genome of *C. jejuni* is relatively small, 1.6–1.8 Mbp, with a GC content around 30% (Champion et al. 2008, Parkhill et al. 2000). There is extensive genetic diversity in *Campylobacter* species, which has implications for strain characterization and studies of epidemiology of strains (Meinersmann et al. 2002, Wassenaar and Newell 2000).

Laboratory diagnostics

Isolation and identification of *Campylobacter* from clinical, food, and environmental samples requires special cultural conditions such as selective media, microaerobic atmosphere, prolonged incubation time, and usually an incubation temperature of 41.5–42.0°C. Enrichment in selective broth before culture on agar plates is usually applied with samples which may contain low numbers or damaged bacteria (Butzler 2004, Corry et al. 1995, Nachamkin et al. 2000). International standardized cultural methods are available for detection, identification and enumeration of *Campylobacter* in food and feeding stuffs (ISO 2006, NMKL 2007). Phenotypic methods or biochemical tests are described for identification of isolates at genus and species level (Nachamkin et al. 2000). However other techniques, mainly molecular methods such as polymerase chain reaction (PCR)-based assays are increasingly used both for detection of *Campylobacter* and for identification of species (Best et al. 2003, Denis et al. 2001, Ridley et al. 2008, Wang et al. 2002).

For strain characterization, also known as typing or subtyping, several methods have been described. The first typing methods, e.g., biotyping, serotyping and phagetyping were based on phenotypic traits. Although the serotyping schemes by Penner (Penner and Hennessy 1980) and Lior (Lior et al. 1982) have been widely used, they have now been replaced almost completely by molecular typing methods. The majority of typing protocols include typing of *C. jejuni* and *C. coli*, however an increasing number of molecular methods are being developed for other *Campylobacter* species. Commonly used methods for *C. jejuni* and *C. coli* are pulsed-field gel electrophoresis (PFGE), amplified fragment length polymorphism analysis (AFLP), restriction fragment length polymorphism (RFLP), ribotyping, multi locus sequence typing (MLST), and others (Dingle et al. 2001, Meinersmann et al. 2002, Nielsen et al. 2000, Wassenaar and Newell 2000). For PFGE, which has been considered the 'gold standard' for genotyping, there are two standard protocols available on internet, Campynet (http://campynet.vetinst.dk/PFGE.html) and PulseNetUSA (www.cdc.gov/PulseNet/protocols/campy_protocol.pdf). PFGE allows great discrimination between different strains, however, the analysis and inter-laboratory comparison of results may be difficult. MLST, which involves sequencing of selected housekeeping genes, has certain advantages by generating sequence data that are easily shared via the PubMLST website (http://pubmlst.org/campylobacter).

Virulence

Although it is well known that *Campylobacter*, like most other enteropathogens, have the ability to adhere, colonize, invade the epithelial cells of the intestinal mucosa and produce toxins, the pathogenesis of *Campylobacter* infection in humans is still poorly understood (Konkel et al. 2001). Studies of *Campylobacter* virulence have included mainly *C. jejuni* and *C. coli*. A number of virulence associated factors, or virulence associated markers have been described; toxins, glycome and other cell surface factors, plasmids, genomic pathogenicity island factors, and other putative markers (Ketley 1997, Poly and Guerry 2008, Young et al. 2007). It is assumed that *C. jejuni* virulence is multifactorial and involves both microbial and host factors such as age, susceptibility to infection and immune status of the host (Janssen et al. 2008). In spite of intensive research, it has so far not been possible to clarify if there are certain virulence factors or combinations thereof that are necessary for generating disease in humans. The mechanisms and interactions between different putative virulence properties and between the bacterium and the human host need to be further investigated. However, such studies are hampered by the lack of a suitable animal disease model mimicking the disease in humans (Newell 2001).

Antimicrobial resistance

C. jejuni and *C. coli* were originally described as highly susceptible to erythromycin, fluoroquinolones, tetracyclines, aminoglycosides and clindamycin (Nachamkin et al. 2000). However in recent years there has been an increase in the proportion of isolates resistant to several antimicrobial agents all over the world (Allos 2001, Engberg et al. 2001, Luangtongkum et al. 2009). Fluoroquinolones were previously considered the drug of choice for treatment of patients, but the rapid increase of resistant strains has impaired the usefulness of this type of drug (Allos 2001). In United States and Canada, approximately 19–47% of human *Campylobacter* strains have been reported as resistant to ciprofloxacin (Luangtongkum et al. 2009). Similar results have been obtained in Europe, where 47% of human isolates were resistant to ciprofloxacin in 2009 (Anon 2011b). Also *Campylobacter* isolated from animals showed high levels of resistance; 46% of *C. jejuni* and 78% of *C. coli* from chicken and 50% of *C. coli* from pigs were resistant to ciprofloxacin in 2009. Erythromycin is nowadays usually considered the optimal drug for treatment of human cases. The resistance to erythromycin is still relatively low in *Campylobacter* isolated from humans and chickens in both USA, Canada and Europe (Anon 2011b, Luangtongkum et al. 2009).

Campylobacteriosis in Humans

Clinical picture

Campylobacter infection in humans is typically a gastrointestinal illness with diarrhea, fever, abdominal pain, malaise, nausea and sometimes vomiting. The symptoms range from mild watery to severely profuse bloody diarrhea which could last for 3 to 5 days (Allos 2001, Butzler 2004). The clinical signs are not distinguishable from those caused by other enteric pathogens such as *Salmonella* or *Shigella*. The incubation time is 2 to 5 days but may vary from 1 to 11 days. In experimental studies, the infectious dose for humans has been demonstrated to be low, as few as 500–800 cfu were reported to result in disease in a study by Black et al. (1988). However, a higher dose is more likely to cause illness as has been reported in later studies (Teunis et al. 2005). Campylobacteriosis is most often a self-limiting disease, although some symptoms may persist for a longer time. The case-fatality is low, usually under 1% as reported from EU and the US (Anon 2011a, http://www.cdc.gov/nczved/divisions/dfbmd/diseases/campylobacter/).

In some patients, usually those who are immunocompromised, *Campylobacter* infection may manifest as a severe illness with bacteraemia and extraintestinal infections which require hospitalization and/or medical treatment with antimicrobials and electrolyte replacement (Allos 2001). The most serious complication of campylobacteriosis is the neurological degenerative disorder Guillain-Barré syndrome which may lead to paralysis and long-term disability (Allos 2001, Blaser and Engberg 2008).

Incidence in humans

It is well documented that *Campylobacter* is one of the most common causes of diarrheal illness worldwide. In the US, it is estimated that more than 2.4 million persons become infected every year with *Campylobacter* (www.cdc.gov). The annual number of confirmed cases in EU is over 190,000, but it is believed that the actual number is around nine million cases per year (http://www.efsa.europa.eu/en/topics/topic/campylobacter.htm). Since 2004, campylobacteriosis is the most frequently reported food borne illness in EU, and was in 2009 almost twice as commonly reported as *Salmonella* (Anon 2011a). Most EU Member States have a notification system in place for campylobacteriosis, however in some countries this is not mandatory but based on voluntary reporting or other systems.

In industrialized countries, campylobacteriosis cases are the most predominant among infants under five and young adults and more often in males than females (Allos 2001, Friedman et al. 2000). The majority of cases are sporadic and usually there are only few outbreaks where a common source of infection can be identified. However outbreaks involving a large number of cases occur, for example those caused by contaminated water (Mentzing 1981, Smith et al. 2006).

In developing countries, campylobacteriosis is endemic and especially in infants it causes significant morbidity, and even mortality (http://www.who.int/topics/en/).

In areas with a colder climate, for example in the Scandinavian countries, there is a seasonal peak in human cases with a significant rise in numbers during the summer months (Friedman et al. 2000).

C. jejuni is the most commonly isolated species from human cases, followed by *C. coli*. The two species together account for more than 95% of all human cases (Lastovica and Allos 2008). A small number of cases are caused by other *Campylobacter* species, i.e., *C. lari*, *C. upsaliensis* or *C. fetus* subsp *fetus*.

Sources of infection

Animals, especially birds, are reservoirs of thermophilic *Campylobacter* spp. Numerous studies have reported on the occurrence and prevalence in wild, domestic and food producing animals (Altekruse et al. 1994, Blaser et al. 1983, Wahlström et al. 2003). The highest prevalences are in poultry, where up to 100% of broiler flocks have been reported positive (Anon 2011a, Jacobs Reitsma 2000). In the Nordic countries, *Campylobacter* prevalence in broilers is relatively low compared to many other European countries. There is also a distinct seasonality, with the highest rates of positive broiler flocks in the summer months (Hansson et al. 2007, Hofshagen and Kruse 2005, Perko-Mäkelä et al. 2002, Wedderkopp et al. 2001). In pigs, *Campylobacter coli* is the species most commonly isolated, and high pig herd prevalences have been found (Anon 2011a, Fosse et al. 2009). Cattle and sheep are also recognized as reservoirs of *Campylobacter* (Stanley and Jones 2003). The reported prevalences vary between studies, and probably reflect differences in sampling strategies and diagnostic procedures but also differences in age distribution, seasonality and geography (Stanley et al. 1998). Dogs and cats are often colonized, especially with *C. upsaliensis* and young individuals are more likely to be carriers (Hald and Madsen 1997, Engvall et al. 2003).

Campylobacter contaminated environment can also constitute source of infection for humans. *Campylobacter* is often isolated from environmental water samples, i.e., surface waters and also drinking water supplies (Jacobs-Reitsma 2000). Although *Campylobacter* spp. do not actually multiply outside the gut, they can survive in water for prolonged periods of time depending on different factors such as temperature and oxidative and osmotic stress (Park 2002).

For identification and establishing the significance of sources, there is a need for reliable strain characterization. Lately, MLST has been increasingly used for source attribution studies, providing information about source of populations of *Campylobacter* strains, rather than matching patient strains with certain animal or food strains. In two recent UK studies, MLST typing and statistical methods were used for estimation of principal sources for human cases. The vast majority of cases were attributed to chicken and cattle. Wild animals and environment were responsible only for a low proportion of cases (Sheppard et al. 2009, Wilson et al. 2008).

Risk factors

Risk factors for human campylobacteriosis include consumption or handling raw or undercooked meat, drinking unpasteurized milk, contaminated water, contact with animals and foreign travel (Friedman et al. 2004, Kapperud et al. 2003, Neimann et al. 2003). Epidemiological studies clearly show that the handling and consumption of contaminated poultry meat is a major source of campylobacteriosis (Nauta et al. 2009, Wingstrand et al. 2006). According to a scientific opinion from the European Food Safety Authority (EFSA) Panel on biological hazards, handling, preparing and consumption of chicken was estimated to account for 20–30% of human cases and 50–80% of the cases could be attributed to the chicken reservoir as a whole (Anon 2010a). Quantitative risk assessment studies have shown that the incidence in humans is related to the numbers of *Campylobacter* in the food (Rosenquist et al. 2003).

Campylobacter in Food and Water

In an extensive literature survey of *Campylobacter* contamination of poultry meat, it was concluded that in most countries all over the world, a majority of retail poultry meats and by-products were contaminated with *Campylobacter* (Suzuki and Yamamoto 2009). Although *Campylobacter* colonizes the gut also of other farm animals than avian species, much lower isolation rates are obtained from samples of beef, pork and lamb (Jacobs-Reitsma 2000, Whyte et al. 2004) (Table 1). The EU Member States submit data on *Campylobacter* in animals and foodstuffs to EFSA in accordance with the 'Zoonosis Directive' (Directive 2003/99/EC). EFSA examines the data and prepares each year a report on the trends and sources of zoonoses and zoonotic agents and food-borne outbreaks in the European Union. In 2009, the results of testing raw meat samples varied considerably between countries (Table 1) but on average, 31% of fresh broiler meat samples were found positive for *Campylobacter.* The corresponding results for fresh pig and bovine meat at retail were 0.6% and 0.5%, respectively (Anon 2011a). However, the number of samples and methods for sampling and laboratory analysis were not harmonized, so the results cannot be regarded as directly comparable between countries (Anon 2011a, Jørgensen et al. 2002, Scherer et al. 2006). A seasonal influence on the prevalence of *Campylobacter* in retail meat could also be expected as has been described in poultry meat in Denmark and Belgium (Boysen et al. 2011, Habib et al. 2008).

Other food categories that are tested in EU for *Campylobacter* are raw milk, dairy products, fruit and vegetables. In 2009, six countries reported results of the proportion of positive raw cow's milk samples to range from 0 to 5.2%. Of dairy products, between 0 and 7% were positive. No fruit or

Table 1. Prevalence of *Campylobacter* in retail meats in different geographical areas (countries/regions).

Country/area (Reference)	No. of samples positive for *Campylobacter* / No. of samples tested (%)				
	Chicken	Turkey	Pork	Beef	Lamb (Goat)
Great Washington D.C., USA (Zhao et al. 2001)	130/184 (70.7%)	25/172 (14.5%)	3/181 (1.7%)	1/182 (0.5%)	nt [a)]
Pakistan (Hussain et al. 2007)	236/492 (48%)	nt	nt	49/451 (10.9%)	24/465 (5.1%)
United Kingdom (Little et al. 2008)	nt	nt	Muscle: 66 /1309 (5.0%) Offal: 24/131 (18.3%)	Muscle:71 /1514 (4.7%) Offal: 6/49 (12.2%)	Muscle: 55 /744 (7.4%) Offal: 59/161 (36.6%)
New Zealand [b)] (Anon 2007)	205/230 (89.1%)	nt	21/230 (9.1%)	Beef: 8/230 (3.5%) Unweaned veal: 9/90 (10%)	16/231 (6.9%)
Ireland (Whyte et al. 2004)	444/890 (49.9%)	33/88 (37.5%)	10/197 (5.1%)	7/221 (3.2%)	31/262 (11.8%)
Japan (Tokumaru et al. 1991)	106/156 (67.9%)	nt	2/94 (2.1%)	0/52 (0%)	nt
Iran (Rahimi et al. 2010)	37/60 (61.7%)	18/50 (36.0%)	nt	nt	3/50 (6.0%) Goat: 17/45 (4.4%)
Ethiopia (Dadi and Asrat 2008)	13/60 (21.7%)	nt	4/47 (8.5%)	14/227 (6.2%)	12/114 (10.5%) Goat: 7/92 (7.6%)
Australia, South Australia	289/310 (93.2%)	nt	nt	nt	nt
New South Wales (Pointon et al. 2008)	484/549 (87.8%)	nt	nt	nt	nt
EU 2009 (range) [c)] (Anon 2011a)	10.8%–90.0%	5.3%–55.2%	0.3%–3.8%	0%–1.8%	nt

a) nt = not tested
b) Raw minced/diced meat
c) Sampling strategies and diagnostic procedures differed between countries

vegetable samples were positive, whereas one out of almost 2800 tested samples of spices and herbs in one country, was positive (Anon 2011a). Studies in other countries have sometimes found higher prevalences of *Campylobacter* positive non-meat samples. So was for example 13 of 127 (10.2%) of raw bulk milk samples positive in a study from Pakistan (Hussain et al. 2007).

Food and water-borne outbreaks

Most cases of campylobacteriosis are sporadic, however small outbreaks occur where a food item is identified as the source of infection. Since subtyping of *Campylobacter* strains is not routinely performed, many food associated outbreaks may pass undetected. Recently, an outbreak associated with consumption of raw peas in the United States was identified and the link between source and cases could be confirmed by PFGE typing. It was established that the peas had been contaminated with faeces from wild birds (Gardner et al. 2011). Otherwise unpasteurized milk is a frequently reported cause of foodborne outbreaks of campylobacteriosis (Friedmann et al. 2000, Heuvelink et al. 2009).

Waterborne outbreaks have been reported from many countries and usually involve a large number of cases (Friedman et al. 2000, Hanninen et al. 2003, Kuusi et al. 2005, Mentzing 1981, Smith et al. 2006). Drinking water supplies may become contaminated in different ways. Heavy rainfall may cause faecal material from pastures to drain into ground water and surface waters may become contaminated by faeces from wild birds. Breakdown in water plants or sewage systems may also cause contamination of the drinking water.

Since 2005, it is mandatory for EU Member States to report investigated food-borne outbreaks. In 2009, 5,550 food-borne outbreaks were reported by 24 Member States. The majority of outbreaks were caused by *Salmonella* (31%) whereas *Campylobacter* accounted for 6% (Anon 2011a). Sixteen of the 333 food-borne outbreaks of campylobacteriosis were classified as verified outbreaks. For verified outbreaks either isolates from the implicated food vehicle or analytical epidemiological evidence (case-control or cohort study) should be provided. For outbreaks with weaker evidence, the term 'possible outbreak' is used. Broiler meat and products thereof and unspecified poultry meat were reported as the implicated foodstuff in seven, bovine meat in two and dairy products in one of the verified *Campylobacter* outbreaks. Three waterborne outbreaks with *Campylobacter* were reported in 2009 in three Member States. One of the outbreaks, involving estimated 500 cases was probably due to contamination of a water pipe installation following heavy rainfall.

Control and Prevention

Surveillance and action plans to reduce *Campylobacter* in broilers

Campylobacter could be monitored and controlled at all levels in the food chain, starting with the primary production. *Campylobacter* spp. are prevalent in the environment and are transmitted horizontally into broiler houses, usually by humans on their clothes or boots. The bacteria can also gain access through ventilation shafts, by water, by contaminated transport crates or be carried by insects (Bull et al. 2006, Hansson et al. 2005, Newell and Fearnley 2003). Broilers are rarely colonized with *Campylobacter* before 2–3 weeks of age. If the bacteria are introduced into a flock, they spread rapidly and colonize all birds so that the within-flock prevalence usually is 100% in a positive flock at slaughter. In the Scandinavian countries, *Campylobacter* spp. in broilers have for many years been in focus for action plans and/or monitoring programmes in order to reduce the prevalence of positive flocks (Hansson et al. 2007, Hofshagen and Kruse 2005, Rosenquist et al. 2009). In UK, the government and industry have together worked out targets to reduce *Campylobacter* in UK produced chickens ("The Joint Government and Industry Target to Reduce *Campylobacter* in UK produced Chickens by 2015", http://www.food.gov.uk/multimedia/pdfs/campytarget.pdf). The New Zealand Food Safety Authority has worked out a risk assessment program in order to reduce the number of human cases ("New Zealand Food Safety Authority's Campylobacter Risk Management Strategy, 2010–2013", http://www.foodsafety.govt.nz/elibrary/industry/Campylobacter_Risk-Comprehensive_Aimed.pdf).

In 2008, EU carried out a baseline survey in which 26 Member States and two other European countries participated. Each country sampled and analysed caecum contents from about 400 slaughter batches according to standardized procedures. At EU level the prevalence of positive batches was 71.2% however the Member State prevalence varied from 2.0% to 100%. In the four Nordic countries, the prevalences of colonized broiler batches were relatively low; from 3.2% to 19% (Anon 2010b). The baseline survey also included analysis of *Campylobacter* in broiler carcasses using the ISO 10272 standard for detection and enumeration of *Campylobacter*. At EU level, the prevalence of *Campylobacter*-contaminated broiler carcasses was 75.8%. The Member State prevalence varied from 4.9% to 100%. Also the counts of *Campylobacter* on the carcasses varied widely between countries (Anon 2010b).

Control measures and risk reducing strategies

In order to prevent transfer of *Campylobacter* from the environment to a broiler flock during rearing, strict biosecurity should be applied. Hygiene measures, i.e., cleaning and disinfection practices and changing clothes and footwear before entering the broiler flocks have proved to reduce the risk of flock colonization (Gibbens et al. 2001, Hansson et al. 2010). However, it is not always possible to prevent *Campylobacter* colonization only by use of hygiene procedures. To prevent flying insects to enter the broiler houses, fly nets have been tested in Denmark and Iceland with promising results (Hald et al. 2004). Other strategies that are currently investigated include vaccination and treatment with competitive exclusion or probiotics products, bacteriophages and bacteriocins (Cox and Pavic 2009, Lin 2009). Another approach is to avoid management practices associated with risk for *Campylobacter* colonization such as partial depopulation of flocks or slaughter at a higher age (Newell and Fearnley 2003, Bouwknegt et al. 2004). In a recent study, biosecurity-based interventions and strategies to reduce *Campylobacter* in broiler farms were thoroughly reviewed by Newell et al. (2011).

At slaughter, faecal material may contaminate the broiler carcass and a correlation between the numbers of *Campylobacter* in intestinal contents and on the carcass has been documented (Rosenquist et al. 2006).

Strategies have been developed to decrease the prevalence of contaminated carcasses and/or reduce the load of *Campylobacter* on the carcasses at the slaughter house. Some of these involve decontamination of carcasses by use of steam and hot water, chilling and freezing procedures (Boysen and Rosenquist 2009), or by use of different chemicals (Cox and Pavic 2009). The measures are effective to some extent; a reduction in numbers is usually obtained by a factor of up to 2 $\log_{10}$ units in numbers of *Campylobacter*. In principle, the use of chemical decontamination is currently not allowed in EU, but this may change. Gamma irradiation of poultry meat is another effective intervention technique which would reduce the risk for consumers substantially. This technology has been approved by the FDA, USDA and the Codex Alimentarius Commission (Cox and Pavic 2009, Havelaar et al. 2007). However, there is a public resistance to buy irradiated food and irradiated meat is only authorized at national level in the EU.

At retail and consumer level, freezing or heat treatment are practical means for preventing cross contamination from poultry meat to other foodstuffs or to people handling the meat. Consumer information campaigns have been successful but are expensive and the effect is usually short lived (Havelaar et al. 2007).

Concluding Remarks

In spite of increased awareness and knowledge about *Campylobacter* epidemiology the number of human campylobacteriosis cases remains high all over the world. The incidence in EU shows a slight upward trend, although there are countries where the incidence has gone down (Anon 2011a).

In order to improve the control of *Campylobacter*, a number of key factors need to be investigated, developed and considered:

- Development of rapid, robust and validated methods for detection, identification and quantification of *Campylobacter* in samples of food and other relevant materials.
- Identification of hitherto unknown or underestimated sources and routes of infection.
- Clarification of mechanisms for virulence and pathogenesis, which virulence associated markers are crucial for generating disease in humans?
- Determine targets and limits for *Campylobacter* in food, based on risk assessment studies.
- Implement harmonized surveillance of *Campylobacter* in poultry and potential risk food.
- Work for a close collaboration between animal/food and public health authorities in order to establish effective and coordinated surveillance and intervention strategies.

References

Allos, B.M. 2001. *Campylobacter jejuni* infections: update on emerging issues and trends. Clin. Infect. Dis. 32: 1201–1206.

Altekruse, S.F., J.M. Hunt, L.K. Tollefson and J.M. Madden. 1994. Food and animal sources of human *Campylobacter jejuni* infection. JAVMA. 204: 57–61.

Anonymous. 2007. ESR Risk Profile: *Campylobacter jejuni/coli* in red meat. ESR Client Report FW0485. Institute of Environmental Science and Research Limited Christchurch Science Centre, New Zealand.

Anonymous. 2010a. EFSA Panel on Biological Hazards (BIOHAZ). Scientific opinion on quantification of the risk posed by broiler meat to human campylobacteriosis in the EU. EFSA Journal. 2010; 8(1): 1437.

Anonymous. 2010b. Analysis of the baseline survey on the prevalence of *Campylobacter* in broiler batches and of *Campylobacter* and *Salmonella* on broiler carcasses in the EU, 2008, Part A: *Campylobacter* and *Salmonella* prevalence estimates. EFSA Journal 2010; 8(03): 1503.

Anonymous. 2011a. European Food Safety Authority, European Centre for Disease Prevention and Control; The European Union Summary Report on Trends and Sources of Zoonoses, Zoonotic Agents and Food-borne Outbreaks in 2009; EFSA Journal 2011; 9(3): 2090.

Anonymous. 2011b. European Union summary report on antimicrobial resistance in zoonotic and indicator bacteria from animals and food in the European Union in 2009. EFSA Journal 2011; 9(7): 2154.

Best, E.L., E.J. Powell, C. Swift, K.A. Grant and J. A. Frost. 2003. Applicability of a rapid duplex real-time PCR assay for speciation of *Campylobacter jejuni* and *Campylobacter coli* directly from culture plates. FEMS Microbiol. Lett. 229: 237–241.

Black, R.E., M.M. Levine, M.L. Clements, T.P. Hughes and M.J. Blaser. 1988. Experimental *Campylobacter jejuni* infection in humans. J. Infect. Dis. 157: 472–479.

Blaser, M.J., D.N. Taylor and R.A. Feldman. 1983. Epidemiology of *Campylobacter jejuni* infections. Epidemiol. Rev. 5: 157–176.

Blaser, M.J. and J. Engberg. 2008. Clinical aspects of *Campylobacter jejuni* and *Campylobacter coli* infections. *In*: I. Nachamkin, C.M. Szymanski and M.J. Blaser (eds.). *Campylobacter*, 3rd ed. ASM Press, Washington, DC. pp. 99–121.

Boysen, L. and H. Rosenquist. 2009. Reduction of thermotolerant *Campylobacter* species on broiler carcasses following physical decontamination at slaughter. J. Food Prot. 72: 497–502.

Boysen, L., H. Vigre and H. Rosenquist. 2011. Seasonal influence on the prevalence of thermotolerant *Campylobacter* in retail broiler meat in Denmark. Food Microbiol. 28: 1028–1032.

Bouwknegt, M., A.W. van de Giessen, W.D. Dam-Deisz, A.H. Havelaar, N.J. Nagelkerke and A.M. Henken. 2004. Risk factors for the presence of *Campylobacter* spp. in Dutch broiler flocks. Prev. Vet. Med. 62: 35–49.

Bull, S.A., V.M. Allen, G. Domingue, F. Jørgensen, J.A. Frost, R. Ure, R. Whyte, D. Tinker, J.E.L. Corry, J. Gillard-King and T.J. Humphrey. 2006. Sources of *Campylobacter* spp colonizing housed broiler flocks during rearing. Appl. Environ. Microbiol. 72: 645–652.

Butzler, J.P. 2004. *Campylobacter*, from obscurity to celebrity. Clin. Microbiol. Infect. 10: 868–876.

Champion, O.L., S. Al-Jaberi, R.A. Stabler and B.W. Wren. 2008. Comparative genomics of *Campylobacter* jejuni. *In*: I. Nachamkin, C.M. Szymanski and M.J. Blaser (eds.). *Campylobacter*, 3rd ed. ASM Press, Washington, DC. pp. 63–71.

Corry, J.E.L., D.E. Post, P. Colin and M.J. Laisney. 1995. Culture media for the isolation of campylobacters. Int. J. Food Microbiol. 26: 43–76.

Cox, J.M. and A. Pavic. 2009. Advances in enteropathogen control in poultry production. J. Appl. Microbiol. 108: 745–755.

Dadi, L. and D. Asrat. 2008. Prevalence and antimicrobial susceptibility profiles of thermotolerant *Campylobacter* strains in retail raw meat products in Ethiopia. Ethiop. J. Health Develop. 22: 195–200.

Denis, M., J. Refregier-Petton, M.J. Laisney, G. Ermel and G. Salvat. 2001. *Campylobacter* contamination in French chicken production from farm to consumers. Use of a PCR assay for detection and identification of *Campylobacter jejuni* and *Camp. coli*. J. Appl. Microbiol. 91: 255–267.

Dingle, K.E., F.M. Colles, D.R.A. Wareing, R. Ure, A.J. Fox, F. E. Bolton, H.J. Bootsma, R.J.L. Willems, R. Urwin and M.C.J. Maiden. 2001. Multilocus sequence typing system for *Campylobacter jejuni*. J. Clin. Microbiol. 39: 14–23.

Directive 2003/99/EC of the European Parliament and of the Council of 17 November 2003 on the monitoring of zoonoses and zoonotic agents, amending Council Decision 90/424/EEC and repealing Council Directive 92/117/EEC. OJ L 325, 12.12.2003, pp. 31–40.

Doyle, M.P and D.J. Roman. 1981. Growth and survival of *Campylobacter fetus* subsp. *jejuni* as a function of temperature and pH. J. Food Protect. 44: 596–601.

Engberg, J., F.M. Aarestrup, D.E. Taylor, P. Gerner-Smidt and I. Nachamkin. 2001. Quinolone and macrolide resistance in *Campylobacter jejuni* and *C. coli*: resistance mechanisms and trends in human isolates. Emerg. Infect. Dis. 7: 24–34.

Engvall, E.O., B. Brändström, L. Andersson, V. Båverud, G. Trowald-Wigh and L. Englund. 2003. Isolation and identification of thermophilic *Campylobacter* species in faecal samples from Swedish dogs. Scand. J. Infect. Dis. 35: 713–718.

Fosse, J., H. Seegers and C. Magras. 2009. Prevalence and risk factors for bacterial food-borne zoonotic hazards in slaughter pigs: a review. Zoonoses Public Health 56: 429–454.

Friedman, C.R., R.M. Hoekstra, M. Samuel, R. Marcus, J. Bender, B. Shiferaw, S. Reddy, S.D. Ahuja, D.L. Helfrick, F. Hardnett, M. Carter, B. Anderson and R.V. Tauxe. 2004. Risk factors for sporadic *Campylobacter* infection in the United States: A case-control study in FoodNet sites. Clin. Infect. Dis. 38 Suppl. 3: S285–96.

Friedman, C.R., J. Neimann, H.C. Wegener and R.V. Tauxe. 2000. Epidemiology of *Campylobacter jejuni* infections in the United States and other industrialized nations. *In*: I. Nachamkin and M.J. Blaser (eds.). *Campylobacter*. 2nd ed. ASM Press, Washington D.C. pp. 121–138.

Gardner, T.J., C. Fitzgerald, C. Xavier, R. Klein, J. Pruckler, S. Stroika and J.B. McLaughlin. 2011. Outbreak of campylobacteriosis associated with consumption of raw peas. Clin. Infect. Dis. 53: 26–32.

Gibbens, J.C., S.J. Pascoe, S.J. Evans, R.H. Davies and A.R. Sayers. 2001. A trial of biosecurity as a means to control *Campylobacter* infection of broiler chickens. Prev. Vet. Med. 48: 85–99.

Habib, I., I. Sampers, M. Uyttendaele, D. Berkvens and L. de Zutter. 2008. Baseline data from a Belgium-wide survey of *Campylobacter* species contamination in chicken meat preparations and considerations for a reliable monitoring program. Appl. Environ. Microbiol. 74: 5483–5489.

Hald, B. and M. Madsen. 1997. Healthy puppies and kittens as carriers of *Campylobacter* spp., with special reference to *Campylobacter upsaliensis*. J. Clin. Microbiol. 35: 3351–3352.

Hald, B., H. Skovgård, D.D. Bang, K. Pedersen, J. Dybdahl, J.B. Jespersen and M. Madsen. 2004. Flies and *Campylobacter* infection of broiler flocks. Emerg. Infect. Dis. 10: 1490–1492.

Hanninen, M.L., H. Haajanen, T. Pummi, K. Wermundsen, M.L. Katila, H. Sarkkinen, I. Miettinen and H. Rautelin. 2003. Detection and typing of *Campylobacter jejuni* and *Campylobacter coli* and analysis of indicator organisms in three waterborne outbreaks in Finland. Appl. Environ. Microbiol. 69: 1391–1396.

Hansson, I., M. Ederoth, L. Andersson, I. Vågsholm and E. Olsson Engvall. 2005. Transmission of *Campylobacter* spp. to chickens during transport to slaughter. J. Appl. Microbiol. 99: 1149–1157.

Hansson, I., L.P. Forshell, P. Gustafsson, S. Boqvist, J. Lindblad, E.O. Engvall, Y. Andersson and I. Vågsholm. 2007. Summary of the Swedish *Campylobacter* program in broilers, 2001 through 2005. J. Food. Protect. 70: 2008–2014.

Hansson, I., E.O. Engvall, I. Vågsholm and A. Nyman. 2010. Risk factors associated with the presence of *Campylobacter*-positive broiler flocks in Sweden. Prev. Vet. Med. 96: 114–21.

Havelaar, A.H., M.J. Mangen, A.A. de Koeijer, M.J. Bogaardt, E.G. Evers, W.F. Jacobs-Reitsma, W. van Pelt, J.A. Wagenaar, G.A. de Wit, H. van der Zee and M.J. Nauta. 2007. Effectiveness and efficiency of controlling *Campylobacter* on broiler chicken meat. Risk. Anal. 27: 831–44.

Heuvelink, A.E., C. van Heerwaarden, A. Zwartkruis-Nahuis, J.J. Tilburg, M.H. Bos, F.G. Heilmann, A. Hofhuis, T. Hoekstra and E. de Boer. 2009. Two outbreaks of campylobacteriosis associated with the consumption of raw cows' milk. Int. J. Food Microbiol. 134: 70–74.

Hofshagen, M. and H. Kruse. 2005. Reduction in flock prevalence of *Campylobacter* spp. in broilers in Norway after implementation of an action plan. J. Food Protect. 68: 2220–2223.

Hussain, I., M.S. Mahmood, M. Akhtar and A. Khan. 2007. Prevalence of *Campylobacter* species in meat, milk and other food commodities in Pakistan. Food Microbiol. 24: 219–222.

[ISO 2006] ISO 10272-1: 2006 (E) "Microbiology of food and animal feeding stuffs–Horizontal method for detection and enumeration of *Campylobacter* spp.—Part 1: Detection method".

ISO 10272- 2:2006 (E) "Microbiology of food and animal feeding stuffs – Horizontal method for detection and enumeration of *Campylobacter* spp.—Part 2: Colony-count technique".

Jacobs-Reitsma, W. 2000. *Campylobacter* in the food supply. *In*: I. Nachamkin and M.J. Blaser (eds.). *Campylobacter*. 2nd ed. ASM Press, Washington D.C. pp. 467–481.

Janssen, R., K.A. Krogfelt, S.A. Cawthraw, W. van Pelt, J.A. Wagenaar and R.J. Owen. 2008. Host-pathogen interactions in *Campylobacter* infections: the host perspective. Clin. Microbiol. Rev. 21: 505–518.

Jørgensen, F., R. Bailey, S. Williams, P. Henderson, D.R.A. Wareing, F.J. Bolton, J.A. Frost, L. Ward and T.J. Humphrey. 2002. Prevalence and numbers of *Salmonella* and *Campylobacter* spp. on raw, whole chickens in relation to sampling methods. Int. J. Food Microbiol. 76: 151–164.

Kapperud, G., G. Espeland, E. Wahl, A. Walde, H. Herikstad, S. Gustavsen, I. Tveit, O. Natås, L. Bevanger and A. Digranes. 2003. Factors associated with increased and decreased risk of *Campylobacter* infection: a prospective case-control study in Norway. Am. J. Epidemiol. 158: 234–242.

Ketley, J.M. 1997. Pathogenesis of enteric infection by *Campylobacter*. Microbiology 143: 5–21.

Konkel, M.E., M.R. Monteville, V. Rivera-Amill and L.A. Joens. 2001. The pathogenesis of *Campylobacter jejuni* mediated enteritis. Curr. Issues Intest. Microbiol. 2: 55–71.

Kuusi, M., J.P. Nuorti, M.L. Hänninen, M. Koskela, V. Jussila, E. Kela, I. Miettinen and P. Ruutu. 2005. A large outbreak of campylobacteriosis associated with municipal water supply in Finland. Epidemiol. Infect. 133: 593–601.

Lastovica, A.J. and B.M. Allos. 2008. Clinical significance of *Campylobacter* and related species other than *Campylobacter jejuni* and *Campylobacter coli*. *In*: I. Nachamkin, C.M. Szymanski and M.J. Blaser (eds.). *Campylobacter*. ASM Press, Washington, DC. pp. 123–149.

Lin, J. 2009. Novel approaches for *Campylobacter* control in poultry. Foodborne Pathog. Dis. 6: 755–765.

Lior, H., D.L. Woodward, J.A. Edgar, L.J. Laroche and P. Gill. 1982. Serotyping of *Campylobacter jejuni* by slide agglutination based on heat-labile antigenic factors. J. Clin. Microbiol. 15: 761–768.

Little, C.L., J.F. Richardsson, R.J. Owen, E. de Pinna and E.J. Threlfall. 2008. *Campylobacter* and *Salmonella* in raw red meats in the United Kingdom: Prevalence, characterization and antimicrobial resistance pattern, 2003–2005. Food Microbiol. 25: 538–543.

Luangtongkum, T., B. Jeon, J. Han, P. Plummer, C.M. Logue and Q. Zhang. 2009. Antibiotic resistance in *Campylobacter*: emergence, transmission and persistence. Future Microbiol. 4: 189–200.

Meinersmann, R.J., C.M. Patton, G.M. Evins, I.K. Wachsmuth and P.I. Fields. 2002. Genetic diversity and relationships of *Campylobacter* species and subspecies. Int. J. Syst. Evol. Microbiol. 52: 1789–1797.

Mentzing, L.-O. 1981. Waterborne outbreaks of *Campylobacter* enteritis in central Sweden. Lancet 2 (8242): 352–354.

Nachamkin, I., J. Engberg and F.M. Aarestrup. 2000. Diagnosis and antimicrobial susceptibility of *Campylobacter* species. *In*: I. Nachamkin and M.J. Blaser (eds.). *Campylobacter* 2nd ed. ASM Press Washington D.C. pp. 45–46.

Nauta, M., A. Hill, H. Rosenquist, S. Brynestad, A. Fetsch, P. van der Logt, A. Fazil, B. Christensen, E. Katsma, B. Borck and A. Havelaar. 2009. A comparison of risk assessments on *Campylobacter* in broiler meat. Int. J. Food Microbiol. 129: 107–123.

Neimann, J., J. Engberg, K. Molbak and H.C. Wegener. 2003. A case-control study of risk factors for sporadic campylobacter infections in Denmark. Epidemiol. Infect. 130: 353–366.

Newell, D.G. 2001. Animal models of *Campylobacter jejuni* colonization and disease and the lessons to be learned from similar *Helicobacter pylori* models. Symp. Ser. Soc. Appl. Microbiol. 30: 57S–67S.

Newell, D.G., K.T. Elvers, D. Dopfer, I. Hansson, P. Jones, J. Gittins, N.J. Stern, R. Davies, I. Connerton, D. Pearson, G. Salvat and V.M. Allen. 2011. Biosecurity-based interventions and strategies to reduce *Campylobacter* spp. on poultry farms. Appl. Environ. Microbiol. 77: 8605–8614.

Newell, D.G. and C. Fearnley. 2003. Sources of *Campylobacter* colonization in broiler chickens. Appl. Environ. Microbiol. 69: 4343–4351.

Nielsen, E.M., J. Engberg, V. Fussing, L. Petersen, C.H. Brogren and S.L.W. On. 2000. Evaluation of phenotypic and genotypic methods for subtyping *Campylobacter jejuni* isolates from humans, poultry and cattle. J. Clin. Microbiol. 38: 3800–3810.

[NMKL 2007] NMKL 119, 3 ed. 2007. Nordic Committee on Food Analyses, 2007. Thermotolerant *Campylobacter*. Detection, semi-quantitative and quantitative determination in foods and drinking water. Method No. 119, 3 Ed. Nordic Committee on Food Analysis.

Park, S.F. 2002. The physiology of *Campylobacter* species and its relevance to their role as foodborne pathogens. Internat J. Food Microbiol. 74: 177–188.

Parkhill, J., B.W. Wren, K. Mungall, J.M. Ketley, C. Churcher, D. Basham, T. Chillingworth, R.M. Davies, T. Feltwell, S. Holroyd, K. Jagels, A.V. Karlyshev, S. Moule, M.J. Pallen, C.W. Penn, M.A. Quail, M.A. Rajandream, K.M. Rutherford, A.H. van Vliet, S. Whitehead and B.G. Barrell. 2000. The genome sequence of the food-borne pathogen *Campylobacter jejuni* reveals hypervariable sequences. Nature 403: 665–668.

Penner, J.L. and J.N. Hennessy. 1980. Passive hemagglutination technique for serotyping *Campylobacter jejuni* subsp. *jejuni* on the basis of heat-stable antigens. J. Clin. Microbiol. 12: 732–737.

Perko-Mäkelä, P., H. Hakkinen, T. Honkanen-Buzalski and M.L. Hänninen. 2002. Prevalence of *Campylobacter* in chicken flocks during the summer of 1999 in Finland. Epidemiol. Infect. 129: 187–192.

Pointon, A., M. Sexton, P. Dowsett, T. Saputra, A. Kiermeier, M. Lorimer, G. Holds, G. Arnold, D. Davos, B. Combs, S. Fabiansson, G. Raven, H. McKenzie, A. Chapman and J. Sumner. 2008. A baseline survey of the microbiological quality of chicken portions and carcasses at retail in two Australian States (2005–2006). J. Food Protect. 71: 1123–1134.

Poly, F. and P. Guerry. 2008. Pathogenesis of *Campylobacter*. Curr. Opin. Gastroenterol. 24: 27–31.

Rahimi, E., H.R. Kazemeini, S. Safaei, K. Allahbakhshi, M. Momeni and M. Riahi. 2010. Detection and identification of *Campylobacter* spp. from retail raw chicken, turkey, sheep and goat meat in Ahvaz, Iran. Afr. J. Microbiol. Res. 4: 1620–1623.

Ridley, A.M., V.M. Allen, M. Sharma, J.A. Harris and D.G. Newell. 2008. Real-time PCR approach for detection of environmental sources of *Campylobacter* strains colonizing broiler flocks. Appl. Environ. Microbiol. 74: 2492–2504.

Rosenquist, H., N.L. Nielsen, H.M. Sommer, B. Norrung and B.B. Christensen. 2003. Quantitative risk assessment of human campylobacteriosis associated with thermophilic *Campylobacter* species in chickens. Int. J. Food Microbiol. 83: 87–103.

Rosenquist, H., L. Boysen, C. Galliano, S. Nordentoft, S. Ethelberg and B. Borck. 2009. Danish strategies to control *Campylobacter* in broilers and broiler meat: facts and effects. Epidemiol. Infect. 137: 1742–1750.

Rosenquist, H., H.M. Sommer, N.L. Nielsen and B.B. Christensen. 2006. The effect of slaughter operations on the contamination of chicken carcasses with thermotolerant *Campylobacter*. Int. J. Food Microbiol. 108: 226–232.

Scherer, K., E. Bartelt, C. Sommerfeld and G. Hildebrandt. 2006. Comparison of different sampling techniques and enumeration methods for the isolation and quantification of *Campylobacter* spp. in raw retail chicken legs. Int. J. Food Microbiol. 108: 115–119.

Sheppard, S.K., J.F. Dallas, N.J.C. Strachan, M. MacRae, N.D. McCarthy, D.J. Wilson, F.J. Gormley, D. Falush, I.D. Ogden, M.C.J. Maiden and K.J. Forbes. 2009. *Campylobacter* genotyping to determine the source of human infection. Clin. Infect. Dis. 48: 1072–1078.

Smith, A., M. Reacher, W. Smerdon, G.K. Adak, G. Nichols and R.M. Chalmers. 2006. Outbreaks of waterborne infectious intestinal disease in England and Wales, 1992–2003. Epidemiol. Infect. 134: 1141–1149.

Stanley, K. and K. Jones. 2003. Cattle and sheep farms as reservoirs of *Campylobacter*. J. Appl. Microbiol. 94: 104S–113S.

Stanley, K.N., J.S. Wallace, J.E. Currie, P.J. Diggle and K. Jones. 1998. The seasonal variation of thermophilic *Campylobacter* in beef cattle, dairy cattle and calves. J. Appl. Microbiol. 85: 472–480.

Suzuki, H. and S. Yamamoto. 2009. *Campylobacter* contamination in retail poultry meats and by-products in the world: a literature survey. J. Vet. Med. Sci. 71: 255–261.

Teunis, P., W. Van den Brandhof, M. Nauta, J. Wagenaar, H. Van den Kerkhof and W. Van Pelt. 2005. A reconsideration of the *Campylobacter* dose-response relation. Epidemiol. Infect. 133: 583–592.

Tokumaru, M., H. Konuma, M. Umesako, S. Konno and K. Shinagawa. 1991. Rates of detection of *Salmonella* and *Campylobacter* in meats in response to the sample size and the infection level of each species. Int. J. Food Microbiol. 13: 41–46.

Wahlström, H., E. Tysén, E. Olsson Engvall, B. Brändström, E. Eriksson, T. Mörner and I. Vågsholm. 2003. Survey of *Campylobacter* species, VTEC O157 and *Salmonella* species in Swedish wildlife. Vet. Rec. 153: 74–80.

Wang, G., C.G. Clark, T.M. Taylor, C. Pucknell, C. Barton, L. Price, D.L. Woodward and F.G. Rodgers. 2002. Colony multiplex PCR assay for identification and differentiation of *Campylobacter jejuni, C. coli, C. lari, C. upsaliensis,* and *C. fetus* subsp. *fetus*. J. Clin. Microbiol. 40: 4744–4747.

Wassenaar, T.M. and D.G. Newell. 2000. Genotyping of *Campylobacter* spp. Appl. Environ. Microbiol. 66: 1–9.

Wedderkopp, A., K.O. Gradel, J.C. Jörgensen and M. Madsen. 2001. Preharvest surveillance of *Campylobacter* and *Salmonella* in Danish broiler flocks: a 2-year study. Int. J. Food Microbiol. 68: 53–59.

Whyte, P., K. McGill, D. Cowley, R.H. Madden, L. Moran, P. Scates, C. Carroll, A. O'Leary, S. Fanning, J.D. Collins, E. McNamara, J.E. Moore and M. Cormican. 2004. Occurrence of *Campylobacter* in retail foods in Ireland. Int. J. Food Microbiol. 95: 111–118.

Wilson, D.J., E. Gabriel, A.J.H. Leatherbarrow, J. Cheesbrough, S. Gee, E. Bolton, A. Fox, P. Fearnhead, C.A. Hart and P.J. Diggle. 2008. Tracing the source of campylobacteriosis. PLoS Genet. 4(9): e1000203.

Wingstrand, A., J. Neimann, J. Engberg, E.M. Nielsen, P. Gerner-Smidt, H.C. Wegener and K. Molbak. 2006. Fresh chicken as main risk factor for campylobacteriosis, Denmark. Emerg. Infect. Dis. 12: 280–285.

Young, K.T., L.M. Davis and V.J. DiRita. 2007. *Campylobacter jejuni*: molecular biology and pathogenesis. Nat. Rev. Microbiol. 5: 665–679.

Zhao, C., B. Ge, J. De Villena, R. Sudler, E. Yeh, S. Zhao, D.G. White, D. Wagner and J. Meng. 2001. Prevalence of *Campylobacter* spp., *Escherichia coli,* and *Salmonella* serovars in retail chicken, turkey, pork and beef from the greater Washington, D.C., area. Appl. Environ. Microbiol. 67: 5431–5436.

CHAPTER 6

Food-borne Pathogenic *Clostridia*

Miia Lindström[#,*] and *Hannu Korkeala*[#]

Clostridia are prevalent in the environment and in the gastrointestinal (GI) tract of animals and humans, thus they are also frequently found in raw food materials. *Clostridia* form resistant spores, which may survive environmental stress and thus persist in the food chain. Obligate anaerobes, *Clostridia* need specific conditions with a low redox potential for growth in foods. Many clostridia produce toxins which cause diseases in humans and animals. *Clostridia* of relevance to human food safety include *Clostridium botulinum* producing a highly potent neurotoxin, which causes a potentially lethal paralysis when ingested, and *Clostridium perfringens* producing enterotoxin, which causes one of the most common GI diseases worldwide. Although *Clostridium difficile* has not been proven to be transmitted through food, its presence in food animals and foods has raised a question of its role as a food pathogen.

Clostridium botulinum—the Old Threat Is Still There

The microorganism

C. botulinum is prevalent in soils (Nevas et al. 2006) and marine environments (Hielm et al. 1998), and is thus frequently present in the GI tract of animals (Dahlenborg et al. 2001, Myllykoski et al. 2006) and raw

[#] Department of Food Hygiene and Environmental Health, Faculty of Veterinary Medicine, P. O. Box 66, 00014 University of Helsinki, Finland.
Emails: miia.lindstrom@helsinki.fi; hannu.korkeala@helsinki.fi
[*] Corresponding author

food materials (Lindström et al. 2006a). Based on the serological properties of the neurotoxins they produce, *C. botulinum* strains are divided into seven toxin types, A–G. Phenotypically and genetically *C. botulinum* strains form four distinct groups of organisms, designated I to IV (Hutson et al. 1993, Keto-Timonen et al. 2006). Groups I and II strains, producing toxin types A, B, E and/or F, are involved with botulism in humans and are discussed here. Group III strains produce types C and D toxins and cause botulism in animals. Type G toxin producing Group IV strains have not been associated with illness and the name of *Clostridium argentinense* has been proposed for this group (Suen et al. 1988). In addition to *C. botulinum*, some strains of *Clostridium butyricum* and *Clostridium baratii* also produce type E and F neurotoxins, respectively.

Group I and II *C. botulinum* strains are physiologically distinct (Lindström and Korkeala 2006). Group I strains are proteolytic, while Group II strains utilize mainly carbohydrates for energy. Group I spores have a high heat resistance (Schmidt 1964, Stumbo 1973, Lynt et al. 1981, Scott and Bernard 1982) and their vegetative cells are mesophilic, being able to grow at a temperature range of 12 to 48°C and having an optimum growth temperature slightly above 37°C (Hinderink et al. 2009). Group II spores have a moderate heat resistance (Lynt et al. 1983, Peck et al. 1993, Lindström et al. 2003) and vegetative growth has been reported in a temperature range of 3 to 40°C (Schmidt et al. 1961, Eklund et al. 1967a, b, Graham et al. 1997, Derman et al. 2011). The inhibitory water-phase NaCl concentration and water activity for Group I strains are 10% and 0.94 and for Group II strains are 5% and 0.97, respectively (Smith and Sugiyama 1988). Growth-limiting pH is 4.6 for Group I and 5.0 for Group II strains (Segner et al. 1966, Smelt et al. 1982).

C. botulinum secretes the neurotoxin during its vegetative growth, with the peak production occurring at late logarithmic growth phase (Bradshaw et al. 2004, Lövenklev et al. 2004, Chen et al. 2008). Botulinum neurotoxins target human and animal motoric nerve ends, blocking acetylcholine release in synapses and thus causing flaccid paralysis. The neurotoxin molecule is accompanied by nontoxic neurotoxin associated proteins which probably protect the neurotoxin and assist in absorption from the GI tract (Fujinaga et al. 1997, Sugawara and Fujinaga 2011, Gu et al. 2012). The maximal potency of the botulinum neurotoxins is achieved after enzymatic cleavage of the toxin molecule. In the proteolytic Group I *C. botulinum*, the cleavage of toxin is achieved by endogenous enzymes, but neurotoxins produced by non-proteolytic Group II strains probably require external proteases, such as the pancreatic trypsin, for their activation (Duff et al. 1956, Das Gupta and Sugiyama 1972, Ohishi and Sakaguchi 1977).

The genes encoding botulinum neurotoxins may be located in the bacterial chromosome (types A, B, E and F), on plasmids (types A, B and G; Eklund et al. 1988, Zhou et al. 1995, Marshall et al. 2007), or on prophages (types C and D; Eklund et al. 1972, Eklund and Poysky 1974). Neurotoxin gene sequencing, binding studies with monoclonal antibodies, and mass spectrometry have revealed five type A neurotoxin subtypes, seven type B, five type F, and eight type E neurotoxin subtypes (Smith et al. 2005, Hill et al. 2007, Chen et al. 2007, Dover et al. 2009, Carter et al. 2010, Macdonald et al. 2011, Kalb et al. 2011, Dover et al. 2013) which show up to 32% amino acid differences (Smith et al. 2005). Although probably possessing little relevance for food safety, the sequence variability is of great concern in the development of vaccines and antitoxins to prevent and treat botulism.

The patient

Botulism is a neuroparalytic disease caused by botulinum toxin blocking neurotransmission and characterized by descending flaccid paralysis. Typical first symptoms include double vision, paralysis of facial muscles, drying of mouth, and difficulty in swallowing and speech. The paralysis proceeds to distal muscles and finally blocks respiratory muscles, which may lead to death unless treated promptly. Botulinum neurotoxins are among the most poisonous natural substances, with an estimated lethal oral dose in humans of 1 μg/kg. Incubation time varies by the toxic dose from a few hours to several days (Arnon et al. 2001). Similarly, the dose of neurotoxin affects the duration of disease: Mild cases with the entire hospitalization period restricted to 1–2 wk (Lindström et al. 2006b) and extremely severe ones requiring mechanical ventilation for several months have been reported (Marcus 2009). Humans of all age groups are at risk. Development of immunity is irrelevant since the naturally occurring doses of neurotoxins in foods causing botulism are substantially higher than those with observed immunogenicity (Berman et al. 2005); nevertheless, recurrence of botulism is extremely rare (Bilusic et al. 2008, Yuan et al. 2011).

Botulism is a rare disease with 150–300 cases and a notification rate of 0.02 per 100, 000 inhabitants (2008) being reported annually in the European Union (Table 1). Not all countries report the number of botulism cases. Worldwide, 52% of all cases were due to type B toxin, and 34% and 12% due to types A and E toxins, respectively, in 1995 (Hatheway 1995). Type F cases are extremely rare (1%; Harvey et al. 2002, Gupta et al. 2005). Outbreaks are usually sporadic and restricted to a few people but also large outbreaks affecting dozens of people or distributed to more than one country have been reported. A food-borne botulism outbreak in Thailand in 2006 affected 209 people, 42 of which developed respiratory failure after having consumed home-prepared bamboo shoots contaminated with type

Table 1. Human botulism outbreaks in the Nordic countries in 1999–2011[a] (European Centre for Disease Prevention and Control).

Yr	Denmark	Finland	Iceland	Norway	Sweden	EU
1999	0	1	2	4	0	247
2000	0	0	3	1	0	269
2001	0	0	0	0	0	181
2002	0	1	0	1	0	305
2003	1	0	0	6	2	241
2004	0	0	0	1	0	210
2005	0	0	0	6	1	207
2006	0	2	0	2	2	155
2007	0	0	0	0	0	171
2008	1	0	0	0	0	157
2009	0	2	0	0	1	185
2010	1	1	0	1	0	NR[b]
2011	NR	2	NR	NR	NR	NR

[a]The figures include all forms of botulism.
[b]NR, not reported.

A neurotoxin (Kongsaengdao et al. 2006). Moreover, a widespread type A outbreak affected six patients in Georgia and Florida, USA, and Ontario, Canada, in 2006 and was traced to commercial carrot juice (Sheth et al. 2008).

Botulism may affect humans in several different forms. The most well known form of human botulism is the classical food poisoning botulism, an intoxication that follows the consumption of food containing preformed neurotoxin. Depending on the amount of toxin ingested, the incubation period typically varies from 12 to 72 hrs, but can be as long as several weeks (Gao et al. 1990). In addition to paralysis, the food-borne form is often manifested by GI symptoms. Mild cases with GI symptoms dominating have been rarely reported but are likely to occur and require attention in the future (Korkeala et al. 1998, Lindström et al. 2006b).

Regions with a particularly high incidence of food-borne botulism include the Republic of Georgia, Poland, China, Russia, Kyrgyzstan, and certain ethnic populations in the North. On the contrary, only a small number of cases have been reported in the Nordic countries in the past decade, with the highest number being reported for Norway (Table 1). The sources of contamination in the Nordic outbreaks included various locally produced fish products (Lindström et al. 2004, 2006b, King et al. 2009), canned garlic of German origin (Krusell and Lohse 2003), or olives of Italian origin (Jalava et al. 2011).

Infant botulism is a consequence of ingested spores of botulinum neurotoxigenic clostridia germinating and growing into neurotoxic cultures in the infantile GI tract with poorly developed intestinal normal bacterial population (Arnon 1986). Due to the requirement of growth from spores *in vivo*, infant botulism is related mainly to the mesophilic Group I *C. botulinum* (Hatheway et al. 1981, Hoffman et al. 1982, Hatheway and McCroskey 1987, Barash and Arnon 2004) or *C. butyricum* (Aureli et al. 1986, McCroskey et al. 1986, Hatheway and McCroskey 1987, Fenicia et al. 2007) and *C. baratii* (Hall et al. 1985, Barash et al. 2005, Keet et al. 2005). The disease generally affects babies under one year. Clinical appearance varies from subclinical to sudden death (Arnon and Chin 1979), with case-fatality rate being less than 2% (Centers for Disease Control and Prevention 1998). Infant botulism is the most common form of botulism in the USA. While infant botulism has been reported to arise from foods contaminated with spores of neurotoxigenic clostridia (Brett et al. 2005), honey being one of the most frequently discussed potential source (Nevas et al. 2002, 2006), household dust is probably another important source of spores (Arnon 1992, Nevas et al. 2005).

Other forms of botulism include wound botulism, a consequence of spore germination into toxic culture in deep wounds (Merson and Dowell 1973), and adult infectious botulism which is etiologically analogous to infant botulism. While wound botulism is increasingly reported in intravenous drug users as a consequence of poor hygiene, adult infectious botulism is extremely rare (Fenicia et al. 2007, Sheppard et al. 2012). A potentially emerging form of botulism is iatrogenic, resulting from therapeutic or cosmetic use of botulinum neurotoxin (Marcus 2009). A particular threat is introduced through availability of counterfeit neurotoxin products with poorly characterized toxin potency (Pickett and Mewies 2009).

Diagnosis of botulism is primarily clinical, which may be challenging due to the extremely rare nature of the disease. While descending flaccid paralysis of variable degree is always present, the different forms of botulism may manifest with a range of other symptoms, such as vomiting and constipation in the food-borne form or abscess formation in wound botulism. The clinical diagnosis is confirmed by the laboratory detection of botulinum neurotoxin in the patient's serum and GI contents, in the suspected source. Detection and isolation of *C. botulinum* strain in the sample materials supports the laboratory diagnosis and should always be aimed at a better understanding of the epidemiology of the disease (Lindström and Korkeala 2006). Laboratory diagnosis is often hampered by delayed sampling after administration of specific and supportive medication, and by the lack of selective microbiological media or reliable *in vitro* neurotoxin tests.

Treatment of all forms of botulism involves administration of neutralizing antitoxin and supportive care. It is essential that antitoxin is administered at an early stage of the disease; only circulating toxin can be neutralized while neurotoxin bound to presynaptic membranes or internalized in motor nerves cannot. Equine serum based antitoxins used in adults have occasionally been reported to cause anaphylaxis (Black and Gunn 1980), thus a skin test prior to use is advised. Due to the pitfalls related to the available antitoxins, active research aimed at novel therapeutics of botulism is ongoing (Capková et al. 2009, Li et al. 2010). Supportive care is primarily targeted to maintain respiration and control of secondary infections, but measures to control blood pressure and body weight may also be required, particularly in severe cases requiring long-term intensive care (Marcus 2009). Debridement of wounds and antimicrobial treatment are often required in cases of wound botulism.

Risk food

Food-borne botulism can be caused by any non-sterilized low-acid food with a pH above 4.6–5.0 and water activity above 0.94–0.97. Prerequisites for intoxication include spore contamination of raw food materials; mild heat treatment allowing spores to survive; conditions allowing spore germination and outgrowth into toxic culture, such as negative redox potential and abused storage temperature; and consumption of food without reheating (Lindström et al. 2006a). As a consequence of frequent occurrence of *C. botulinum* spores in the environment, most raw food materials can be contaminated and should be considered as potential vehicles of *C. botulinum* spores. The prevalence of *C. botulinum* in most raw food materials is in the range of 1 to 10%, but also as high as 40% prevalence has been reported for certain fish species (Hyytiä et al. 1998). The reported contamination levels are usually below 1,000 spores/kg of raw material (Lindström et al. 2009).

To eliminate *C. botulinum* spores by heat treatment, generally a sterilization process at >100°C is required for Group I spores, while pasteurization processes at 70 to 90°C suffice for Group II spores. To control the risk of Group I spores in the canning industry, processes are often targeted to eliminate the risk by a factor of 10^{12} (12D process, 'botulinum cook'). Such products are considered to be commercially sterile and can be stored at room temperature for months or years. Problems arise from failure to control the time and temperature during processing, a particular problem in home canning, and leaving viable spores in the products. Bulging of lids indicates gas formation and thus outgrowth from survived spores, and should be a warning sign of increased botulism risk related to the entire batch of cans. Psychrotrophic Group II spores form a particular challenge in minimally heated refrigerated packaged foods of extended durability (REPFED), such as vacuum-packaged hot-smoked fish and most ready-to-eat meals. Recommendations to target heat treatments of REPFED products to control the risk posed by Group II spores by a factor of 10^6 (6D process) have been set (Food Standards Agency 2008) but are often ignored in favor of sensory properties, particularly in hot-smoking of fish (Lindström et al. 2003). Consequently, vacuum-packaged hot-smoked fish products are frequently positive for *C. botulinum* spores (Hyytiä et al. 1997, Korkeala et al. 1998, Lindström et al. 2006b, King et al. 2009).

Spore germination and outgrowth into toxic culture is prevented by combined effect of intrinsic (salt and preservatives acting towards a decreased water activity, redox potential, and low pH) and extrinsic factors (storage temperature and time), depending on product type (Food Standards Agency 2008). While the safety of traditional low-acid canned foods relies merely on the heat treatment killing practically all *C. botulinum* spores, control of Group II spores in the complex variety of ready meals, their components and compartments is a challenging task, particularly as the maximum storage temperatures of such foods can be up to 6–8°C. The use of salt and preservatives would be effective (Hyytiä et al. 1997, Keto-Timonen et al. 2012) but is practically limited due to consumer trends towards natural and healthy foods. The cold chain is challenged by the increasing distribution chain, extended storage times and uneducated consumer behavior.

Botulinum neurotoxin is a heat-labile toxin and can be eliminated by heating at 85°C for 5 min (Siegel 1993). However, not all REPFEDs are reheated and those intended for consumption as hot meals are often advised to be reheated in a microwave oven. Microwave heating does not ensure even distribution of heat, thus a brief heating may be insufficient to eliminate the neurotoxin in all parts of the product. The neurotoxin is odorless and tasteless and thereby does not possess warning signs to the consumer. Moreover, while the proteolytic *C. botulinum* Group I strains produce off-odors and off-flavors during their growth causing spoilage of the product and thus rejection by the consumer, Group II strains are nonproteolytic and thereby often unassociated with sensory spoilage of the product. Therefore the sensory quality of a lethal food product may be flawless.

Considering the theoretically high botulism risk related to REPFEDs, the number of botulism outbreaks caused by these products is surprisingly small. This suggests that the *C. botulinum* contamination level in many food raw materials is low, the hurdle concept in control of spore survival, germination and outgrowth is mainly efficient, and/or food-borne botulism and cases with mild symptoms are under-diagnosed.

Most frequently reported vehicles for botulism include home-prepared foods, such as cured meats, canned vegetables, and fermented fish products (Hauschild and Dodds 1993). Canned vegetables typically involve Group I *C. botulinum*, while Group II organisms are related to marine foods or preserved meat with mild heat processing (Lindström et al. 2006a). Of particular relevance to Northern Europe are vacuum-packaged hot-smoked fish products which have caused many outbreaks. Outbreaks due to dairy produce are limited in number (Lindström et al. 2010). Commercial foods are increasingly involved in food-borne botulism and such outbreaks may be large and cause significant economic losses to the food industry (O'Mahony et al. 1990, Weber et al. 1993, Korkeala et al. 1998, Aureli et al. 2000, Lindström et al. 2006b, Sheth et al. 2008, Jalava et al. 2011). An example of a large type E outbreak due to commercial product faseikh (uneviscerated, salted mullet fish) affected 91 people, with 18 dying, in Cairo in 1991 (Weber et al. 1993). The only foodstuffs associated with infant botulism are honey (Aureli et al. 2002), which often carries high numbers of *C. botulinum* spores (Arnon 1992, Hauschild and Dodds 1993, Nevas et al. 2002, 2006) and contaminated infant milk powder (Brett et al. 2005).

Concluding Remarks

Although botulism and *C. botulinum* have been known for more than a century, the ever-growing selection of food products associated with botulism outbreaks and the different manifestations of the disease are an important reminder of the emerging nature of the risk this pathogen possesses in modern society. Development of novel control strategies requires a thorough understanding of the physiology of *C. botulinum* strains at cellular and genetic levels. This has been hampered by the lack of suitable genetic manipulation tools until the very recent years when a series of gene knockout and over-expression tools became available for *Clostridia* (Heap et al. 2007, 2009). First reports on genetic mechanisms in *C. botulinum* focused on neurotoxin production (Bradshaw et al. 2010, Cooksley et al. 2010, Marshall et al. 2010, Zhang et al. 2013), response mechanisms at low (Söderholm et al. 2011, Lindström et al. 2012) and high temperature (Selby et al. 2011), and sporulation (Kirk et al. 2012).

Enterotoxigenic *Clostridium perfringens* type A

The microorganism

C. perfringens food poisoning is caused by *C. perfringens* enterotoxin (CPE) encoded by *cpe* and produced by less than 5% of *C. perfringens* type A strains (Smedley et al. 2004). *C. perfringens* strains are divided into five types, A to E, according to the major toxins (include alpha, beta, epsilon and iota toxins) they produce. Most *C. perfringens* strains also produce a range of other toxins or potential virulence factors, such as the CPE. While all toxin types have been reported to be virulent, type A and C strains alone have been associated with human disease. Those associated with CPE include food poisoning (Knox and Mac Donald 1943), antibiotic-associated diarrhea (AAD, Borriello et al. 1984), sporadic diarrhea (SD, Brett et al. 1992), and sudden infant death syndrome (SIDS, Murrell et al. 1987).

C. perfringens strains are generally able to grow at the temperature range of 15–54°C (de Jong et al. 2004, Juneja et al. 2006), the optimum being above 40°C. Growth is very rapid at optimal conditions with the generation time being as short as 7 min. Optimal conditions for *C. perfringens* food poisonings arise when food contaminated with *cpe*-positive *C. perfringens* spores is slowly chilled, or held or served at a temperature range of 20 to 50°C, allowing germination and rapid growth of *C. perfringens* (Li and McClane 2006b). After ingestion of large numbers of vegetative *C. perfringens* cells, they sporulate in the intestinal lumen and produce CPE (McClane 2001).

The enterotoxin gene can be located in the bacterial chromosome or on large plasmids (Brynestad et al. 1997, Miyamoto et al. 2002, Cornillot et al. 1995), with the genetic arrangement in the chromosomal and plasmid-borne *cpe* loci being different (Brynestad et al. 1997). Novel *cpe* genotypes associated with unidentified genetic structures have also been identified (Heikinheimo et al. 2006, Li et al. 2007). The *cpe* expression is linked to sporulation (Harry et al. 2009). In the human ileum, CPE acts as a pore-forming toxin causing morphological damage and loss of transport (Fernández Miyakawa et al. 2005). Depending on the amount of CPE, cellular levels of calcium may increase rapidly and the cells die via either apoptotic or oncotic pathway (Smedley and McClane 2004). The condition manifests with abdominal cramps and heavy diarrhea (Skjelkvåle and Uemura 1977).

Genetic analysis of *cpe*-positive *C. perfringens* strains reveals that chromosomal *cpe*-carrying strains form a genetically homogeneous cluster distinct from other *C. perfringens* strains (Keto-Timonen et al. 2006, Deguchi et al. 2009). In contrast, the plasmid-borne *cpe*-positive strains are diverse and share genetic similarities with *cpe*-negative *C. perfringens* strains (Keto-Timonen et al. 2006, Deguchi et al. 2009). Considering the different genetics, the range of diseases they cause, and different resistance to environmental factors, it is evident that the chromosomal and plasmid-borne *cpe*-carrying *C. perfringens* type A strains are epidemiologically distinct. A comparative genomic hybridization analysis with DNA microarrays addressed partly distinct energy, fatty acid and phospholipid metabolism between chromosomal and plasmid-borne *cpe* genotypes, suggesting that the plasmid-borne *cpe*-carrying strains are associated with intestinal environment, whereas the chromosomal *cpe* genotypes, although present in small numbers in the gut, are adapted to other environments (Lahti et al. 2012).

The patient

Abdominal cramps and diarrhea appear within 8–12 hr of ingestion of *C. perfringens* cells, followed by recovery within 24 hrs. Fatalities are rare but possible in elderly or debilitated humans (Smith 1998, Bos et al. 2005). The chromosomal *cpe*-positive strains alone were long associated with food poisonings (Sparks et al. 2001). This was explained by the greater stress resistance of the chromosomal *cpe*-positive strains than plasmid-borne *cpe*-positive strains (Sarker et al. 2000, Li and McClane 2006a, b, Grant et al. 2008). However, recent research has confirmed plasmid-borne *cpe*-positive strains are also present in foods (Tanaka et al. 2003, 2007, Grant et al. 2008, Lahti et al. 2008a) and cause food poisoning (Lahti et al. 2008a).

Together with campylobacteria, salmonellae, and noroviruses, *C. perfringens* type A is among the most common causes of food-borne illness in the United States (Scallan et al. 2011). *C. perfringens* food poisonings became common after World War II upon establishment of mass catering. Still today, *C. perfringens* food poisonings are mainly associated with mass catering due to challenges in rapid chilling or heating of large amounts of food. Due to the relatively mild and self-limiting symptoms, *C. perfringens* food poisonings are probably highly under-diagnosed (Table 2). Most outbreaks have been associated with foods of animal origin.

In addition to food poisonings with acute onset and implicated source, *cpe*-positive *C. perfringens* type A has also been reported to cause up to 15% of AAD and sporadic diarrhea SD cases in humans (Modi and Wilcox 2001, Asha and Wilcox 2002). These conditions have been considered to be caused by plasmid-borne *cpe*-positive *C. perfringens* and non-food-related (Collie et al. 1998, McClane and Chakrabarti 2004, Raju and Sarker 2005). However, based on current knowledge on plasmid-borne *cpe*-positive *C. perfringens* strains being able to cause food poisonings (Lahti et al. 2008), we suggest that some cases of CPE-mediated AAD and SD may also be transmitted through food (Lindström et al. 2011).

Table 2. *Clostridium perfringens* food poisonings (number of outbreaks/cases) in the Nordic countries in 2003–2009. Data for Iceland are not available.

Yr	Denmark	Finland	Norway	Sweden
2003	NR[a]	2/20	NR	3/81
2004	NR	2/63	NR	1/19
2005	4/109	2/29	NR	7/214
2006	5/134	0	NR	1/20
2007	5/166	2/NR	5/47	1/17
2008	1/2	1/NR	4/8	0
2009	NR/66	4/91	1/33	NR

[a]NR, not reported.

Risk food

The CPE-mediated food poisoning outbreaks typically involve a large number of victims and are associated with temperature-abused meat or poultry dishes. Thus meat animals have been considered to be the main reservoir of *cpe*-positive *C. perfringens*. However, a Finnish study found a large number of *cpe*-positive *C. perfringens* in the GI tract of healthy humans, suggesting that food chain workers serve as a reservoir for *cpe*-positive *C. perfringens* and therefore play a role in contamination of foods (Heikinheimo et al. 2006). Unsuccessful attempts to isolate *cpe*-positive *C. perfringens* in animal fecal samples support this theory (Lindström et al. 2011). Thus it appears that any foods supporting the growth of *C. perfringens* can act as vehicles for food poisoning.

Concluding Remarks

Although self-limiting and hardly fatal, *C. perfringens* food poisoning is a very common illness in industrialized countries all over the world. Recent research has proposed several novel characteristics in the epidemiology of *C. perfringens* food poisoning. Both chromosomal and plasmid-borne *cpe*-positive *C. perfringens* type A strains cause food poisonings, thus also part of AAD and SD caused by plasmid-borne *cpe*-positive strains may be transmitted through foods. The chromosomal and plasmid-borne *cpe*-positive strains are different in their genetic background and physiology, thus they are likely to reside in different, as-yet poorly understood reservoirs. Humans serve as a rich reservoir of *cpe*-positive *C. perfringens* and may thus play a greater role in the epidemiology of *C. perfringens* food poisoning than previously understood. Further research is warranted to address these questions.

Clostridium difficile—Emerging Food Pathogen?

Apart from soil and aquatic environments, *C. difficile* is frequently found in hospital environments and hospitalized patients, and in the GI tract of healthy animals and humans (Costa et al. 2012, Janezic et al. 2012). In both animals and humans *C. difficile* strains may colonize the GI tract if the gut microbes are disturbed, e.g., by antimicrobial treatments, and cause a potentially life-threatening diarrhea through production of two major toxins, the enterotoxin TcdA and the cytotoxin TcdB (Burdon et al. 1981, Alfa et al. 2000). Both toxins are glucosyltransferases that target and inactivate the Rho family of GTPases (Voth and Ballard 2005).

Since the beginning of the millennium, *C. difficile* has emerged as the major cause of healthcare-associated infections. An average of 5,000

C. difficile-associated disease (CDAD) cases per yr were reported in Finland in 2008–2011, while the disease was not reported prior to 2007. *C. difficile* causes an increasing number of deaths among hospitalized patients with prolonged or wide-spectrum antibiotic treatment, elderly, and immune compromised individuals. Of particular concern is the emergence of strains representing the PCR ribotype 027, which produce significantly increased amounts of toxins and are resistant to fluoroquinolones (Kuijper et al. 2006). These strains have been detected in the past 5–10 yr in most European countries. Hospital-acquired infection may be spread through patients and healthcare staff or through hospital air (Roberts et al. 2008).

Apart from hospital origin, CDAD cases are increasingly reported among outpatients and persons with no apparent healthcare contacts or predisposing factors (Gould and Limbago 2010). Therefore the question arises whether CDAD is a zoonosis and if food could serve as a vehicle for virulent *C. difficile* strains. Recent research suggests a high prevalence of *C. difficile* in meat animals but also in various meats and meat products in the USA and Canada, although the prevalence in Europe seems to be relatively low (Table 3), perhaps due to different slaughter technique (Rupnik 2010) or diagnostics. Further concern arises from the presence of PCR ribotype

Table 3. Prevalence of *Clostridium difficile* in foods.

Product	Country	No. of positive samples/No. of samples studied (%)	PCR ribotype 027 present	Reference
Ground beef	USA	13/26 (50)	Yes	Songer et al. 2009
	Canada	11/53 (21)	No	Rodriguez-Palacios et al. 2007
	Canada	10/149 (7)	No	Rodriguez-Palacios et al. 2009
	Canada	14/115 (12)	Yes	Weese et al. 2009
	France	2/105 (2)	No	Bouttier et al. 2010
	Austria	3/100 (3)	No	Jöbstl et al. 2010
Ground veal	Canada	1/7 (14)	No	Rodriguez-Palacios et al. 2007
Ground pork	USA	3/7 (43)	Yes	Songer et al. 2009
	Canada	14/115 (12)	Yes	Weese et al. 2009
Ground turkey	USA	4/9 (44)	No	Songer et al. 2009
Various ground meats	Sweden	2/82 (2)	NR[a]	Von Abercron et al. 2009
Veal chops	Canada	3/65 (5)	No	Rodriguez-Palacios et al. 2009
Sausages	USA	17/46 (37)	Yes	Songer et al. 2009
Salad	UK	3/40 (8)	No	Bakri et al. 2009
Vegetables	UK	7/300 (2)	NR	al Saif and Brazier 1996
Raw milk	Austria	0/50	No	Jöbstl et al. 2010

[a]NR, not reported.

027 isolates in foods (Table 3), as well as from significant overlap of genotypes being detected from food animals, foods and humans (Gould and Limbago 2010, Janezic et al. 2012). Although no evidence of food-borne CDAD infection is available, it seems plausible that exposure to *C. difficile* through food may explain at least part of the community-acquired CDAD cases (Rupnik 2007, Weese et al. 2009). More research efforts are required to reveal the role of other factors, such as the household environment and pet animals, in transmission of community-acquired CDAD, infectious dose, and predisposing factors other than healthcare and age-related.

References

al Saif, N. and J.S. Brazier. 1996. The distribution of *Clostridium difficile* in the environment of South Wales. J. Med. Microbiol. 45: 133–137.

Alfa, M.J., A. Kabani, D. Lyerly, S. Moncrief, L.M. Neville, A. Al-Barrak, G.K. Harding, B. Dyck, K. Olekson and J.M. Embil. 2000. Characterization of a toxin A-negative, toxin B-positive strain of *Clostridium difficile* responsible for a nosocomial outbreak of *Clostridium difficile*-associated diarrhea. J. Clin. Microbiol. 38: 2706–2714.

Arnon, S.S. 1986. Infant botulism: anticipating the second decade. J. Infect. Dis. 154: 201–206.

Arnon, S.S. 1992. Infant botulism. *In:* R.D. Feigen and J.D. Gerry (eds.). Textbook of Pediatric Infectious Diseases. Saunders Press, Philadelphia, USA. pp. 1095–1102.

Arnon, S.S. and J. Chin. 1979. The clinical spectrum of infant botulism. Rev. Infect. Dis. 1: 614–624.

Arnon, S.S., R. Schechter, T.V. Inglesby, D.A. Henderson, J.G. Bartlett, M.S. Ascher, E. Eitzen, A.D. Fine, J. Hauer, M. Layton, S. Lillibridge, M.T. Osterholm, T. O'Toole, G. Parker, T.M. Perl, P.K. Russell, D.L. Swerdlow, K. Tonat and Working Group on Civilian Biodefense. 2001. Botulinum toxin as a biological weapon: medical and public health management. Review. J. Am. Med. Assoc. 285: 1059–70.

Asha, N.J. and M.H. Wilcox. 2002. Laboratory diagnosis of *Clostridium perfringens* antibiotic-associated diarrhoea. J. Med. Microbiol. 51: 891–894.

Aureli, P., L. Fenicia, B. Pasolini, M. Gianfranceschi, L.M. McCroskey and C.L. Hatheway. 1986. Two cases of type E infant botulism caused by neurotoxigenic *Clostridium butyricum* in Italy. J. Infect. Dis. 154: 201–206.

Aureli, P., G. Franciosa and L. Fenicia. 2002. Infant botulism and honey in Europe: a commentary. Pediatr. Infect. Dis. J. 21: 866–868.

Aureli, P., M. Di Cunto, A. Maffei, G. De Chiara, G. Franciosa, L. Accorinti, A.M. Gambardella and D. Greco. 2000. An outbreak in Italy of botulism associated with a dessert made with mascarpone cream cheese. Eur. J. Epidemiol. 16: 913–918.

Bakri, M.M., D.J. Brown, J.P. Butcher and A.D. Sutherland. 2009. *Clostridium difficile* in ready-to-eat salads, Scotland. Emerg. Infect. Dis. 15: 817–818.

Barash, J.R. and S.S. Arnon. 2004. Dual toxin-producing strain of *Clostridium botulinum* type Bf isolated from a California patient with infant botulism. J. Clin. Microbiol. 42: 1713–1715.

Barash, J.R., T.W. Tang and S.S. Arnon. 2005. First case of infant botulism caused by *Clostridium baratii* type F in California. J. Clin. Microbiol. 43: 4280–4282.

Berman, B., L. Seeberger and R. Kumar. 2005. Long-term safety, efficacy, dosing, and development of resistance with botulinum toxin type B in cervical dystonia. Mov. Disord. 20: 233–237.

Bilusic, M., J. Pattathil, M. Brescia, W. McHugh, M. Zaboski and B. Schanzer. 2008. Recurrent bulbar paralysis caused by botulinum toxin type B. Clin. Infect. Dis. 46:e72–74.

Black, R.E. and R.A. Gunn. 1980. Hypersensitivity reactions associated with botulinal antitoxin. Am. J. Med. 69: 567–570.

Borriello, S.P., H.E. Larson, A.R. Welch, F. Barclay, M.F. Stringer and B.A. Bartholomew. 1984. Enterotoxigenic *Clostridium perfringens*: A possible cause of antibiotic-associated diarrhoea. Lancet 1: 305–307.

Bos, J., L. Smithee, B. McClane, R.F. Distefano, F. Uzal, J.G. Songer, S. Mallonee and J.M. Crutcher. 2005. Fatal necrotizing colitis following a foodborne outbreak of enterotoxigenic *Clostridium perfringens* type A infection. Clin. Infect. Dis. 40: e78–83.

Bouttier, S., M.C. Barc, B. Felix, S. Lambert, A. Collignon and F. Barbut. 2010. *Clostridium difficile* in ground meat, France. Emerg. Infect. Dis. 16: 733–735.

Bradshaw, M., K.M. Marshall, J.T. Heap, W.H. Tepp, N.P. Minton and E.A. Johnson. 2010. Construction of a nontoxigenic *Clostridium botulinum* strain for food challenge studies. Appl. Environ. Microbiol. 76: 387–393.

Bradshaw, M., S.S. Dineen, N.D. Maks and E.A. Johnson. 2004. Regulation of neurotoxin complex expression in *Clostridium botulinum* strains 62A, Hall A-hyper and NCTC 2916. Anaerobe 10: 321–333.

Brett, M.M., J. McLauchlin, A. Harris, S. O'Brien, N. Black, R.J. Forsyth, D. Roberts and F.J. Bolton. 2005. A case of infant botulism with a possible link to infant formula milk powder: evidence for the presence of more than one strain of *Clostridium botulinum* in clinical specimens and food. J. Med. Microbiol. 54: 769–776.

Brett, M.M., J.C. Rodhouse, T.J. Donovan, G.M. Tebbutt and D.N. Hutchinson. 1992. Detection of *Clostridium perfringens* and its enterotoxin in cases of sporadic diarrhoea. J. Clin. Pathol. 45: 609–611.

Brynestad, S., B. Synstad and P.E. Granum. 1997. The *Clostridium perfringens* enterotoxin gene is on a transposable element in type A human food poisoning strains. Microbiol. 143: 2109–2115.

Burdon, D.W., R.H. George, G.A. Mogg, Y. Arabi, H. Thompson, M. Johnson, J. Alexander-Williams and M.R. Keighley. 1981. Faecal toxin and severity of antibiotic-associated pseudomembranous colitis. J. Clin. Pathol. 34: 548–551.

Capková, K., N.T. Salzameda and K.D. Janda. 2009. Investigations into small molecule non-peptidic inhibitors of the botulinum neurotoxins. Review. Toxicon 54: 575–582.

Carter, A.T., D.R. Mason, K.A. Grant, G. Franciosa, P. Aureli and M.W. Peck. 2010. Further characterization of proteolytic *Clostridium botulinum* type A5 reveals that neurotoxin formation is unaffected by loss of the *cntR* (*botR*) promoter sigma factor binding site. J. Clin. Microbiol. 48: 1012–1013.

Centers for Disease Control and Prevention. 1998. Botulism in the United States 1899–1996. Handbook for Epidemiologists, Clinicians, and Laboratory Workers U.S. Department of Health, Education, and Welfare, CDC, Atlanta, GA.

Chen, Y., H. Korkeala, J. Aarnikunnas and M. Lindström. 2007. Sequencing the botulinum neurotoxin and related genes in *Clostridium botulinum* type E strains reveals *orfx3* and a novel type E neurotoxin subtype. J. Bacteriol. 189: 8643–8650.

Chen, Y., H. Korkeala, J. Lindén and M. Lindström. 2008. Quantitative real-time reverse transcription-PCR analysis reveals stable and prolonged neurotoxin cluster gene activity in a *Clostridium botulinum* type E strain at refrigeration temperature. Appl. Environ. Microbiol. 74: 6132–6137.

Collie, R.E., J.F. Kokai-Kun and B.A. McClane. 1998. Phenotypic characterization of enterotoxigenic *Clostridium perfringens* isolates from non-foodborne human gastrointestinal diseases. Anaerobe 4: 69–79.

Cornillot, E., B. Saintjoanis, G. Daube, S. Katayama, P.E. Granum, B. Canard and S.T. Cole. 1995. The enterotoxin gene *cpe* of *Clostridium perfringens* can be chromosomal or plasmid-borne. Mol. Microbiol. 15: 639–647.

Cooksley, C.M., I.J. Davis, K. Winzer, W.C. Chan, M.W. Peck and N.P. Minton. 2010. Regulation of neurotoxin production and sporulation by a Putative *agrBD* signaling system in proteolytic *Clostridium botulinum*. Appl. Environ. Microbiol. 76: 4448–4460.

Costa, M.C., R. Reid-Smith, S. Gow, S.J. Hannon, C. Booker, J. Rousseau, K.M. Benedict, P.S. Morley and J.S. Weese. 2012. Prevalence and molecular characterization of *Clostridium difficile* isolated from feedlot beef cattle upon arrival and mid-feeding period. BMC Vet. Res. 8: 38.

Dahlenborg, M., E. Borch and P. Rådström. 2001. Development of a combined selection and enrichment PCR procedure for *Clostridium botulinum* types B, E, and F and its use to determine prevalence in faecal samples from slaughtered pigs. Appl. Environ. Microbiol. 67: 4781–4788.

Das Gupta, B.R. and H. Sugiyama. 1972. Role of a protease in natural activation of *Clostridium botulinum* neurotoxin. Infect. Immun. 6: 587–590.

de Jong, A.E., F.M. Rombouts and R.R. Beumer. 2004. Behavior of *Clostridium perfringens* at low temperatures. Int. J. Food Microbiol. 97: 71–80.

Deguchi, A., K. Miyamoto, T. Kuwahara, Y. Miki, I. Kaneko, J. Li, B.A. McClane and S. Akimoto. 2009. Genetic characterization of type A enterotoxigenic *Clostridium perfringens* strains. PLoS One 4: e5598.

Derman, Y., M. Lindström, K. Selby and H. Korkeala. 2011. Growth of Group II *Clostridium botulinum* strains at extreme temperatures. J. Food Protect. 74: 1797–1804.

Dover, N., J.R. Barash and S.S. Arnon. 2009. Novel *Clostridium botulinum* toxin gene arrangement with subtype A5 and partial subtype B3 botulinum neurotoxin genes. J. Clin. Microbiol. 47: 2349–2350.

Dover, N., J.R. Barash, K.K. Hill, K.W. Davenport, H. Teshima, G. Xie and S.S. Arnon. 2013. *Clostridium botulinum* strain Af84 contains three neurotoxin gene clusters: *Bont/A2, bont/F4* and *bont/F5*. PLoS One 8(4): e61205. doi:10.1371/journal.pone.0061205.

Duff, J.T., G.G. Wright and A. Yarinsky. 1956. Activation of *Clostridium botulinum* type E toxin by trypsin. J. Bacteriol. 72: 455–460.

Eklund, M.W. and F. Poysky. 1974. Interconversion of type C and D strains of *Clostridium botulinum* by specific bacteriophages. Appl. Microbiol. 27: 251–258.

Eklund, M.W., F.T. Poysky and S.M. Reed. 1972. Bacteriophage and the toxigenicity of *Clostridium botulinum* type D. Nature. 235: 16–17.

Eklund, M.W., F.T. Poysky and D.I. Wieler. 1967a. Characteristics of *Clostridium botulinum* type F isolated from the Pacific Coast of the United States. Appl. Microbiol. 15: 1316–1323.

Eklund, M.W., F.T. Poysky, L.M. Mseitif and M.S. Strom. 1988. Evidence for plasmid-mediated toxin and bacteriocin production in *Clostridium botulinum* type G. Appl. Environ. Microbiol. 54: 1405–1408.

Eklund, M.W., D.I. Wieler and F.T. Poysky. 1967b. Outgrowth and toxin production of nonproteolytic type B *Clostridium botulinum* at 3.3 to 5.6°C. J. Bacteriol. 93: 1461–1462.

Fenicia, L., F. Anniballi and P. Aureli. 2007. Intestinal toxemia botulism in Italy, 1984–2005. Eur. J. Clin. Microbiol. Infect. Dis. 26: 385–394.

Fernández Miyakawa, M.E., V. Pistone Creydt, F.A. Uzal, B.A. McClane and C. Ibarra. 2005. *Clostridium perfringens* enterotoxin damages the human intestine *in vitro*. Infect. Immun. 73: 8407–8410.

Food Standards Agency. 2008. Food Standards Agency guidance on the safety and shelf-life of vacuum and modified atmosphere packed chilled foods with respect to non-proteolytic *Clostridium botulinum*. Food Standards Agency, United Kingdom. Available at: http://food.gov.uk/multimedia/pdfs/publication/vacpacguide.pdf.

Fujinaga, Y., K. Inoue, S. Watanabe, Y. K.okota, Y. Hirai, E. Nagamachi and K. Oguma. 1997. The haemagglutinin of *Clostridium botulinum* type C progenitor toxin plays an essential role in binding of toxin to the epithelial cells of guinea pig small intestine, leading to the efficient absorption of the toxin. Microbiol. 143: 3841–3847.

Gao, Q.Y., Y.F. Huang, J.G. Wu, H.D. Liu and H.Q. Xia. 1990. A review of botulism in China. Biomed. Environ. Sci. 3: 326–336.

Gould, L.H. and B. Limbago. 2010. *Clostridium difficile* in food and domestic animals: A new foodborne pathogen? Clin. Infect. Dis. 51: 577–582.

Graham, A.F., D.R. Mason, F.J. Maxwell and M.W. Peck. 1997. Effect of pH and NaCl on growth from spores of nonproteolytic Clostridium botulinum at chill temperature. Lett. Appl. Microbiol. 24: 95–100.

Grant, K.A., S. Kenyon, I. Nwafor, J. Plowman, C. Ohai, R. Halford-Maw, M.W. Peck and J. McLauchlin. 2008. The identification and characterization of *Clostridium perfringens* by real-time PCR, location of enterotoxin gene, and heat resistance. Foodborne Pathog. Dis. 5: 629–639.

Gu, S., S. Rumpel, J. Zhou, J. Strotmeier, H. Bigalke, K. Perry, C.B. Shoemaker, A. Rummel and R. Jin. 2012. Botulinum neurotoxin is shielded by NTNHA in an interlocked complex. Science 335: 977–981.

Gupta, A., C.J. Sumner, M. Castor, S. Maslanka and J. Sobel. 2005. Adult botulism type F in the United States, 1981–2002. Neurol. 65: 1694–1700.

Hall, J.D., L.M. McCroskey, B.J. Pincomb and C.L. Hatheway. 1985. Isolation of an organism resembling *Clostridium barati* which produces type F botulinal toxin from an infant with botulism. J. Clin. Microbiol. 21: 654–655.

Harry, K.H., R. Zhou L. Kroos and S.B. Melville. 2009. Sporulation and enterotoxin (CPE) synthesis are controlled by the sporulation-specific sigma factors SigE and SigK in *Clostridium perfringens*. J Bacteriol. 2009: 2728–2742.

Harvey, S.M., J. Sturgeon and D.E. Dassey. 2002. Botulism due to *Clostridium baratii* type F toxin. J. Clin. Microbiol. 40: 2260–2262.

Hatheway, C.L. 1995. Botulism: The present status of the disease. Curr. Top. Microbiol. Immunol. 195: 55–75.

Hatheway, C.L. and L.M. McCroskey. 1987. Examination of feces and serum for diagnosis of infant botulism in 336 patients. J. Clin. Microbiol. 25: 2334–2338.

Hatheway, C.L., L.M. McCroskey, G.L. Lombard and V.R. Dowell Jr. 1981. Atypical toxin variant of *Clostridium botulinum* type B associated with infant botulism. J. Clin. Microbiol. 14: 607–611.

Hauschild, A.H.W. and K.L. Dodds. 1993. *Clostridium botulinum*. Ecology and control in foods. Marcel Dekker, New York, USA.

Heap, J.T., O.J. Pennington, S.T. Cartman, G.P. Carter and N.P. Minton. 2007. The ClosTron: a universal gene knock-out system for the genus *Clostridium*. J. Microbiol. Methods. 70: 452–464.

Heap, J.T., O.J. Pennington, S.T. Cartman and N.P. Minton. 2009. A modular system for *Clostridium* shuttle plasmids. J. Microbiol. Methods. 78: 79–85.

Heikinheimo, A., M. Lindström, P.E. Granum and H. Korkeala. 2006. Humans as reservoir for enterotoxin gene-carrying *Clostridium perfringens* type A. Emerg. Infect. Dis. 12: 1724–1729.

Hielm, S., E. Hyytiä, A.B. Andersin and H. Korkeala. 1998. A high prevalence of *Clostridium botulinum* type E in Finnish freshwater and Baltic Sea sediment samples. J. Appl. Microbiol. 84: 133–137.

Hill, K.K., T.J. Smith, C.H. Helma, L.O. Ticknor, B.T. Foley, R.T. Svensson, J.L. Brown, E.A. Johnson, L.A. Smith, R.T. Okinaka, P.J. Jackson and J.D. Marks. 2007. Genetic diversity among botulinum neurotoxin-producing clostridial strains. J. Bacteriol. 189: 818–832.

Hinderink, K., M. Lindström and H. Korkeala. 2009. Group I *Clostridium botulinum* strains show significant variation in growth at low and high temperatures. J. Food Protect. 72: 375–383.

Hoffman, R.E., B.J. Pincomb, M.R. Skeels and M.J. Burkhart. 1982. Type F infant botulism. Am. J. Dis. Child. 136: 270–271.

Hutson, R.A., D.E. Thompson and M.D. Collins. 1993. Genetic interrelationships of saccharolytic *Clostridium botulinum* types B, E and F and related clostridia as revealed by small-subunit rRNA gene sequences. FEMS Microbiol. Lett. 108: 103–110.

Hyytiä, E., S. Eerola, S. Hielm and H. Korkeala. 1997. Sodium nitrite and potassium nitrate in control of nonproteolytic *Clostridium botulinum* outgrowth and toxigenesis in vacuum-packed cold-smoked rainbow trout. Int. J. Food Microbiol. 37: 63–72.

Hyytiä, E., S. Hielm and H. Korkeala. 1998. Prevalence of *Clostridium botulinum* type E in Finnish fish and fishery products. Epidemiol. Infect. 120: 245–250.

Jalava, K., K. Selby, A. Pihlajasaari, E. Kolho, E. Dahlsten, N. Forss, T. Bäcklund, H. Korkeala, T. Honkanen-Buzalski, T. Hulkko, Y. Derman, A. Järvinen, H. Kotilainen, L. Kultanen, P. Ruutu, O. Lyytikäinen and M. Lindström. 2011. Two cases of food-borne botulism in Finland caused by conserved olives, October 2011. Euro Surveill. 16: 20034.

Janezic, S., M. Ocepek, V. Zidaric and M. Rupnik. 2012. *Clostridium difficile* genotypes other than ribotype 078 that are prevalent among human, animal and environmental isolates. BMC Microbiol. 12: 48.

Juneja, V.K., L. Huang and H.H. Thippareddi. 2006. Predictive model for growth of *Clostridium perfringens* in cooked cured pork. Int. J. Food Microbiol. 110: 85–92.

Jöbstl, M., S. Heuberger, A. Indra, R. Nepf, J. Köfer and M. Wagner. 2010. *Clostridium difficile* in raw products of animal origin. Int. J. Food Microbiol. 138: 172–175.

Kalb, S.R., W.I. Santana, I.N. Geren, C. Garcia-Rodriguez, J. Lou, T.J. Smith, J.D. Marks, L.A. Smith, J.L. Pirkle and J.R. Barr. 2011. Extraction and inhibition of enzymatic activity of botulinum neurotoxins /B1, /B2, /B3, /B4, and /B5 by a panel of monoclonal anti-BoNT/B antibodies. BMC Biochem. 12: 58.

Keet, C.A., C.K. Fox, M. Margeta, E. Marco, A.L. Shane, S.J. Dearmond, J.B. Strober and S.P. Miller. 2005. Infant botulism, type F, presenting at 54 hours of life. Pediatr. Neurol. 32: 193–196.

Keto-Timonen, R., A. Heikinheimo, E. Eerola and H. Korkeala. 2006. Identification of *Clostridium* species and DNA fingerprinting of *Clostridium perfringens* by amplified fragment length polymorphism analysis. J. Clin. Microbiol. 44: 4057–4065.

Keto-Timonen, R., M. Lindström, E. Puolanne, M. Niemistö and H. Korkeala. 2012. Inhibition of toxigenesis of group Group II (nonproteolytic) *Clostridium botulinum* type B in meat products by reduced level of nitrite. J. Food Protect. 75: 1346–1349.

King, L. A., T. Niskanen, M. Junnikkala, E. Moilainen, M. Lindström, H. Korkeala, T. Korhonen, M. Popoff, C. Mazuet, H. Callon, N. Pihier, F. Peloux, C. Ichai, H. Quintard, P. Dellamonica, E. Cua, M. Lasfargue, F. Pierre and H. de Valk. 2009. Botulism and hot-smoked whitefish: a family cluster of type E botulism in France, September 2009. Euro Surveill. 14: 193947.

Kirk, D.G., E. Dahlsten, Z. Zhang, H. Korkeala and M. Lindström. 2012. Involvement of *Clostridium botulinum* ATCC 3502 sigma factor K in early-stage sporulation. Appl. Environ. Microbiol. 78: 4590–4596.

Knox, R. and E.K. MacDonald. 1943. Outbreaks of food poisoning in certain Leicester institutions. Med. Off. 69: 21–22.

Kongsaengdao, S., K. Samintarapanya, S. Rusmeechan, A. Wongsa, C. Pothirat, C. Permpikul, S. Pongpakdee, W. Puavilai, P. Kateruttanakul, U. Phengtham, K. Panjapornpon, J. Janma, K. Piyavechviratana, P. Sithinamsuwan, A. Deesomchok, S. Tongyoo, W. Vilaichone, K. Boonyapisit, S. Mayotarn, B. Piya-Isragul, A. Rattanaphon, P. Intalapaporn, P. Dusitanond, P. Harnsomburana, W. Laowittawas, P. Chairangsaris, J. Suwantamee, W. Wongmek, R. Ratanarat, A. Poompichate, H. Panyadilok, N. Sutcharitchan, A. Chuesuwan, P. Oranrigsupau, C. Sutthapas, S. Tanprawate, J. Lorsuwansiri, N. Phattana and Thai Botulism Study Group. 2006. An outbreak of botulism in Thailand: clinical manifestations and management of severe respiratory failure. Clin. Infect. Dis. 43: 1247–1256.

Korkeala, H., G. Stengel, E. Hyytiä, B. Vogelsang, A. Bohl, H. Wihlman, P. Pakkala and S. Hielm. 1998. Type E botulism associated with vacuum-packaged hot-smoked whitefish. Int. J. Food Microbiol. 43: 1–5.

Krusell, L. and N. Lohse. 2003. A case of human botulism in Denmark after consumption of garlic in chilli oil dressing produced in Germany. Euro Surveill. 7: 2163.

Kuijper, E.J., B. Coignard, P. Tüll and ESCMID Study Group for *Clostridium difficile*. 2006. Emergence of *Clostridium difficile*-associated disease in North America and Europe. Clin. Microbiol. Infect. 12: 2–18.

Lahti, P., A. Heikinheimo, T. Johansson and H. Korkeala. 2008. *Clostridium perfringens* type A isolates carrying plasmid-borne enterotoxin gene genotypes IS1151-*cpe* or IS1470-like-*cpe* are a common cause of food poisonings. J. Clin. Microbiol. 46: 371–373.

Lahti, P., M. Lindström, P. Somervuo, A. Heikinheimo and H. Korkeala. 2012. Comparative genomic hybridization analysis shows different epidemiology of chromosomal and plasmid-borne *cpe*-carrying *Clostridium perfringens* type A. PLoS One 7(10): e46162. doi:10.1371/journal.pone.0046162.

Li, B., N.P. Peet, M.M. Butler, J.C. Burnett, D.T. Moir and T.L. Bowlin. 2010. Small molecule inhibitors as countermeasures for botulinum neurotoxin intoxication. Molecules. 16: 202–220.

Li, J. and B.A. McClane. 2006a. Further comparison of temperature effects on growth and survival of *Clostridium perfringens* type A isolates carrying a chromosomal or plasmid-borne enterotoxin gene. Appl. Environ. Microbiol. 72: 4561–4568.

Li, J. and B.A. McClane. 2006b. Comparative effects of osmotic, sodium nitrite-induced, and pH-induced stress on growth and survival of *Clostridium perfringens* type A isolates carrying chromosomal or plasmid-borne enterotoxin genes. App. Environ. Microbiol. 72: 7620–7625.

Li, J., K. Miyamoto and B.A. McClane. 2007. Comparison of virulence plasmids among *Clostridium perfringens* type E isolates. Infect. Immun. 75: 1811–1819.

Lindström, M. and H. Korkeala. 2006. Laboratory diagnostics of botulism. Clin. Microbiol. Rev. 19: 298–314.

Lindström, M., E. Dahlsten, H. Söderholm, K. Selby. P. Somervuo, J.T. Heap, N.P. Minton and H. Korkeala. 2012. Involvement of two-component system CBO0366/CBO0365 in the cold shock response and growth of group I (proteolytic) *Clostridium botulinum* ATCC 3502 at low temperatures. Appl. Environ. Microbiol. 78, 13, s. 5466–5470.

Lindström, M., S. Hielm, M. Nevas, S. Tuisku and H. Korkeala. 2004. Proteolytic *Clostridium botulinum* type B in the gastric content of patient with type E botulism due to whitefish eggs. Foodborne Path. Dis. 1: 53–57.

Lindström, M., M. Fredriksson-Ahomaa and H. Korkeala. 2009. Molecular epidemiology of Group I and II *Clostridium botulinum*. *In:* G. Gottschalk and H. Brüggemann (eds.). *Clostridia*. Molecular Biology in the Post-Genomic Era. Caister Academic Press, Norfolk, UK. pp. 103–130.

Lindström, M., A. Heikinheimo, P. Lahti and H. Korkeala. 2011. Novel insights into the epidemiology of *Clostridium perfringens* type A food poisoning. Food Microbiol. 28: 192–198.

Lindström, M., K. Kiviniemi and H. Korkeala. 2006a. Hazard and control of group II non-proteolytic *Clostridium botulinum* in modern food processing. Review. Int. J. Food Microbiol. 108: 92–104.

Lindström, M., J. Myllykoski, S. Sivelä and H. Korkeala. 2010. *Clostridium botulinum* in cattle and dairy products. Crit. Rev. Food Sci. Nutr. 50: 281–304.

Lindström, M., M. Nevas, S. Hielm, L. Lähteenmäki, M.W. Peck and H. Korkeala. 2003. Thermal inactivation of nonproteolytic *Clostridium botulinum* type E spores in model fish media and in vacuum-packaged hot-smoked vacuum-packaged fish products. Appl. Environ. Microbiol. 69: 4029–4035.

Lindström, M., M. Vuorela, K. Hinderink, H. Korkeala, E. Dahlsten, M. Raahenmaa and M. Kuusi. 2006b. Botulism associated with vacuum-packed smoked whitefish in Finland, June–July 2006. Euro Surveill. 11: E060720.3.

Lövenklev, M., E. Holst, E. Borch and P. Rådström. 2004. Relative neurotoxin gene expression in *Clostridium botulinum* type B, determined using quantitative reverse transcription-PCR. Appl. Environ. Microbiol. 70: 2919–2927.

Lynt, R.K., D.A. Kautter and H.M. Solomon. 1981. Heat resistance of proteolytic *Clostridium botulinum* type F in phosphate buffer and crabmeat. J. Food Sci. 47: 204.

Lynt, R.K., D.A. Kautter and H.M. Solomon. 1983. Effect of delayed germination by heat-damaged spores on estimates of heat resistance of *Clostridium botulinum* types E and F. J. Food Sci. 48: 226–229.

Marcus, S.M. 2009. Reflections on the care of a patient severely poisoned by 'rogue' botulinum toxin and rendered paralysed for a protracted hospital stay. The Botulinum J. 1: 318–339.

Marshall, K.M., M. Bradshaw, S. Pellett and E.A. Johnson. 2007. Plasmid encoded neurotoxin genes in *Clostridium botulinum* serotype A subtypes. Biochem. Biophys. Res. Commun. 361: 49–54.

Marshall, K.M., M. Bradshaw and E.A. Johnson. 2010. Conjugative botulinum neurotoxin-encoding plasmids in *Clostridium botulinum*. PLoS One. 5:e11087.

McClane, B.A. 2001. *Clostridium perfringens*. *In:* M.P. Doyle and L.R. Beuchat (eds.). Food Microbiology: Fundamentals and Frontiers, 2nd ed. ASM Press, Washington DC, USA. pp. 351–372.

McClane, B.A. and G. Chakrabarti. 2004. New insights into the cytotoxic mechanisms of *Clostridium perfringens* enterotoxin. Anaerobe. 10: 107–114.

McCroskey, L.M., C.L. Hatheway, L. Fenicia, B. Pasolini and P. Aureli. 1986. Characterization of an organism that produces type E botulinal toxin but which resembles *Clostridium butyricum* from the faeces of an infant with type E botulism. J. Clin. Microbiol. 23: 201–202.

Macdonald, T.E., C.H. Helma, Y. Shou, Y.E. Valdez, L.O. Ticknor, B.T. Foley, S.W. Davis, G.E. Hannett, C.D. Kelly-Cirino, J.R. Barash, S.S. Arnon, M. Lindström, H. Korkeala, L.A. Smith, T.J. Smith and K.K. Hill. 2011. Analysis of *Clostridium botulinum* serotype E strains by using multilocus sequence typing, amplified fragment length polymorphism, variable-number tandem-repeat analysis, and botulinum neurotoxin gene sequencing. Appl. Environ. Microbiol. 77: 8625–8634.

Merson, M.H. and V.R. Dowell Jr. 1973. Epidemiologic, clinical and laboratory aspects of wound botulism. N. Engl. J. Med. 289: 1005–1010.

Miyamoto, K., G. Chakrabarti, Y. Morino and B.A. McClane. 2002. Organization of the plasmid *cpe* locus in *Clostridium perfringens* type A isolates. Infect. Immun. 70: 4261–4272.

Modi, N. and M.H. Wilcox. 2001. Evidence for antibiotic induced *Clostridium perfringens* diarrhoea. J. Clin. Pathol. 54: 748–751.

Murrell, T.G., B.G. Ingham, J.R. Moss and W.B. Taylor. 1987. A hypothesis concerning *Clostridium perfringens* type A enterotoxin CPE and sudden infant death syndrome SIDS. Med. Hypotheses 22: 401–413.

Myllykoski, J., M. Nevas, M. Lindström and H. Korkeala. 2006. The detection and prevalence of *Clostridium botulinum* in pig intestinal samples. Int. J. Food Microbiol. 110: 172–177.

Nevas, M., S. Hielm, M. Lindström, H. Horn, K. Koivulehto and H. Korkeala. 2002. High prevalence of *Clostridium botulinum* types A and B in honey samples detected by polymerase chain reaction. Int. J. Food Microbiol. 72: 45–52.

Nevas, M., M. Lindström, A. Hörman, R. Keto-Timonen and H. Korkeala. 2006. *Clostridium botulinum* in the production environment of honey. Environ. Microbiol. 8: 1085–1094.

Nevas, M., M. Lindström, A. Virtanen, S. Hielm, M. Kuusi, S.S. Arnon, E. Vuori and H. Korkeala. 2005. Infant botulism acquired from household dust presenting as sudden infant death syndrome. J. Clin. Microbiol. 43: 511–513.

O'Mahony, M., E. Mitchell, R.J. Gilbert, D.N. Hutchinson, N.T. Begg, J.C. Rodhouse and J.E. Morris. 1990. An outbreak of foodborne botulism associated with contaminated hazelnut yoghurt. Epidemiol. Infect. 1990: 389–395.

Ohishi, I. and G. Sakaguchi. 1977. Activation of botulinum toxins in the absence of nicking. Infect. Immun. 17: 402–407.

Peck, M.W., D.A. Fairbairn and B.M. Lund. 1993. Heat-resistance of spores of non-proteolytic *Clostridium botulinum* estimated on medium containing lysozyme. Lett. Appl. Microbiol. 16: 126–131.

Pickett, A. and M. Mewies. 2009. Serious issues relating to the clinical use of unlicensed botulinum toxin products. J. Am. Acad. Dermatol. 61: 149–150.

Raju, D. and M.R. Sarker. 2005. Comparison of the levels of heat resistance of wild-type, *cpe* knockout, and *cpe* plasmid-cured *Clostridium perfringens* type A strains. Appl. Environ. Microbiol. 71: 7618–7620.

Roberts K., C.F. Smith, A.M. Snelling, K.G. Kerr, K.R. Banfield, P.A. Sleigh and C.B. Beggs. 2008. Aerial dissemination of *Clostridium difficile* spores. BMC Infect. Dis. 8: 7.

Rodriguez-Palacios, A., R.J. Reid-Smith, H.R. Staempfli, D. Daignault, N. Janecko, B.P. Avery, H. Martin, A.D. Thomspon, L.C. McDonald, B. Limbago and J.S. Weese. 2009. Possible seasonality of *Clostridium difficile* in retail meat, Canada. Emerg. Infect. Dis. 15: 802–805.

Rodriguez-Palacios, A., H.R. Staempfli, T. Duffield and J.S. Weese. 2007. *Clostridium difficile* in retail ground meat, Canada. Emerg. Infect. Dis. 13: 485–487.

Rupnik, M. 2007. Is *Clostridium difficile*-associated infection a potentially zoonotic and foodborne disease? Clin. Microbiol. Infect. 13: 457–459.

Rupnik, M. 2010. *Clostridium difficile*: (Re)emergence of zoonotic potential. Clin. Infect. Dis. 51: 583–584.

Sarker, M.R., R.P. Shivers, S.G. Sparks, V.K. Juneja and B.A. McClane. 2000. Comparative experiments to examine the effects of heating on vegetative cells and spores of *Clostridium perfringens* isolates carrying plasmid genes versus chromosomal enterotoxin genes. Appl. Environ. Microbiol. 66: 3234–3240.

Scallan, E., R.M. Hoekstra, F.J. Angulo, R.V. Tauxe, M.A. Widdowson, S.L. Roy, J.L. Jones and P.M. Griffin. 2011. Foodborne illness acquired in the United States—major pathogens. Emerg. Infect. Dis. 17: 7–15.

Schmidt, C.F. 1964. Spores of *C. botulinum*: Formation, resistance, germination. *In:* K.H. Lewis and K. Cassel Jr. (eds.). Botulism. U. S. Department of Health, Education, and Welfare, Public Health Service, Cincinnati, Ohio, USA. pp. 69–82.

Schmidt, C.F., R.V. Lechowich and J.F. Folinazzo. 1961. Growth and toxin production by type E *Clostridium botulinum* below 40°F. J. Food Sci. 26: 626–630.

Scott, V.N. and D.T. Bernard. 1982. Heat resistance of spores of non-proteolytic type B *Clostridium botulinum*. J. Food Protect. 45: 909–912.

Segner, W.P., C.F. Schmidt and J.K. Boltz. 1966. Effect of sodium chloride and pH on the outgrowth of spores of type E *Clostridium botulinum* at optimal and suboptimal temperatures. Appl. Microbiol. 14: 49–54.

Selby, K., M. Lindström, P. Somervuo, J.T. Heap, N.P. Minton and H. Korkeala. 2011. Important role of class I heat shock genes *hrcA* and *dnaK* in the heat shock response and the response to pH and NaCl stress of group I *Clostridium botulinum* strain ATCC 3502. Appl. Environ. Microbiol. 77: 2823–2830.

Sheth, A.N., P. Wiersma, D. Atrubin, V. Dubey, D. Zink, G. Skinner, F. Doerr, P. Juliao, G. Gonzalez, C. Burnett, C. Drenzek, C. Shuler, J. Austin, A. Ellis, S. Maslanka and J. Sobel. 2008. International outbreak of severe botulism with prolonged toxemia caused by commercial carrot juice. Clin. Infect. Dis. 47: 1245–1251.

Sheppard, Y.D., D. Middleton, Y. Whitfield, F. Tyndel, S. Haider, J. Spiegelman, R.H. Swartz, M.P. Nelder, S.L. Baker, L. Landry, R. Maceachern, S. Deamond, L. Ross, G. Peters, M. Baird, D. Rose, G. Sanders and J.W. Austin. 2012. Intestinal toxemia botulism in 3 adults, Ontario, Canada, 2006–2008. Emerg. Infect. Dis. 18: 1–6.

Siegel, L.S. 1993. Destruction of botulinum toxins in food and water. *In:* A.H.W. Hauschild and K.L. Dodds (eds.). *Clostridium botulinum*. Ecology and control in foods. Marcel Dekker, New York, USA. pp. 323–342.

Skjelkvåle, R. and T. Uemura. 1977. Experimental diarrhea in human volunteers following oral administration of *Clostridium perfringens* enterotoxin. J. Appl. Bacteriol. 43: 281–286.

Smedley, J.G. and B.A. McClane. 2004. Fine mapping of the N-terminal cytotoxicity region of *Clostridium perfringens* enterotoxin by site-directed mutagenesis. Infect. Immun. 72: 6914–6923.

Smedley, J.G., D.J. Fisher, S. Sayeed, G. Chakrabarti and B.A. McClane. 2004. The enteric toxins of *Clostridium perfringens*. Rev. Physiol. Bioch. Pharmacol. 152: 183–204.

Smelt, J.P., G.J. Raatjes, J.S. Crowther and C.T. Verrips. 1982. Growth and toxin formation by *Clostridium botulinum* at low pH values. J. Appl. Bacteriol. 52: 75–82.

Smith, J.L. 1998. Foodborne illness in the elderly. J. Food Protect. 61: 1229–1239.

Smith, T.J., J. Lou, I.N. Geren, C.M. Forsyth, R. Tsai, S.L. Laporte, W.H. Tepp, M. Bradshaw, E.A. Johnson, L.A. Smith and J.D. Marks. 2005. Sequence variation within botulinum neurotoxin serotypes impacts antibody binding and neutralization. Infect. Immun. 73: 5450–5457.

Smith, L.D.S. and H. Sugiyama. 1988. Botulism. The Organism, Its Toxins, the Disease, 2nd ed. Charles C. Thomas Publisher, Springfield, Illinois, USA.

Songer, J.G., H.T. Trinh, G.E. Killgore, A.D. Thompson, L.C. McDonald and B.M. Limbago. 2009. *Clostridium difficile* in retail meat products, USA, 2007. Emerg. Infect. Dis. 15: 819–821.

Sparks, S.G., R.J. Carman, M.R. Sarker and B.A. McClane. 2001. Genotyping of enterotoxigenic *Clostridium perfringens* fecal isolates associated with antibiotic-associated diarrhea and food poisoning in North America. J. Clin. Microbiol. 39: 883–888.

Stumbo, C.R. 1973. Thermobacteriology in food processing, 2nd ed. Academic Press, New York, USA.

Suen, J.C., C.L. Hatheway, A.G. Steigerwalt and D.J. Brenner. 1988. *Clostridium argentinense*, sp. nov: A genetically homogenous group composed of all strains of *Clostridium botulinum* type G and some nontoxigenic strains previously identified as *Clostridium subterminale* or *Clostridium hastiforme*. Int. J. Syst. Bacteriol. 38: 375.

Sugawara, Y. and Y. Fujinaga. 2011. The botulinum toxin complex meets E-cadherin on the way to its destination. Cell Adh. Migr. 5: 34–36.

Söderholm, H., M. Lindström, P. Somervuo, J. Heap, N.P. Minton, J. Lindén and H. Korkeala. 2011. *cspB* encodes a major cold shock protein in *Clostridium botulinum* ATCC 3502. Int. J. Food Microbiol. 146: 23–30.

Tanaka, D., J. Isobe, S. Hosorogi, K. Kimata, M. Shimizu, K. Katori, Y. Gyobu, Y. Nagai, T. Yamagishi, T. Karasawa and S. Nakamura. 2003. An outbreak of food-borne gastroenteritis caused by *Clostridium perfringens* carrying the *cpe* gene on a plasmid. Jpn. J. Infect. Dis. 56: 137–139.

Tanaka, D., K. Kimata, M. Shimizu, J. Isobe, M. Watahiki, T. Karasawa, T. Yamagishi, S. Kuramoto, T. Serikawa, F. Ishiguro, M. Yamada, K. Yamaoka, M. Tokoro, T. Fukao, M. Matsumoto, R. Hiramatsu, C. Monma and Y. Nagai. 2007. Genotyping of *Clostridium perfringens* isolates collected from food poisoning outbreaks and healthy individuals in Japan based on the *cpe* locus. Jpn. J. Infect. Dis. 60: 68–69.

Von Abercron, S.M., F. Karlsson, G.T. Wigh, M. Wierup and K. Krovacek. 2009. Low occurrence of *Clostridium difficile* in retail ground meat in Sweden. J. Food Protect. 72: 1732–1734.

Voth, D.E. and J.D. Ballard. 2005. *Clostridium difficile* toxins: Mechanism of action and role in disease. Clin. Microbiol. Rev. 18: 247–263.

Weber, J.T., R.G. Hibbs Jr., A. Darwish, B. Mishu, A.L. Corwin, M. Rakha, C.L. Hatheway, S. el Sharkawy, S.A. el-Rahim, M.F. al-Hamd, J.E. Sarn, P.A. Blake and R.V. Tauxe. 1993. A massive outbreak of type E botulism associated with traditional salted fish in Cairo. J. Infect. Dis. 167: 451–454.

Weese, J.S., B.P. Avery, J. Rousseau and R.J. Reid-Smith. 2009. Detection and enumeration of *Clostridium difficile* spores in retail beef and pork. Appl. Environ. Microbiol 75: 5009–5011.

Zhang, Z., H. Korkeala, E. Dahlsten, E. Sahala, J.T. Heap, N.P. Minton and M. Lindström. 2013. Two-component signal transduction system CBO0787/CBO0786 represses transcription from botulinum neurotoxin promoters in *Clostridium botulinum* ATCC 3502. PLoS Pathog 9(3): e1003252. doi:10.1371/journal.ppat.1003252.

Zhou, Y., H. Sugiyama, H. Nakano and E.A. Johnson. 1995. The genes for the *Clostridium botulinum* type G toxin complex are on a plasmid. Infect. Immun. 63: 2087–2091.

Yuan, J., G. Inami, J. Mohle-Boetani and D.J. Vugia. 2011. Recurrent wound botulism among injection drug users in California. Clin. Infect. Dis. 52: 862–866.

CHAPTER 7

Diarrheagenic *E. coli* with Emphasis on Shiga Toxin-producing *E. coli*[†]

Peter Gerner-Smidt[#]

Introduction

Escherichia coli is one of the microorganisms most frequently encountered in human clinical specimens and specimens from animals, food and water. Most strains of *E. coli* are low pathogenic or not pathogenic at all. Since they are part of the normal intestinal flora in humans and animals, their presence in food and water is an indication of faecal contamination. However, a number of strains are human pathogens and *E. coli* is a common cause of extraintestinal infections such as urinary tract infections and septicemia. During the past 50 years *E. coli* has increasingly been recognized as a cause of infectious gastroenteritis. At least six pathogroups of such diarrheagenic *E. coli* (DEC) have been described (Kaper et al. 2004): enteropathogenic *E. coli* (EPEC), which is associated with infantile diarrhea; enterotoxigenic *E. coli* (ETEC), which is a frequent cause of watery diarrhea in travelers

[#] Enteric Diseases Laboratory Branch, Centers for Disease Control & Prevention, MS-CO-3, 1600 Clifton Rd, 30333, Atlanta GA, USA.
Email: plg5@cdc.gov

[†] Disclaimer: The findings and conclusions in this presentation have not been formally disseminated by the Centers for Disease Control and Prevention and should not be construed to represent any agency determination or policy.

to third world countries; enteroinvasive *E. coli* (EIEC), which is related to *Shigella* and cause diarrhea in some third world countries; in fact, *Shigella* may be considered a variant of EIEC since this genus cannot be differentiated genetically from *E. coli* (Lukjancenko et al. 2010, Parsot 2005); entero aggregative *E. coli* (EAEC), which have also been described as a cause of diarrhea in third world countries; diffusely adherent *E. coli* (DAEC), the pathogenicity of which is still debated; and finally Shiga toxin-producing *E. coli* (STEC), which is the subject of this chapter. These pathogroups were identified before molecular methods were available for routine detection in clinical laboratories by their ability to adhere to cell cultures in specific patterns (EPEC, STEC, EAEC, DAEC), production of enterotoxins (ETEC) or cytotoxins (STEC) or by the use of the animal tests such as the Sereny test for EIEC (and *Shigella*) (Nataro and Kaper 1998). Most pathogroups cannot be differentiated from commensal *E. coli* by traditional cultural methods but a number of serotypes are correlated with each pathogroup. However, traditional methods to identify DEC are used only in clinical laboratories with specific interest in these pathogens and for this reason infections caused by DEC are rarely diagnosed. This is changing now with the increasing use of immunoassays and molecular methods such as PCR. A number of virulence factors and virulence associated genes have been identified in each DEC pathogroup and as molecular methods are becoming commercially available and implemented in the clinical laboratory routine, our knowledge of DEC will increase greatly. The DEC pathogroup that is most commonly diagnosed in clinical laboratories is STEC and for this reason, we know more about the epidemiology and virulence of STEC than any other DEC pathogroup. However, our understanding of this pathogroup is still incomplete.

Classification and Detection of STEC

STEC is characterized by its ability to produce a cytotoxin that destroys endothelial cells. Its activity was first demonstrated on cultures of vero-cells, a kidney cell line from the African green monkey, and the toxin is therefore also known as verocytotoxin (Konowalchuk et al. 1977). For a long time STEC was called verocytotoxin producing *E. coli* (VTEC). However, it was shown that the toxin was structurally and functionally related closely to Shiga toxin (Stx) produced by strains of *Shigella dysenteriae* type 1 and it was proposed to rename verocytotoxin to Shiga toxin (Calderwood et al. 1996). Today both names are used interchangeably. The name enterohemorrhagic *E. coli* (EHEC) is used by some scientists to denote a subset of STEC that produce severe illness in humans. However, many variants of EHEC definitions exist; none include all human pathogenic strains or exclude all that do not cause human disease in order to avoid confusion, this expression should not be used. The term HUSEC has been proposed for STEC that are associated with acute kidney failure, hemolytic uremic syndrome (HUS) (Mellmann et al. 2008).

Serotyping is the basic and first line tool to group STEC. Traditionally two major groups are considered: serogroup O157 and non-O157 serogroups. Numerous STEC serotypes of which at least 150 have been associated with human illness, have been described (Gould et al. 2009). The serotype O157:H7 is the prototype of human pathogenic STEC. Shortly after the first description of STEC in 1977, this serotype was identified as cause of outbreaks and HUS (Karmali et al. 1983, Wells et al. 1983). Strains of this serotype with the exception of the "German clone" discussed later do not or only slowly ferment sorbitol at variance with most other *E. coli* (Wells et al. 1983). This trait is used to identify rapidly O157 strains growing on sorbitol McConkey agar (March and Ratnam 1986) and other sorbitol containing selective indicator media, which will differentiate STEC O157 from most other *E. coli* by the color of the colonies. This method was for many years the only practical way to identify STEC in diagnostic laboratories and for that reason most of our knowledge about STEC relates to this serotype. However, with the development of enzyme immuno-assays (EIA's), and molecular diagnostic methods like colony hybridization and PCR, STEC is now increasingly being diagnosed by the presence of the toxin itself or the genes encoding it. In United States (U.S.), the clinical laboratories are recommended to test all diarrheal stools for STEC using a combination of culture for O157 on sorbitol McConkey agar or similar media, and detection of Stx, e.g., by EIA (Gould et al. 2009). This dual approach is recommended because EIA alone is not sufficiently sensitive to detect all O157 STEC and in order to save time diagnosing aggressive O157 infections. For the detection of outbreaks, it is very important that the causative organism is isolated from specimens that test positive by EIA or molecular methods and further characterized by serotyping and/or molecular subtyping, e.g., pulsed field gel-electrophoresis (PFGE), like it is done in PulseNet USA (Gerner-Smidt et al. 2006, Swaminathan et al. 2001) and PulseNet International (Swaminathan et al. 2006), the U.S. and international molecular subtyping networks for surveillance of foodborne infections.

Methods that detect Stx or the genes encoding them (*stx*) are not sufficiently specific in detecting human pathogenic STEC in animals and food because not all STEC in animals cause disease in man. Instead, surrogate methods targeting the most common STEC serogroups associated with human illness, e.g., O26, O45, O103, O111, O121 and O145 by immunomagnetic separation (IMS) of enrichment broth cultures, are used. These six serotypes cause approx. 80% of all illness caused by non-O157 STEC in the U.S. and have, since 2012 been considered adulterants in meat by the United States Department of Agriculture (USDA). A four stage approach to detect and isolate the organisms was proposed (LQAD 2012): first an enrichment broth is tested by PCR for the presence of two virulence genes present in these serotypes: *eae* and *stx*; next, if the genes are detected, another PCR is performed to detect genes specific for the six serotypes; if positive, then the organisms are cultured using IMS; and finally single colonies are confirmed positive for *eae* and *stx* by PCR, and biochemically and serologically confirmed as *E. coli* belonging to one of the six serotypes. The European reference laboratory for VTEC at Isituto Superiore di Sanita in Rome has also published PCR methods for the detection of virulent STEC including assays targeting the serogroups O157, O55 and two *eae* negative serogroups O91 and O113 (http://www.iss.it/vtec/work/cont.php?id=152&tipo=3&lang=2) that are common in Europe.

Unfortunately, these surrogate methods will not capture all strains that are pathogenic to humans and are therefore suboptimal. Ideally the methods should capture characteristics that are common to all human pathogenic STEC. However, what makes a strain pathogenic to humans is not fully understood.

Clinical Presentation and Epidemiology

STEC is transmitted to humans via the faecal-oral route through ingestion of contaminated food or water, contact with animals that may carry the organism without being sick, or through contact with a sick person. Since most STEC infections are transmitted through food and water, they are potentially preventable. The infectious dose is thought to be low; as low as 50–100 organisms may cause disease. All ages are susceptible to STEC infection but the incidence is highest in young children (Brooks et al. 2005, Mead and Griffin 1998). The incubation period is usually less than one week but may be longer as was seen in the German outbreak of O104:H4 infections in 2011 (Frank et al. 2011). In the U.S., disease caused by non-O157 serotypes is more often acquired during travel abroad than O157 disease (Hadler et al. 2011).

Ruminants, especially cattle but also sheep and goats, are the main reservoir for STEC O157 but the serotype has also been found in horses, pigs, cats, dogs, wildlife and poultry. Although less is known about the reservoir of other STEC serotypes, non-O157 STEC has been found in the same animals as O157. STEC has also been found in meat from virtually any meat animal, dairy products, fish, shellfish, fruit and produce with higher rates of non-O157 STEC compared to STEC O157 (Mathusa et al. 2010, Pennington 2010, Pihkala et al. 2012). It is difficult to compare the prevalence of STEC in different animals and foods from different studies because there is no general consensus on which method to use and the methods differ in their performance characteristics. Recently, it was shown that different plating media may each select for growth strains belonging to different serotypes and virulence profiles (Quiñones et al. 2012).

When STEC is isolated from a diarrheal stool from a person, the cause of the patient's disease is usually considered established. However, not all STEC are pathogenic to man and not all pathogenic strains cause disease in all individuals. The clinical spectrum (Fig. 1) varies from an asymptomatic infection, to watery diarrhea with no or low grade fever, to serious bloody diarrhea, a sign of hemorrhagic colitis; most patients with watery or bloody diarrhea recover spontaneously within approx. one week (Mead and Griffin 1998) but a small percent develop HUS, a condition characterized by microangiopathic hemolytic anemia, reduced number of circulating platelets and kidney failure and sometimes neurological symptoms (Pennington 2010), e.g., cognitive impairment or aphasia and epileptic seizures (Magnus et al. 2012). The risk of HUS is dependent on microbial as well as host factors; it is highest among infants and young children and the condition has a case-fatality rate of approx. 5% despite treatment. Patients recovering from the most severe form of HUS, oliguric or anuric HUS, which requires dialysis, carry a high risk of chronic kidney failure showing as hypertension and proteinuria, which may develop more than a year following initial recovery (Rosales et al. 2012). STEC infections also carry a significant mortality among the elderly albeit usually not associated with HUS. Treatment of the infection is supportive and early re- or overhydration has been shown to prevent progression to oliguric and anuric HUS of infections caused by the O157 serotype (Ake et al. 2005). Since the virulence factors associated with HUS progression are also found in non-O157 strains, aggressive hydration therapy is also likely to be beneficial for the treatment of infections caused by non-O157 serotypes. Antimicrobials are contraindicated since their use seems to precipitate the development of HUS (Wong et al. 2012), although bacteriostatic antimicrobials that act by inhibiting the protein synthesis *in vitro* has been shown to decrease the production of Stx (Bielaszewska et al. 2012, Pedersen et al. 2008). Carriage of STEC is often short, a few weeks, and cultures for STEC in HUS patients

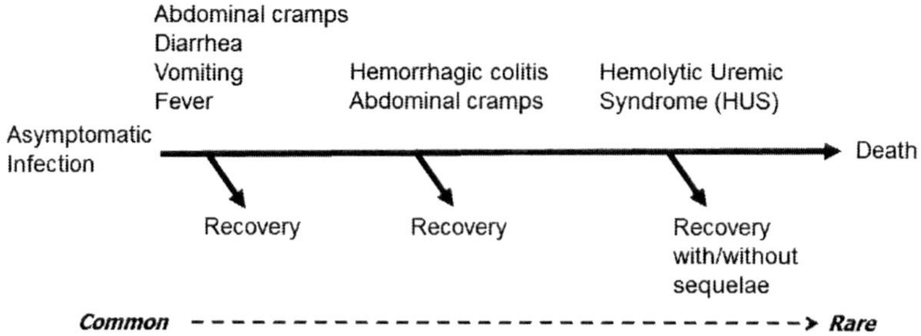

Figure 1. Progression of clinical symptoms of STEC infection.

are often negative because carriage has ceased when the kidney failure develops approx. one week after onset of diarrhea. Although chronic carriers have not been identified, post-diarrheal carriage for several months may occur (Jensen et al. 2005) may be associated with specific strains, e.g., the German O104:H4 outbreak strain (Nitschke et al. 2012). Treatment of asymptomatic carriage seems to be an efficient way to terminate carriage (Jensen et al. 2005). Like other DEC, STEC infections are under diagnosed in most countries compared to infections caused by traditional diarrheal pathogens like *Salmonella, Shigella, Yersinia* and *Campylobacter* because not all stools are being tested for this pathogen. STEC is an important public health problem because of the severity of the disease and because it is a common cause of outbreaks of food or waterborne disease. In the U.S., an estimated 173,905 foodborne STEC illnesses occur every year, leading to 2,409 hospitalizations and 20 deaths (Scallan et al. 2011). STEC infections are notifiable in most countries. In the U.S. the disease has been nationally notifiable since 2000 through reporting to the National Notifiable Diseases Surveillance System (NNDSS) by health care providers. Suspect, not culture confirmed, infections may be reported through this system. Laboratory confirmed illness is monitored passively in the Laboratory Enteric Illness Surveillance system (LEDS) and actively in FoodNet. FoodNet, the Foodborne Diseases Active Surveillance Network, monitors laboratory confirmed STEC infections at 10 sentinel sites covering approx. 15% of the population in the U.S. The network performs epidemiological studies on the burden of foodborne illness (Scallan et al. 2011) and risk factors for sporadic illness, e.g., through case-control studies (Fullerton et al. 2012), and conducts laboratory surveys ensuring completeness of the reported data and monitors the trends in diagnosed illness (Centers for Disease and Prevention 2011). The incidence of O157 illness has decreased in the last five years to less than 1 case per 100,000 in the population, which is approximately the same as the reported incidence of non-O157 infections, which is rising, most likely because an increasing number of clinical laboratories today are looking for them. Outbreaks of foodborne illnesses are reported by the public health departments to CDC through the Foodborne and Diarrheal Outbreaks Surveillance System (FDOSS) and information about these are available in the Foodborne Outbreak On-line Database (FOOD): http://www.cdc.gov/foodborneoutbreaks. Three-hundred and thirty seven outbreaks of STEC O157 were reported in United States during the period 1998–2009. Similarly 15 outbreaks of non-O157 infections were reported. Most outbreaks were small and local, associated with restaurants and private gatherings but 44 involved patients in multiple states. The database contains only information about outbreaks that are confirmed or suspected to be foodborne. Foodborne sources most commonly identified for serogroup O157 were beef, especially ground beef, leafy greens (lettuce and spinach), sprouts and raw milk. Non-foodborne sources include contaminated drinking and recreational water, contact with animals, person-person spread and rarely laboratory acquired infections (Rangel et al. 2005). Vehicles for outbreaks caused by non-O157 serogroups are the same as those caused by STEC O157 (Mathusa et al. 2010, Pihkala et al. 2012). With efficient epidemiological follow-up of outbreaks, new vehicles for STEC outbreaks have been identified recently including raw cookie dough (Neil et al. 2012) and hazel nuts (Miller et al. 2012). The recognition of novel vehicles for outbreaks emphasizes the need for an open minded approach to investigation of outbreaks not leaving out any possibilities. More multi-state outbreaks have been detected in recent years, likely to be associated with improved molecular surveillance by PulseNet USA (Gerner-Smidt et al. 2006, Swaminathan et al. 2001). PulseNet USA is a network of more than 80 public health, state agriculture and food regulatory agency laboratories who perform high-discriminatory molecular subtyping by pulsed field gel electrophoresis (PFGE) (www.pulsenetinternational.org/protocols) in real-time on isolates of foodborne pathogens including STEC received from the clinical diagnostic laboratories, and from food and animals as part of outbreak investigations, projects or routine regulatory food inspections. PFGE is supplemented with another complementary high discriminatory subtyping method, multi locus variable number of tandem repeats (VNTR) analysis (MLVA) when a common PFGE pattern is involved in a cluster/outbreak or to confirm PFGE findings (Hyytia-Trees et al. 2006). A MLVA protocol for non-O157 STEC has been described but it has not been validated extensively (Lobersli et al. 2012). When PulseNet detects a cluster of STEC patients infected with isolates with indistinguishable PFGE fingerprints at a rate that is higher than the historical background, an epidemiological investigation is initiated. Often this follow-up consists only of a review of the data submitted along with the isolates and monitoring of the cluster to see how it evolves. If the number of STEC case-patients continues to increase, a full-fledged outbreak investigation is initiated with patient interviews and follow-up with inspection and microbiological sampling of suspected food, production premises, or outbreak venue, e.g., a petting zoo, a restaurant, etc. PulseNet detects between 40 and 50 STEC clusters every year and at any given time approx. five clusters are being monitored/investigated. The number of uploads of PFGE patterns to PulseNet by year from 1996–2010 is shown for STEC O157 and non-O157 in Fig. 2. The number of uploads of non-O157 patterns has increased dramatically in recent years and has exceeded O157 since 2010. This is not likely a reflection of an increase in infections caused by non-O157 serotypes but as mentioned before, rather caused by increasing use of methods that detect Stx or *stx* by the diagnostic laboratories.

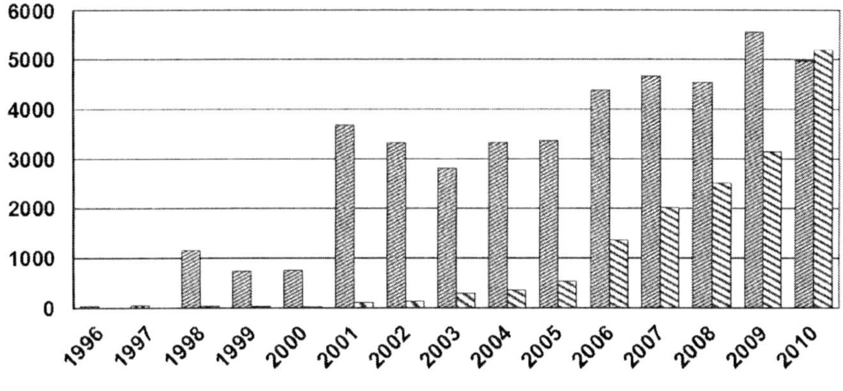

Figure 2. Annual uploads of PFGE profiles from clinical isolates of STEC O157 and non-O157 to the PulseNet USA national *E. coli* database 1996–2010.

The societal costs of STEC disease in the U.S. has been estimated conservatively to approximately US$ 700 million annually (Scharff 2012).

Today non-O157 serotypes seem to outnumber O157 strains as cause of diarrhea in most countries. In general O157 strains are more virulent than non-O157 strains and for that reason this serotype remains the predominant cause of HUS. However, as will be discussed later some non-O157 strains may be as virulent as the most virulent O157 strains.

Virulence of STEC

In an attempt to classify STEC based on their association with HUS and outbreak potential a system of five sero-pathotypes has been proposed (Karmali et al. 2003): type A contains STEC O157 which is the serogroup most commonly associated with HUS and a frequent cause of outbreaks; type B contains serogroups that have been associated with outbreaks and HUS but less frequently than O157, e.g., O26, O45, O103, O111, O121 and O145; type C contains serogroups that rarely cause outbreaks but are associated with HUS, e.g., O91, O104, and O113; type D contains strains that rarely cause outbreaks and are not associated with HUS; and finally type E contains strains that are not considered human pathogens. Although useful for a general grouping of STEC, it classifies all strains belonging to a particular serotype as being equally virulent, which is not the case. Considerable virulence heterogeneity exists in many serotypes, e.g., most strains of serogroup O103 do not cause HUS, nonetheless an O103 strain (O103:H25) caused an outbreak in Norway in 2006 associated with consumption of a fermented mutton sausage with an extremely high HUS-rate of 10 out of

16 reported cases (Schimmer et al. 2008); this strain was just as virulent as the most virulent O157 strains ever described. As described below, even within STEC O157 there are large differences between the virulence of different genetic lineages. The STEC populations are also dynamic and the pathogenic potential of a serotype may change and therefore move it to a different seropathotype. For example the serogroup O104 could after the devastating outbreak in Germany in the summer of 2011 (Frank et al. 2011) be reclassified as seropathotype B; before then it was classified as seropathotype C.

Thus, the serogroups or serotype alone does not define the virulence or outbreak potential of an STEC strain. The best predictor(s) of virulence in STEC is still incompletely understood.

Stx is a critical virulence factor. It is encoded by a lambdoid prophage and belongs to the AB_5 toxin family. The B subunit binds to globotriaosylceramide (Gb3) on the host cell. Following this, the A subunit is internalized and activated to a N-glycosidase which binds to the ribosome, disrupting protein synthesis causing cell death (Karmali 2004, Nakao and Takeda 2000). Stx exists in two types, Stx1 and Stx2, each with several subtypes and numerous variants. Stx1 and Stx2 are distinct immunologically and neutralizing antibodies raised against each will not cross react with the other; similarly there are no DNA-DNA cross hybridization between *stx1* and *stx2,* the genes encoding the Shiga toxins. Subtypes of Stx cross react immunologically with Stx of the same type but significant differences exist in their biological activity, receptor binding and serological reactivity. The nomenclature of Stx subtypes was for a long time not standardized and extremely confusing (Table 1) with different names for the same subtype; recently, a standardized system has been proposed greatly simplifying the Stx nomenclature (Scheutz et al. 2012). Four Stx1 subtypes, Stx1, Stx1a, Stx1c and Stx1d have been described; Stx1 or Stx is the Shiga toxin produced by *Shigella dysenteriae* type 1; the other three are found in STEC. Similarly, seven subtypes of Stx2 have been described, Stx2a-g. Strains producing Stx2e are associated with edema disease in pigs and do not seem to be associated with human illness. More than 12 amino acid variants of Stx1 and more than 80 variants of Stx2 have been described so far. Although related to the other Stx2 subtypes, the Stx2f subtype is so different that it will not be detected by many immunological and molecular assays for the detection of Stx2 (Prager et al. 2009, Schmidt et al. 2000). For this reason, infections with strains containing this subtype may be under diagnosed compared to strains producing the other Stx2 subtypes. Stx2 is in general more toxic than Stx1 and strains that express Stx2 are more likely to cause severe disease than strains expressing Stx1-only (Boerlin et al. 1999, Ethelberg et al. 2004). HUS and hemorrhagic colitis is further associated with specific Stx2

Table 1 Current nomenclature for Shiga toxin subtypes and previous designations for genes and subtypes.*

Toxin type	Toxin subtype (gene)	Previous gene designation(s)	Previous toxin designation(s)
Stx1	Stx (*stx*)	*Stx*	Stx1
	Stx1a (*stx1a*)	*stx1,slt-1*	SLT-1, VT1
	Stx1c (*stx1c*)	stx_{1OX3}, *stx1c*	Stx(1c)
	Stx1d (*stx1d*)	*stx1d*	
Stx2	Stx2a (*stx2a*)	*slt-II, stx2*	SLT-II, SLT-II/O48, Stx2vhd
	Stx2b (*stx2b*)	*stx2, stx2b,* Stx_{2vO111}, *stx2d*	SLT-II/OX3a/VT2d-OX3a, Stx2d-OX3a, SLT-ii/O111/VT2d-O111, Stx2d-O111, Stx2d/VT2-Ount
	Stx2c (*stx2c*)	*stx2, stx2c,*$stx_{2vOX393}$ *stx2v-ha,* stx_{2vhd}	SLT-II/OX3b/2 / VT2d-OX3/2, Stx2v/VT2v, SLTIIc, *Stx2*-OX3b
	Stx2d (*stx2d*)	stx2vha, stx_{2d1}/stx2d1, stx2ha, $stx2d_a$, stx2v-ha, stx2vhb, stx_{2d2}/stx2d2, stx2hb, $stx2d_b$, stx2v-hb, $stx2_{c2}$, stx2vhc (stx2d3)	SLT-IIvh/SL-IIvha, Stx2vha/VT2vha, stx2vh-a/VT2vh-a, stx2d1/VT2d1, VT2v-a, VT2v-b, stx2vhb/vVT2vhb, stxvh-b/VT2vh-b, SL-IIvhb, Stx2c2, Stx2vhc, Stx2cf
	Stx2e (*stx2e*)	*slt*-IIv, *slt*-IIva, *slt*-IIe, *slt*-IIera, *stx2e*-ONT-26725-97 (92), *stx2e*	SLT-IIv, SLT-IIva, SLT-IIe/VT2e, VT2vp, VT2vp1, Stx2era
	Stx2f (*stx2f*)	*stx2f,* stx_{2ev}/*stx2ev*, *slt*-IIvhc, *slt*-IId	Stx2ev/VT2ev/VTev, Stx2vp2/VT2vp2, Stx2va/Stx2v, SLTIIvhc, SLT-IId/VT2d, SLTIIva
	Stx2g (*stx2g*)	*stx2g*	

*: from Scheutz et al. 2012.

subtypes with the phylogenetically closely related Stx2a, Stx2c and Stx2d containing strains being most virulent (Bielaszewska et al. 2006a, Persson et al. 2007, Scheutz et al. 2012).

An STEC may contain and express more than one Stx-gene. Thus, STEC may be classified as producing Stx1-only, Stx1 and Stx2, or Stx2-only. Additionally, strains may contain more copies of the same Stx-gene in its genome. An outbreak caused by an O157:NM (non-motile) strain that contained two copies of the same Stx2 gene has been described (Bielaszewska et al. 2006b). One of the genes was stably integrated into the chromosome at the site usually associated with insertion of this gene whereas the other gene was not stably integrated into the chromosome leading to differences in the PFGE patterns from different patients involved and even of different colony picks of a pure culture, confusing the outbreak investigation.

STEC contains a number of other confirmed or putative virulence factors. The vast majority of these are located on mobile genetic elements, e.g., pathogenicity islands (PAI's) and prophages in the chromosome or on plasmids. The pathogenicity island, locus for enterocyte effacement (LEE) contains genes encoding an adhesin, intimin (*eae*), its receptor (*tir*) and effectors of a type III secretion system (T3SS). LEE is besides in STEC also found in enteropathogenic *E. coli* (EPEC). The LEE associated virulence genes enables the bacterium to adhere to the enterocytes in the intestine and are responsible for the so-called attaching and effacing lesions of the endothelium of the intestine and the so called localized adherence pattern that is characteristic for STEC and EPEC in cell cultures (Nataro and Kaper 1998). LEE is associated with severe disease in STEC but is not essential for virulence since a subset of strains associated with severe disease do not contain these virulence factors, e.g., some Stx2d containing strains (Bielaszewska et al. 2006b). Several different variants of the genes located on LEE exist and at least some of these seem to affect the virulence of the strains to humans, e.g., it has been shown that the predominant *tir* variant in bovine isolates of STEC O157 contains a SNP that is rarely present human isolates (Bono et al. 2007). Some important virulence factors are located outside the LEE and include among other T3SS effectors (non-LEE effectors, *nle*) present on different PAI's and a number of plasmid encoded factors: enterohemolysin (*ehxA*), STEC autoagglutinating adhesin (*saa*) and another AB_5 toxin, subtilase (*sub*). The number of virulence associated genes in an STEC strain seems to be positively correlated with increasing virulence (Buvens and Pierard 2012, Coombes et al. 2008, Wickham et al. 2006).

Genomics and Phylogenetics

The genomes of STEC strains are larger, 5.3–5.7Mb, than other sequenced *E. coli* strains, 4.4–5.2Mb. Compared to other *E. coli* strains, STEC in general contain more tRNA genes, prophages, other inserted sequences, e.g., PAI's and plasmids (Ogura et al. 2009). STEC is very diverse phylogenetically; serogroups belong to different lineages, e.g., O157 belongs to phylogroup D, whereas O26, O103 and O111 belong to group B1 and less virulent serotypes are spread out among phylogroups A, B1, B2, and D (Girardeau et al. 2005).

STEC O157 strains also split in different sublineages and clades with markedly different virulence (Manning et al. 2008). STEC O157 is assumed to have arisen from a presumed EPEC strain of O55:H7 through stepwise evolution to a sorbitol fermenting intermediate STEC O157:H7 through acquisition of a Stx2 phage and a change of the serogroup. This intermediate then split into two groups: one that lost its motility and became the virulent sorbitol fermenting O157 clone, which is known as the "German clone" since it was first detected in that country and has spread from there (Karch and Bielaszewska 2001), and another which lost the capacity to ferment sorbitol and became the ancestor of all known sorbitol negative O157:H7 and O157:H-clones (Feng et al. 1998, Jenke et al. 2012, Leopold et al. 2009).

Bono et al. (Bono et al. 2012) recently identified 160 nucleotide polymorphisms specific to the cattle reservoir in a comprehensive study of human and bovine STEC O157 indicating that some STEC O157 isolates, which are common in cattle, may not cause disease in humans at all. The reservoir for sorbitol fermenting STEC O157 is not known. These strains belong to the lineage, which is the closest genetic relative to the serotype O55:H7, the progenitor of STEC O157. This serotype belongs to the EPEC pathogroup and has a human reservoir. This could indicate that STEC 157 acquired the ability to colonize and infect ruminants when it lost its ability to ferment sorbitol and that some strains have evolved further and lost the potential to infect humans as exemplified by the aforementioned bovine specific lineage (Bono et al. 2012).

Hybrid *E. coli*

Many *E. coli* strains and pathotypes contain traits that are hallmarks of different pathogroups, incl. DEC. This is hardly surprising since most DEC virulence genes are located on mobile genetic elements. For example, long before anything was known about the virulence factors of DEC such hybrid strains were causing devastating outbreaks of bacillary dysentery. The etiology of this disease is a good example of a DEC hybrid strain: *Shigella* has the genetic backbone of *E. coli* and shares the invasion armamentarium with EIEC (Parsot 2005, van den Beld and Reubsaet 2012), and *Shigella dysenteriae* type 1, the cause of bacillary dysentery, produce Stx, i.e., the organism is a hybrid between an EIEC and an STEC.

A recent example of a virulent DEC hybrid, is the enteroaggregative Shiga toxin-producing O104:H4 strain that caused a large outbreak of HUS associated with sprout consumption in Germany and France in 2011 (Frank et al. 2011, King et al. 2012). A little less than 4,000 cases of patients were reported, approx. 850 presenting with HUS and 54 died in this outbreak. The virulence of the outbreak strain was initially a mystery to the microbiologists investigating the outbreak: it was *eae* and *saa* negative, i.e., lacking the adherence mechanisms usually associated with virulent STEC, and produced *stx2a*; although *stx2a* is associated with severe illness, a non-adherent strain had never before been associated with outbreaks of severe STEC illness; however, the strain was quickly identified as enteroaggregative, confirming its capacity to adhere to enterocytes thereby providing a plausible explanation for its virulence (Bielaszewska et al. 2011, Scheutz et al. 2011). Whole genome sequencing rapidly followed allowing for generating hypotheses about the origin of the outbreak strain (Mellmann et al. 2011, Rasko et al. 2011). One study suggested that it originated from a common O104:H4 STEC/EAEC progenitor, which by stepwise gain and loss of chromosomal and plasmid-encoded virulence factors developed into the outbreak strain (Mellmann et al. 2011), whereas the other study indicated that it originated from horizontal gene transfer from an already existing EAEC strain (Rasko et al. 2011). More sequencing studies of EAEC and STEC O104:H4 strains unrelated to the outbreak are required to clarify this question. Before the outbreak, a few sporadic isolates of similar strains associated with HUS but not with outbreaks had been kept in reference collections in different countries (ECDC and EFSA 2011) and a similar strain had possibly caused an outbreak of HUS in the Republic of Georgia in 2009 (Scheutz et al. 2011). Following the outbreak in Germany and France, another cluster of O104 illnesses with HUS was detected in French travelers returning from Turkey without any evidence of any outbreak in the host population (Jourdan-da Silva et al. 2012). The natural reservoir for O104:H4 has not been determined and it is not clear if this clone will remain a major public health threat, i.e., cause outbreaks in the future.

Many EPEC and STEC strains also seem to have evolved from common ancestors. As mentioned above STEC O157 seems to have evolved from an EPEC O55 ancestor. The same is probably true for many non-O157 strains which share a number of LEE and non-LEE effectors with EPEC strains (Afset et al. 2006, Coombes et al. 2008, Wickham et al. 2006).

Some less virulent STEC strains have been shown to contain the gene encoding the heat stable enterotoxin of ETEC (*StIa*) (Prager et al. 2011).

Finally, virulence genes for EAEC seem to be present in a not insignificant number of *E. coli* strains associated with urinary tract infections (Abe et al. 2008).

Concluding Remarks

In brief, STEC is a diverse pathogroup of *E. coli;* most strains are likely to be harmless to humans but a number of strains are among the most deadly foodborne pathogens. Because of this and their ability to cause outbreaks, STEC is a major public health problem. Our knowledge about STEC is still very biased towards O157 since this is the only serotype for which a simple culture method is available. Ruminants seems to be the main reservoir for most STEC and beef remains the most commonly identified vehicle for infection although other sources are important, too. Our understanding of the epidemiology of non-O157 STEC will improve with the increasing use of diagnostic methods that detect the Shiga toxins or the genes encoding them. Huge gaps exists in our understanding of the sources and virulence of human pathogenic STEC and a simple diagnostic method to detect these strains in animals and food is badly needed to improve our understanding of STEC epidemiology.

References

Abe, C.M., F.A. Salvador, I.N. Falsetti, M.A. Vieira, J. Blanco, J.E. Blanco, M. Blanco, A.M. Machado, W.P. Elias, R.T. Hernandes and T.A. Gomes. 2008. Uropathogenic *Escherichia coli* (UPEC) strains may carry virulence properties of diarrhoeagenic *E. coli*. FEMS Immunol. Med. Microbiol. 52: 397–406.

Afset, J.E., G. Bruant, R. Brousseau, J. Harel, E. Anderssen, L. Bevanger and K. Bergh. 2006. Identification of virulence genes linked with diarrhea due to atypical enteropathogenic *Escherichia coli* by DNA microarray analysis and PCR. J. Clin. Microbiol. 44: 3703–3711.

Ake, J.A., S. Jelacic, M.A. Ciol, S.L. Watkins, K.F. Murray, D.L. Christie, E.J. Klein and P.I. Tarr. 2005. Relative nephroprotection during *Escherichia coli* O157:H7 infections: association with intravenous volume expansion. Pediatrics 115: e673–e680.

Bielaszewska, M., A.W. Friedrich, T. Aldick, R. Schurk-Bulgrin and H. Karch. 2006a. Shiga toxin activatable by intestinal mucus in *Escherichia coli* isolated from humans: predictor for a severe clinical outcome. Clin. Infect. Dis. 43: 1160–1167.

Bielaszewska, M., R. Prager, W. Zhang, A.W. Friedrich, A. Mellmann, H. Tschape and H. Karch. 2006b. Chromosomal dynamism in progeny of outbreak-related sorbitol-fermenting enterohemorrhagic *Escherichia coli* O157:NM. Appl. Environ. Microbiol. 72: 1900–1909.

Bielaszewska, M., A. Mellmann, W. Zhang, R. Kock, A. Fruth, A. Bauwens, G. Peters and H. Karch. 2011. Characterisation of the *Escherichia coli* strain associated with an outbreak of haemolytic uraemic syndrome in Germany, 2011: a microbiological study. Lancet Infect. Dis. 11: 671–676.

Bielaszewska, M., E.A. Idelevich, W. Zhang, A. Bauwens, F. Schaumburg, A. Mellmann, G. Peters and H. Karch. 2012. Effects of antibiotics on Shiga toxin 2 production and bacteriophage induction by epidemic *Escherichia coli* O104:H4 strain. Antimicrob. Agents Chemother. 56: 3277–3282.

Boerlin, P., S.A. McEwen, F. Boerlin-Petzold, J.B. Wilson, R.P. Johnson and C.L. Gyles. 1999. Associations between virulence factors of Shiga toxin-producing *Escherichia coli* and disease in humans. J. Clin. Microbiol. 37: 497–503.

Bono, J.L., J.E. Keen, M.L. Clawson, L.M. Durso, M.P. Heaton and W.W. Laegreid. 2007. Association of *Escherichia coli* O157:H7 tir polymorphisms with human infection. BMC Infect. Dis. 7: 98.

Bono, J.L., T.P. Smith, J.E. Keen, G.P. Harhay, T.G. McDaneld, R.E. Mandrell, W.K. Jung, T.E. Besser, P. Gerner-Smidt, M. Bielaszewska, H. Karch and M.L. Clawson. 2012. Phylogeny of Shiga toxin-producing *Escherichia coli* O157 Isolated from cattle and clinically ill humans. Mol. Biol. Evol. 29: 2047–2062.

Brooks, J.T., E.G. Sowers, J.G. Wells, K.D. Greene, P.M. Griffin, R.M. Hoekstra and N.A. Strockbine. 2005. Non-O157 Shiga toxin-producing *Escherichia coli* infections in the United States, 1983–2002. J. Infect. Dis. 192: 1422–1429.

Buvens, G. and D. Pierard. 2012. Virulence profiling and disease association of verocytotoxin-producing *Escherichia coli* O157 and non-O157 isolates in Belgium. Foodborne Pathog. Dis. 9: 530–535.

Calderwood, S.B., D.W.K. Acheson, G.T. Keusch, T.J. Barrett, P.M. Griffin, N.A. Strockbine, B. Swaminathan, J.B. Kaper, M.M. Levine, B.S. Kaplan, H. Karch, A. O'Brien, T.G. Obrig, P.I. Tarr and I.K. Wachsmuth. 1996. Proposed new nomenclature for Shiga-like toxin (verotoxin) family. ASM News 62: 118–119.

Centers for Disease C, Prevention. 2011. Vital signs: incidence and trends of infection with pathogens transmitted commonly through food—foodborne diseases active surveillance network, 10 U.S. sites, 1996–2010. MMWR Morb. Mortal. Wkly Rep. 60: 749–755.

Coombes, B.K., M.E. Wickham, M. Mascarenhas, S. Gruenheid, B.B. Finlay and M.A. Karmali. 2008. Molecular analysis as an aid to assess the public health risk of non-O157 Shiga toxin-producing *Escherichia coli* strains. Appl. Environ. Microbiol. 74: 2153–2160.

ECDC, EFSA. 2011. Shiga toxin/verotoxin-producing *Escherichia coli* in humans, food and animals in the EU/EEA, with special reference to the German outbreak strain STEC O104. Stockholm: ECDC.

Ethelberg, S., K.E. Olsen, F. Scheutz, C. Jensen, P. Schiellerup, J. Enberg, A.M. Petersen, B. Olesen, P. Gerner-Smidt and K. Molbak. 2004. Virulence factors for hemolytic uremic syndrome, Denmark. Emerg Infect. Dis. 10: 842–847.

Feng, P., K.A. Lampel, H. Karch and T.S. Whittam. 1998. Genotypic and phenotypic changes in the emergence of *Escherichia coli* O157:H7. J. Infect. Dis. 177: 1750–1753.

Frank, C., D. Werber, J.P. Cramer, M. Askar, M. Faber, M. an der Heiden, H. Bernard, A. Fruth, R. Prager, A. Spode, M. Wadl, A. Zoufaly, S. Jordan, M.J. Kemper, P. Follin, L. Muller, L.A. King, B. Rosner, U. Buchholz, K. Stark, G. Krause and Team HUSI. 2011. Epidemic profile of Shiga-toxin-producing *Escherichia coli* O104:H4 outbreak in Germany. N. Engl. J. Med. 365: 1771–1780.

Fullerton, K.E., E. Scallan, M.D. Kirk, B.E. Mahon, F.J. Angulo, H. de Valk, W. van Pelt, C. Gauci, A.M. Hauri, S. Majowicz, S.J. O'Brien and International Collaboration Working G. 2012. Case-control studies of sporadic enteric infections: a review and discussion of studies conducted internationally from 1990 to 2009. Foodborne Pathog. Dis. 9: 281–292.

Gerner-Smidt, P., K. Hise, J. Kincaid, S. Hunter, S. Rolando, E. Hyytia-Trees, E.M. Ribot and B. Swaminathan. 2006. PulseNet USA: a five-year update. Foodborne Pathog. Dis. 3: 9–19.

Girardeau, J.P., A. Dalmasso, Y. Bertin, C. Ducrot, S. Bord, V. Livrelli, C. Vernozy-Rozand and C. Martin. 2005. Association of virulence genotype with phylogenetic background in comparison to different seropathotypes of Shiga toxin-producing *Escherichia coli* isolates. J. Clin. Microbiol. 43: 6098–6107.

Gould, L.H., C. Bopp, N. Strockbine, R. Atkinson, V. Baselski, B. Body, R. Carey, C. Crandall, S. Hurd, R. Kaplan, M. Neill, S. Shea, P. Somsel, M. Tobin-D'Angelo, P.M. Griffin, P. Gerner-Smidt and Centers for Disease C, Prevention. 2009. Recommendations for diagnosis of shiga toxin—producing *Escherichia coli* infections by clinical laboratories. MMWR Recomm. Rep. 58: 1–14.

Hadler, J.L., P. Clogher, S. Hurd, Q. Phan, M. Mandour, K. Bemis and R. Marcus. 2011. Ten-year trends and risk factors for non-O157 Shiga toxin-producing *Escherichia coli* found through Shiga toxin testing, Connecticut, 2000–2009. Clin. Infect. Dis. 53: 269–276.

Hyytia-Trees, E., S.C. Smole, P.A. Fields, B. Swaminathan and E. Ribot. 2006. Second generation subtyping: a proposed PulseNet protocol for multiple-locus variable-number tandem repeat analysis of Shiga toxin-producing *Escherichia coli* O157 (STEC O157). Foodborne Pathog. Dis. 3: 118–131.

Jenke, C., S.R. Leopold, T. Weniger, J. Rothganger, D. Harmsen, H. Karch and A. Mellmann. 2012. Identification of intermediate in evolutionary model of enterohemorrhagic *Escherichia coli* O157. Emerg. Infect. Dis. 18: 582–588.

Jensen, C., P. Schiellerup, K. Olsen, F. Scheutz, E. Petersen, P. Gerner-Smidt and K. Molbak. 2005. Antimicrobial treatment of asymptomatic carriers of verocytotoxin-producing *Escherichia coli*: an empiric study. Scand. J. Infect. Dis. 37: 61–63.

Jourdan-da Silva, N., M. Watrin, F.X. Weill, L.A. King, M. Gouali, A. Mailles, D. van Cauteren, M. Bataille, S. Guettier, C. Castrale, P. Henry, P. Mariani, V. Vaillant and H. de Valk. 2012. Outbreak of haemolytic uraemic syndrome due to Shiga toxin-producing *Escherichia coli* O104:H4 among French tourists returning from Turkey, September 2011. Euro Surveill 17: pii 20065.

Kaper, J.B., J.P. Nataro and H.L. Mobley. 2004. Pathogenic *Escherichia coli*. Nat. Rev. Microbiol. 2: 123–140.

Karch, H., M. Bielaszewska. 2001. Sorbitol-fermenting Shiga toxin-producing *Escherichia coli* O157:H(–) strains: epidemiology, phenotypic and molecular characteristics, and microbiological diagnosis. J. Clin. Microbiol. 39: 2043–2049.

Karmali, M.A. 2004. Infection by Shiga toxin-producing *Escherichia coli*: an overview. Mol. Biotechnol. 26: 117–122.

Karmali, M.A., B.T. Steele, M. Petric and C. Lim. 1983. Sporadic cases of haemolytic-uraemic syndrome associated with faecal cytotoxin and cytotoxin-producing *Escherichia coli* in stools. Lancet 1: 619–620.

Karmali, M.A., M. Mascarenhas, S. Shen, K. Ziebell, S. Johnson, R. Reid-Smith, J. Isaac-Renton, C. Clark, K. Rahn and J.B. Kaper. 2003. Association of genomic O island 122 of *Escherichia coli* EDL 933 with verocytotoxin-producing *Escherichia coli* seropathotypes that are linked to epidemic and/or serious disease. J. Clin. Microbiol. 41: 4930–4940.

King, L.A., F. Nogareda, F.X. Weill, P. Mariani-Kurkdjian, E. Loukiadis, G. Gault, N. Jourdan-DaSilva, E. Bingen, M. Mace, D. Thevenot, N. Ong, C. Castor, H. Noel, D. Van Cauteren, M. Charron, V. Vaillant, B. Aldabe, V. Goulet, G. Delmas, E. Couturier, Y. Le Strat, C. Combe, Y. Delmas, F. Terrier, B. Vendrely, P. Rolland and H. de Valk. 2012. Outbreak of Shiga toxin-producing *Escherichia coli* O104:H4 associated with organic fenugreek sprouts, France, June 2011. Clin. Infect. Dis. 54: 1588–1594.

Konowalchuk, J., J.I. Speirs and S. Stavric. 1977. Vero response to a cytotoxin of *Escherichia coli*. Infect. Immun. 18: 775–779.

Leopold, S.R., V. Magrini, N.J. Holt, N. Shaikh, E.R. Mardis, J. Cagno, Y. Ogura, A. Iguchi, T. Hayashi, A. Mellmann, H. Karch, T.E. Besser, S.A. Sawyer, T.S. Whittam and P.I. Tarr. 2009. A precise reconstruction of the emergence and constrained radiations of *Escherichia coli* O157 portrayed by backbone concatenomic analysis. Proc. Natl. Acad. Sci. USA. 13: 13.

Lobersli, I., K. Haugum and B.A. Lindstedt. 2012. Rapid and high resolution genotyping of all *Escherichia coli* serotypes using 10 genomic repeat-containing loci. J. Microbiol. Methods 88: 134–139.

LQAD. 2012. Detection and isolation of non-O157 Shiga toxin-producing *Escherichia coli* (STEC) from meat products. Laboratory Guidebook: United States Department of Agriculture, Food Safety and Inspection Service, Office of Public Health Science. Athens, GA.

Lukjancenko, O., T.M. Wassenaar and D.W. Ussery. 2010. Comparison of 61 sequenced *Escherichia coli* genomes. Microb. Ecol. 60: 708–720.

Magnus, T., J. Rother, O. Simova, M. Meier-Cillien, J. Repenthin, F. Moller, J. Gbadamosi, U. Panzer, M. Wengenroth, C. Hagel, S. Kluge, R.K. Stahl, K. Wegscheider, P. Urban, B. Eckert, M. Glatzel, J. Fiehler and C. Gerloff. 2012. The neurological syndrome in adults during the 2011 northern German *E. coli* serotype O104:H4 outbreak. Brain 135: 1850–1859.

Manning, S.D., A.S. Motiwala, A.C. Springman, W. Qi, D.W. Lacher, L.M. Ouellette, J.M. Mladonicky, P. Somsel, J.T. Rudrik, S.E. Dietrich, W. Zhang, B. Swaminatha, D. Alland and T.S. Whittam. 2008. Variation in virulence among clades of *Escherichia coli* O157:H7 associated with disease outbreaks. Proc. Natl. Acad. Sci. USA 105: 4868–4873.

March, S.B. and S. Ratnam. 1986. Sorbitol-MacConkey medium for detection of *Escherichia coli* O157:H7 associated with hemorrhagic colitis. J. Clin. Microbiol. 23: 869–872.

Mathusa, E.C., Y. Chen, E. Enache and L. Hontz. 2010. Non-O157 Shiga toxin-producing *Escherichia coli* in foods. J. Food Prot. 73: 1721–1736.

Mead, P.S. and P.M. Griffin. 1998. *Escherichia coli* O157:H7. Lancet 352: 1207–1212.

Mellmann, A., M. Bielaszewska, R. Kock, A.W. Friedrich, A. Fruth, B. Middendorf, D. Harmsen, M.A. Schmidt and H. Karch. 2008. Analysis of collection of hemolytic uremic syndrome-associated enterohemorrhagic *Escherichia coli*. Emerg. Infect. Dis. 14: 1287–1290.

Mellmann, A., D. Harmsen, C.A. Cummings, E.B. Zentz, S.R. Leopold, A. Rico, K. Prior, R. Szczepanowski, Y. Ji, W. Zhang, S.F. McLaughlin, J.K. Henkhaus, B. Leopold,

M. Bielaszewska, R. Prager, P.M. Brzoska, R.L. Moore, S. Guenther, J.M. Rothberg and H. Karch. 2011. Prospective genomic characterization of the German enterohemorrhagic *Escherichia coli* O104:H4 outbreak by rapid next generation sequencing technology. PLoS ONE 6: e22751.

Miller, B.D., C.E. Rigdon, J. Ball, J.M. Rounds, R.F. Klos, B.M. Brennan, K.D. Arends, P. Kennelly, C. Hedberg and K.E. Smith. 2012. Use of traceback methods to confirm the source of a multistate *Escherichia coli* O157:H7 outbreak due to in-shell hazelnuts. J. Food Prot. 75: 320–327.

Nakao, H. and T. Takeda. 2000. *Escherichia coli* Shiga toxin. J. Nat. Toxins. 9: 299–313.

Nataro, J.P. and J.B. Kaper. 1998. Diarrheagenic *Escherichia coli* Clin. Microbiol. Rev. 11: 142–201.

Neil, K.P., G. Biggerstaff, J.K. MacDonald, E. Trees, C. Medus, K.A. Musser, S.G. Stroika, D. Zink and M.J. Sotir. 2012. A novel vehicle for transmission of *Escherichia coli* O157:H7 to humans: multistate outbreak of *E. coli* O157:H7 infections associated with consumption of ready-to-bake commercial prepackaged cookie dough—United States, 2009. Clin. Infect. Dis. 54: 511–518.

Nitschke, M., F. Sayk, C. Hartel, R.T. Roseland, S. Hauswaldt, J. Steinhoff, K. Fellermann, I. Derad, P. Wellhoner, J. Buning, B. Tiemer, A. Katalinic, J. Rupp, H. Lehnert, W. Solbach and J.K. Knobloch. 2012. Association between azithromycin therapy and duration of bacterial shedding among patients with Shiga toxin-producing enteroaggregative *Escherichia coli* O104:H4. Jama 307: 1046–1052.

Ogura, Y., T. Ooka, A. Iguchi, H. Toh, M. Asadulghani, K. Oshima, T. Kodama, H. Abe, K. Nakayama, K. Kurokawa, T. Tobe, M. Hattori and T. Hayashi. 2009. Comparative genomics reveal the mechanism of the parallel evolution of O157 and non-O157 enterohemorrhagic *Escherichia coli*. Proc. Natl. Acad. Sci. USA 106: 17939–17944.

Parsot, C. 2005. *Shigella* spp. and enteroinvasive *Escherichia coli* pathogenicity factors. FEMS Microbiol. Lett. 252: 11–18.

Pedersen, M.G., C. Hansen, E. Riise, S. Persson and K.E. Olsen. 2008. Subtype-specific suppression of Shiga toxin 2 released from *Escherichia coli* upon exposure to protein synthesis inhibitors. J. Clin. Microbiol. 46: 2987–2991.

Pennington, H. 2010. *Escherichia coli* O157. Lancet 376: 1428–1435.

Persson, S., K.E. Olsen, S. Ethelberg and F. Scheutz. 2007. Subtyping method for *Escherichia coli* Shiga toxin (verocytotoxin) 2 variants and correlations to clinical manifestations. J. Clin. Microbiol. 45: 2020–2024.

Pihkala, N., N. Bauer, D. Eblen, P. Evans, R. Johnson, J. Webb, C. Williams and the FSIS non-O157 working group. 2012. Risk profile for pathogenic non-O157 Shiga toxin-producing *Escherichia coli* (non-O157 STEC). United States Department of Agriculture, Food Safety and Inspection Service, Office of Policy and Program Development, Office of Public Health Science. Washington D.C.

Prager, R., A. Fruth, U. Busch and E. Tietze. 2011. Comparative analysis of virulence genes, genetic diversity, and phylogeny of Shiga toxin 2g and heat-stable enterotoxin STIa encoding *Escherichia coli* isolates from humans, animals, and environmental sources. Int. J. Med. Microbiol. 301: 181–191.

Prager, R., A. Fruth, U. Siewert, U. Strutz and H. Tschape. 2009. *Escherichia coli* encoding Shiga toxin 2f as an emerging human pathogen. Int. J. Med. Microbiol. 299: 343– 353.

Quiñones, B., M.B. Cooley, M.S.Swimley, D. Carychao, K. Nguyen, R. Patel, E.R. Atwill, M. Jay-Russell and R.E. Mandrell. 2012. Selective recovery by different culture methods of Shiga toxin-producing *Escherichia coli* genotypes from a major produce production region in California. Paper presented at ASM General Meeting, San Francisco, California.

Rangel, J.M., P.H. Sparling, C. Crowe, P.M. Griffin and D.L. Swerdlow. 2005. Epidemiology of *Escherichia coli* O157:H7 outbreaks, United States, 1982–2002. Emerg. Infect. Dis. 11: 603–609.

Rasko, D.A., D.R. Webster, J.W. Sahl, A. Bashir, N. Boisen, F. Scheutz, E.E. Paxinos, R. Sebr, C.S. Chin, D. Iliopoulos, A. Klammer, P. Peluso, L. Lee, A.O. Kislyuk, J. Bullard,

A. Kasarskis, S. Wang, J. Eid, D. Rank, J.C. Redman, S.R. Steyert, J. Frimodt-Moller, C. Struve, A.M. Petersen, K.A. Krogfelt, J.P. Nataro, E.E. Schadt and M.K. Waldor. 2011. Origins of the *E. coli* strain causing an outbreak of hemolytic-uremic syndrome in Germany. N. Engl. J. Med. 365: 709– 717.

Rosales, A., J. Hofer, L.B. Zimmerhackl, T.C. Jungraithmayr, M. Riedl, T. Giner, A. Strasak, D. Orth-Holler, R. Wurzner, H. Karch and German-Austrian HUSSG. 2012. Need for long-term follow-up in enterohemorrhagic *Escherichia coli*-associated hemolytic uremic syndrome due to late-emerging sequelae. Clin. Infect. Dis. 54: 1413–1421.

Scallan, E., R.M. Hoekstra, F.J. Angulo, R.V. Tauxe, M.A. Widdowson, S.L. Roy, J.L. Jones, P.M. Griffin, A. Cronquist, S. Hanna, J. Hatch, T. Hayes, K. Larson, C. Nicholson, K. Wymore, M. Tobin-D'Angelo, N. Strockbine, P. Snippes, R. Atkinson and L.H. Gould. 2011. Foodborne illness acquired in the United States-major pathogens. Emerg. Infect. Dis. 17: 7–15.

Scharff, R.L. 2012. Economic burden from health losses due to foodborne illness in the United States. J. Food. Prot. 75: 123–131.

Scheutz, F., E.M. Nielsen, J. Frimodt–Moller, N. Boisen, S. Morabito, R. Tozzoli, J.P. Nataro and A. Caprioli. 2011. Characteristics of the enteroaggregative Shiga toxin/verotoxin-producing *Escherichia coli* O104:H4 strain causing the outbreak of haemolytic uraemic syndrome in Germany, May to June 2011. Euro. Surveill. 16: pii: 19889.

Scheutz, F., L.D. Teel, L. Beutin, D. Pierard, G. Buvens, H. Karch, A. Mellmann, A. Caprioli, R. Tozzoli, S. Morabito, N.A. Strockbine, A.R. Melton-Celsa, M. Sanchez, S. Persson and A.D. O'Brien. 2012. Multicenter evaluation of a sequence-based protocol for subtyping shiga toxins and standardizing stx nomenclature. J. Clin. Microbiol. 50: 2951–2963.

Schimmer, B., K. Nygard, H.M. Eriksen, J. Lassen, B.A. Lindstedt, L.T. Brandal, G. Kapperud and P. Aavitsland. 2008. Outbreak of haemolytic uraemic syndrome in Norway caused by stx2-positive *Escherichia coli* O103:H25 traced to cured mutton sausages. BMC Infect. Dis. 8: 41.

Schmidt, H., J. Scheef, S. Morabito, A. Caprioli, L.H. Wieler and H. Karch. 2000. A new Shiga toxin 2 variant (Stx2f) from *Escherichia coli* isolated from pigeons. Appl. Environ. Microbiol. 66: 1205–1208.

Swaminathan, B., T.J. Barrett, S.B. Hunte, R.V. Tauxe, and the CDC PulseNet task force. 2001. PulseNet: the molecular subtyping network for foodborne bacterial disease surveillance, United States Emerg. Infect. Dis. 7 382–389.

Swaminathan, B., P. Gerner-Smidt, L.K. Ng, S. Lukinmaa, K.M. Kam, S. Rolando, E.P. Gutierrez and N. Binsztein. 2006. Building PulseNet International: an interconnected system of laboratory networks to facilitate timely public health recognition and response to foodborne disease outbreaks and emerging foodborne diseases. Foodborne Pathog. Dis 3: 36–50.

van den Beld, M.J., F.A. Reubsaet. 2012. Differentiation between *Shigella*, enteroinvasive *Escherichia coli* (EIEC) and noninvasive *Escherichia coli*. Eur. J. Clin. Microbiol. Infect. Dis. 31: 899–904.

Wells, J.G., B.R. Davis, I.K. Wachsmuth, L.W. Riley, R.S. Remis, R. Sokolow and G.K. Morris. 1983. Laboratory investigation of hemorrhagic colitis outbreaks associated with a rare *Escherichia coli* serotype. J. Clin. Microbiol. 18: 512–520.

Wickham, ME., C. Lupp, M. Mascarenhas, A. Vazquez, B.K. Coombes, N.F. Brown, B.A. Coburn, W. Deng, J.L. Puente, M.A. Karmali and B.B. Finlay. 2006. Bacterial genetic determinants of non-O157 STEC outbreaks and hemolytic-uremic syndrome after infection. J. Infect. Dis. 194: 819–827.

Wong, C.S., J.C. Mooney, J.R. Brandt, A.O. Staples, S. Jelacic, D.R. Boster, S.L. Watkins and P.I. Tarr. 2012. Risk factors for the hemolytic uremic syndrome in children infected with *Escherichia coli* O157:H7: a multivariable analysis. Clin. Infect. Dis. 55: 33–41.

CHAPTER 8

Lessons from the *Escherichia coli* O103 Outbreak in Norway 2006

Ivar Vågsholm[#]

Introduction

During the winter of 2006 there was an outbreak caused by an aggressive Verocytotoxin-producing[1] *Escherichia coli (E. coli)* O103:H25, in Norway. The outbreak lasted from January to April and affected 18 persons, mostly young children, ten of whom developed haemolytic uremic syndrome (HUS) and one child died. A case-control study found minced meat being the likely food source of the outbreak. On receiving this information, the risk managers initiated an urgent recall of minced meat products from the market. However, a number of cases appeared after the recall. Continued investigations found that the outbreak was caused by contaminated cold-smoked fermented mutton sausage. The same *E. coli* O103:H25 bacteria were isolated from both sausages, from mutton as well as from the patients. A contentious point was the conflicting advice given to the public during the outbreak; this week discard the minced meat you have in the fridge,

[#] Department of Biomedical Sciences and Veterinary Public Health, SLU, Box 7028, S-750 07 UPPSALA, Sweden.
Email: ivar.vagsholm@slu.se

[1] Note that STEC (Shiga like toxin producing *E. coli*) and VTEC (Verocytotoxin producing *E. coli*) are synonyms

next week discard all minced meat both in the fridge and freezer, and some weeks later finally throw away all fermented sausages made of mutton. In May 2006, the Norwegian Government decided to appoint a committee to evaluate of the handling of the outbreak by the authorities and the industry. The mandate included making recommendations to prevent future food-borne disease outbreaks and to be better prepared when outbreaks occurred in future. This article is based on this report (Anonymous 2007) and the author's reflections as a member of the committee.

Epidemiological Intelligence

Epidemiological intelligence

The public health authorities and industry should actively follow and analyse international and national developments both in terms of emerging serotypes, risk foods and transmission pathways for pathogenic *E. coli* as well as other food borne pathogens. Based on the analysis the risk assessment and management strategies for control of food borne pathogens should be updated regularly.

Monitoring and surveillance

Monitoring of VTEC—There is a need for strategic monitoring of VTEC in the food chain. One way of doing this would be to carry out national baseline studies at regular intervals, e.g., every 3rd year, for VTEC in cattle, sheep and goats, based on the diagnostic method with the best sensitivity, which up to now (2011) have been faeces samples. Moreover, the information provided to health professionals on the diagnosis and management of HUS and the methods used by primary microbiology laboratories for the detection of Verocytotoxin-producing *E. coli* (VTEC) should be updated regularly.

Early warning system—the key to successful risk management are timely actions based on early warnings. The traditional warning system is based on notification to the national public health and food safety authorities of predetermined food pathogens based on clinical symptoms, confirmatory laboratory investigations. Such a notification system might identify an outbreak too late, in particular if the outbreak is not clustered in time and space, e.g., same city and same week. In this outbreak there were scattered cases all over the country which is typical of modern food chains with widely dispersed food distribution. Earlier warning will enable public health and food safety authorities to investigate a possible outbreak and if needed, implementing interim control measures. Consequently, clinical syndromes such as HUS associated with gastro-intestinal symptoms should be made compulsorily notifiable. In this outbreak with the benefit of hindsight, detection and interim risk management response to the outbreak might have been feasible possibly 2–3 weeks earlier.

Outbreak Investigations and Control

Open mind—When investigating outbreaks caused by food borne pathogens, one should be aware of the fact that they may be caused by a wide range of contaminated foods in addition to meat and meat products, including vegetables, sprouted seeds (Buckholz et al. 2011) and drinking-water and also by direct contact with infected farm animals, bathing in contaminated water and by person-to-person contact so as to plan their investigations accordingly. It was also suggested to use Bradford-Hills criteria for causation as guidance for assessing the results of the investigations (Bradford-Hill 1965).

Collaboration risk managers and assessors—In dealing with outbreaks of foodborne diseases, close co-operation between risk managers and risk assessors is needed. Although it is important to separate risk assessment from risk management, this should not prevent risk assessors providing advice on risk management matters and possible control options. However, risk managers are always responsible for final risk management decisions.

Task force—When outbreaks of foodborne disease occur, a task force, with representatives from public health, veterinary, food safety authorities and scientific support organisations as well as experts from the food industry, should be established to ensure that all available information is used in the outbreak investigation and control. One typical task is critical review of the results from epidemiological and microbiological studies during an outbreak and of control actions including, e.g., the active follow-up of decisions to recall contaminated/suspect foods from the market.

Results from epidemiological studies—Case control studies are good for indicating the putative sources of a foodborne outbreak rapidly. In this outbreak, the hypothetical source could be indicated after 2–3 days. However, there are several methodological problems such as: information and confounding biases; the statistical problem of multiple comparisons (EFSA 2011); recall bias—the difficulties cases and controls (or their parents) have in remembering their cost histories several weeks ago. When making inferences, it is important to include all available information, i.e., the industry knowledge about the food chain, relevant descriptive information such as age and sex distributions, and whether the outbreak is a point source or continuous source epidemic. In this outbreak the descriptive epidemiology indicated a protracted point source over several weeks with wide geographical distribution. However, these pieces of information did not easily fit the hypothesis of minced meat being the source; since, as noted by the industry, minced meat has a shelf life of days and different distribution patterns. Targeted confirmative studies based on genetic fingerprinting of foodborne pathogens are also needed as soon as possible during outbreaks. These considerations should be kept in mind when the results are interpreted and considered as basis for risk management decisions and therefore external experts should be invited to review the results and inferences critically.

Food Safety

A priority should be to ensure a uniform and comprehensive food control throughout the industry in particular:

Carcass contamination—To reduce the risk of contamination of meat with *E. coli* O103 as well as other pathogenic organisms present in the gut or skin of the slaughtered animals, further measures are needed to improve the slaughter hygiene. This will include measures to improve the cleanliness of animals presented for slaughter and regular monitoring of faecal contamination of carcases. For sheep, this should include shearing just before slaughter. In general, a dirty animal, regardless of species should not be admitted for slaughter.

Safety standard of the fermentation process including validation—The fermentation process should be quantified and validated as a food safety hurdle, it is not sufficient to rely on historical absence of evidence. The reduction of pathogenic microorganisms achieved during the sausage production process should be documented. The manufacturers of cold-smoked fermented sausage products need to monitor the hygienic standard of all the ingredients in them and only meat from slaughterhouses with a good food hygiene record should be used. Moreover, they should introduce effective HACCP-based in-house control programmes to ensure the safety of the final product. Microbiological analysis of random samples can be used over a period of time to verify that the process is working.

Traceability—The food industry should develop their traceability systems to ensure rapid tracing and if needed, the possibility for a rapid recall from the market of unsafe foods. The food safety authority needs to monitor the recall of unsafe foodstuff from the market, regardless of whether the decision is its own or that of a food business operator.

Risk Communication

Communicating uncertainty—The communication of risks and their management are probably the most important and difficult part of handling an outbreak. There is a need for coordination between all involved, both food safety authorities and industry, as also the need for listening and addressing concerns of all stakeholders including consumers. It is important to give clear answers, but the hard part is to communicate the bounds of knowledge and uncertainty. A better communication of uncertainties and lack of knowledge between risk assessors and management is crucial for the successful management of an outbreak. Although there is a need to give clear messages to consumers, the food industry, retailers and others concerned, it is important that conclusions derived from case-control studies concerning the connection between specific foodstuffs and outbreaks of food-borne disease are balanced and that uncertainties associated with the conclusions are clearly communicated. In particular when there are only a few cases and controls, this is important.

Collaboration—In dealing with outbreaks of foodborne diseases, it is important that there is close co-operation between risk managers and risk assessors. Both have risk communication responsibilities. Although it is important to separate risk assessment from risk management so that risk managers do not influence risk assessors, this should not prevent risk assessors providing advice on risk management matters. However, risk managers are always responsible for final risk management decisions.

Final Words

It is important to note that all involved did their best and made tremendous efforts under very difficult circumstances to contain the outbreak for which they should be commended and acknowledged—still lessons were learned for future outbreaks.

References

Anonymous, 2007. *E. coli*-outbreak, Evaluation of government and industry risk management winter/spring 2006. Report from the government-appointed royal inquiry. Published by the Department of Agriculture and Food, and the Department of Health Oslo. 214 pages. In Norwegian with an English summary. (http://www.regjeringen.no/upload/kilde/hod/prm/2006/0113/ddd/pdfv/301864-e-coli-rapport-151206.pdf)

Bradford-Hill, A. 1965. "The Environment and Disease: Association or Causation?". Proceedings of the Royal Society of Medicine. 58: 295–300.

Buchholz, U., H. Bernard, D. Werbe, M.M. Böhmer, C. Remschmidt, H. Wilking, Y. Deleré, M. an der Heiden, C. Adlhoch, J. Dreesman, J. Ehlers, S. Ethelberg, M. Faber, C. Frank, G. Fricke, M. Greiner, M. Höhle, S. Ivarsson, U. Jark, M. Kirchner, J. Koch, G. Krause, P. Luber, B. Rosner, K. Stark and M. Kühne. 2011. German outbreak of Escherichia coli O104:H4 associated with sprouts. N. Engl. J. Med. 365: 1763–70.

EFSA Scientific Committee; Statistical Significance and Biological Relevance. EFSA Journal 2011; 9(9): 2372. 17 pp.

CHAPTER 9

Enterocococci

Krina Trivedi[#]

Enterococci

Enterococci are Gram positive ovoid or spherical cells; occurring in pairs, clusters or in short-chains. They do not form endospores, some of them are motile.

Scientific classification

Kingdom: Bacteria
Division: *Firmicutes*
Class: Cocci
Order: *Lactobacillales*
Family: *Enterococcaceae*
Genus: *Enterococcus*

This genus previously belonged to the Lancefield's serological group D streptococci; later they were sub-divided into three separate genera: *Streptococcus*, *Lactococcus* and *Enterococcus* based on modern classification techniques and serological studies (Schleifer and Klipper-Balz 1984, Devriese et al. 1993). The pathogenic species remained in the genus *Streptococcus* with the exception of *Streptococcus thermophilus*, they were separated from the non pathogenic and technically important species of the new genera *Lactococcus*

[#] Department of Milk Hygiene and Technology, Veterinary and Pharmaceutical Sciences University Brno, Palackeho 1/3, 612 42 Brno, Czech Republic.
Email: trivedikrina@gmail.com

(Deveriese and Pot 1995). The "faecal streptococci" associated with the gastrointestinal tract of humans and animals, with some fermented foods and with a range of other habitats, constitute the new genus *Enterococcus* (Franz et al. 2003). Enterococci are widely distributed in various habitat (soil, water, plants), followed by a recession in vertebrates, in food and clinical materials (Sedláček 2007).

Identification of bacteria

There are over 100 modifications of selective media for the isolation of enterococci from various specimens and due to the heterogeneity in the composition of the media it is impossible to recommend one universal medium (Domig et al. 2003). Several media for the isolation, enumeration and identification of enterococci have been reviewed, and it has been concluded that there are no 'ideal' media available, because most display drawbacks in terms of selectivity and recovery. As a result, the parallel use of two media, one highly, the other moderately selective, may be a reasonable way to obtain acceptable results from any food habitat.

For further confirmatory examination genotypic methods have been very necessary. In particular, 16S and 23S rRNA targeted probes proved to be successful in identifying *Enterococcus* species (Klein 2003). Other methods for intra species differentiation include protein fingerprinting, polymerase chain reaction (PCR)-based typing methods such as random amplified polymorphic DNA (RAPD), pulsed field gel electrophoresis (PFGE), and restriction enzyme analysis (Klein 2003).

Hazardous effect

Enterococcal cultures have been featured in dairy fermentations for decades, and isolates with histories of safe use are being promoted as probiotic cultures. The role of enterococci as probiotics is still controversial because they are also known as agents involved in several infections. The differences between an enterococcal pathogen and an apparently safe food originating strain are unclear. The molecular taxonomy of enterococci does not lead to a distinction between these two groups of strains. So, when present in food they can be viewed as potential pathogens very occasionally associated with outbreaks of food borne disease.

Enterococci belong to a range of micro-organisms that can decarboxylase free amino acids by substrate specific decarboxylases and creating biogenic amines (Trivedi et al. 2009). The consumption of food containing high levels of biogenic amines can have toxicological effects. Food intoxication caused by ingestion of biogenic amines determines a number of symptoms including headache, vomiting, increased blood pressure and even allergic reactions (Giraffa 2002). The ability to produce biogenic amines in cheese and fermented sausages has been reported for bacteria of the genus *Enterococcus* (Giraffa 2002).

Biogenic amines

The biogenic amines (BA) are low molecular weight; organic bases frequently found in many foods and can occur in high concentrations in fermented products obtained from raw materials characterized by high protein contents (Halász et al. 1994). They are found in different kinds of foods, such as fishery products, cheeses, wine beer, dry sausages and other fermented foods (Bover-Cid and Holzapfel 1999). BA occurs in various tissues and they play an important role in cell regeneration and differentiation. However, higher intake could have toxicological effects especially on allergic individuals as well as those populations who are consuming drugs with an action of the amine oxidase inhibitors (Komprda et al. 2008). The capability of BA formation has been described for several groups of micro-organisms, mainly *Enterobacteriaceae*, *Pseudomonas* spp., enterococci and some other lactic acid bacteria. It is very difficult to remove amines from food products once they are formed, even with the heat treatment such as autoclaving (Zaman et al. 2010). In foods, the presence of biogenic amines is indicative of undesired microbial activity. In addition, the occurrence of relatively high levels of certain biogenic amine has been reported as indicators of a deterioration process and/or defective elaboration. In foods, biogenic amines are mainly generated by decarboxylation of the corresponding amino acids through substrate specific enzymes of the microorganisms present in this environment (ten Brink et al. 1990), particularly histidine, tyrosine, lysine, ornithine and arginine (Silla Santos 1996).

In particular, *E. faecalis*, and *E. faecium* have been found to be prominent in the production of tyramine. Species other than like *E. durans*, *E. casseliflavus*, *E. mundtii* and *E. hirae* have also been reported for the production of tyramine (Trivedi et al. 2009). The ability to form histamine, tyramine and putrescine by the genera *Enterococcus* and *Lactobacillus* is not uniformly shared.

Impact of biogenic amines

Tyramine: Tyramine is the most important biogenic amine in foods, both in terms of the frequency with which it is detected and the concentrations it reaches. In addition, it is also one of the most biologically active of all biogenic amines (Ladero et al. 2010). It is formed by the action of tyrosine decarboxylase by the enterococci associated with the foods (Komprda et al. 2008). Tyramine is broken down in the mammalian organism to ketones, NH_3, and H_2O_2 in the mitochondrion of neurons, hepatocytes and enterocytes by the oxidative deamination catalyzed by the enzyme monoamine oxidase (MAO) (Komprda et al. 2008). Higher intake of tyramine in sensitive individuals could lead to serious health effects such as high blood pressure, brain haemorrhage, hypertension, migraine, and sometimes heart failure (Komprda et al. 2008). Foods in which tyramine could be present in higher toxicological levels can be associated with fermented vegetables (Špička et al. 2002), alcoholic beverages (Kalač et al. 2002), fermented dry sausages (Komprda et al. 2008) and especially ripened cheeses. Ripened cheeses are considered to be most important food borne source of biogenic amines (Stratton et al. 1991).

Histamine: Histamine is considered as the most active amine and is related to almost all food amines poisoning incidences (Zaman et al. 2010). Excessive formation of histamine is usually observed only in tissues rich in the precursor amino acid or in fermented foods, and it may cause adverse health effects due to its vasoactivity. High histamine levels can induce migraine, headaches, vertigo, nausea, vomiting, hypotension, arythmia, anaphylaxia. Histamine excites the smooth muscles of the uterus and gastrointestinal tract, often enhancing the release of acid secretion from the gastric mucosa, provoking dysmenorrhea, cramps, stomach ache and diarrhea (Russo et al. 2010). Scombroid fish poisoning, also known as histaminosis is a food-borne illness caused by consumption of fish containing high concentrations of histamine (Russo et al. 2010).

Other biogenic amines: Putrescine, although it seems to have a lower pharmacological activity, hampers the detoxification of histamine and tyramine. Moreover, the presence of the amine putrescine is investigated in meat products since it can react with nitrite to form nitrosopyrrolidine, a heterocyclic carcinogenic nitrosamine (Landete et al. 2007). Cadaverine is a BA not considered as toxic, although it can potentiate the toxicity of histamine (Landete et al. 2007). The amounts of histamine, putrescine and cadaverine usually increase during spoilage of fish or meat (Pircher et al. 2007).

Legislation: Commission regulation (EC) No 2073/2005 of 15 November 2005 on microbiological criteria for foodstuffs has set the critical levels for histamine not to exceed 200 mg/kg or 100 mg/kg according to whether the product has undergone enzyme maturation treatment in brine or not.

There are no set regulations for tyramine by European Union but there are certain European countries such as Slovak Republic, their Nutritional Codex has set the limitation of 200 mg/kg of tyramine in cheeses (Karovičova and Kohajdova 2003).

Virulence genes

The enterococci are recognized as the causative agents of a number of non food borne clinical infections, such as bacteraemia and endocarditis, and in recent years there has been increasing concern over the number of emerging vancomycin-resistant enterococci strains (VREs). Enterococci are also noted for their capacity to exchange genetic information by conjugation (Clewell 1990), and these processes are known to take place in the gastrointestinal tract (Eaton and Gasson 2001). As well as transmissible antibiotic resistance plasmids, virulence factors such as hemolysin-cytolysin production and the capacity for adhesion are known to be transmissible by highly efficient gene transfer mechanisms (Wirth 1994).

Much progress has been made in determining virulence factors from clinical enterococcal isolates using molecular biological techniques. As a result, we now have the opportunity to study the presence of virulence factors among food isolates. Enterococci may carry various genes directly or indirectly contributing to virulence. Genes encoding virulence factors such as aggregation substances, endocarditis antigen, gelatinase, enterococcal surface protein, hyaluronidase and adhesion collagen protein have been described in enterococci isolated from foodstuffs (Trivedi et al. 2011).

Impact of main virulence genes: Strains encoding virulence factors in their genome are potentially dangerous for the host and may also transfer their virulence genes to susceptible recipients in the intestinal microflora.

*Aggregation substances (asa*1*)*: Strains without a specific means of attachment to the surface of the gastrointestinal tract are considered avirulent and are eliminated by the usual intestinal motility (Jett et al. 1994). Aggregation substance (AS) and the enterococcal surface protein are two important adhesins of enterococci, which mediate attachment.

Aggregation substances encoded by *asa*1 is a pheromone-inducible protein that enables the conjugative transfer of sex pheromone gene-containing plasmids through the clumping of one cell to another (Galli et al. 1990). As a virulence factor, aggregation substance increases bacterial adherence to renal tubular cells and heart endocardial cells (Jett et al. 1998), enhances internalization by intestinal epithelial cells (Jett et al. 1998) and has been shown to increase the valvular vegetation mass in an animal model of endocarditis (Chow et al. 1993).

Enterococcal surface protein (esp): The enterococcal surface protein, encoded by the chromosomal *esp*, has an interesting structure that includes a central core consisting of distinct tandem repeat units. This central repeat region serves as a retractable arm, extending the N terminal globular domain through the cell wall to the surface, which might facilitate immune evasion in case of immune deficiency (Dupré et al. 2003). Enterococcal surface protein is associated with increased virulence, colonization and persistence in the urinary tract and biofilm formation (Dupré et al. 2003). A variant *esp* gene has recently been identified as a marker of highly prevalent vancomycin-resistant *E. faecium* (VREF) clones among hospitalized patients (Willems et al. 2001). However, the *esp* gene has also been detected in vancomycin-susceptible *E. faecium* (VSEF) isolates (Woodford et al. 2001).

*Gelatinase (gel*E*)*: Gelatinase, encoded by the chromosomal *gel*E, is an extracellular zinc endopeptidase that hydrolyzes collagen, gelatin, and small peptides (Su et al. 1991) and it has been reported to increase the severity of endocarditis in an animal model (Vankerckhoven et al. 2004).

*Cytolysin (cyl*A*)*: The production of cytolysin has also been shown to significantly worsen the severity of endocarditis (Chow et al. 1993) and endophthalmitis (Vankerckhoven et al. 2004) in animal models as well as to contribute to the severity of enterococcal disease in humans (Vankerckhoven et al. 2004). Cytolysin genes are carried on a plasmid or are integrated into the bacterial chromosome (Jett et al. 1994). Cytolysin consists of two components, lysin (L) and activator (A). The cytolysin operon consists of five genes, of which *cyl*L1, *cyl*L2, *cyl*M, and *cyl*B are relevant to the expression of component L, whereas *cyl*A is necessary for the expression of component A (Vankerckhoven et al. 2004).

Hyaluronidase (hyl): Recently, another virulence factor, hyaluronidase, was described in *E. faecium* (Rice et al. 2003). The *E. faecium* hyaluronidase, encoded by the chromosomal *hyl*, shows homology of the hyaluronidases previously described in *Streptococcus pyogenes*, *Staphylococcus aureus*, and *Streptococcus pneumoniae*, which are believed to contribute to invasion of the nasopharynx and pneumococcal pneumonia (Vankerckhoven et al. 2004).

Accessory colonization factor (*ace*): Another cell surface protein, which exhibits strong similarities to the *S. aureus* collagen binding protein Cna, has recently been identified (Rich et al. 1999). This *E. faecalis* specific surface component, mediates binding to certain collagens (Rich et al. 1999) and may play a role in the pathogenesis of endocarditis (Nallapareddy et al. 2000).

Endocarditis antigen (*efa*A): Endocarditis antigen encoded by *efa*A, a serum -inducible surface protein that shows extensive similarities to several adhesins of streptococci (Lowe et al. 1995), is a putative endocarditis antigen and demonstrated a potential biological role in a mouse peritonitis model (Singh et al. 1998).

Antibiotic resistance

Enterococci are low grade pathogens but their intrinsic resistance to several antibiotics and their acquisition of resistance to the few antibiotics available for treatment in clinical therapy, such as glycopeptides, has led to difficulties in and a search for new drugs and therapeutic options. Unlike acquired resistance and virulence traits, which are usually transposon or plasmid encoded. Intrinsic resistance is based in chromosomal genes, which are typically non transferable. Penicillin, ampicillin, and vancomycin are among the few antibiotics that show consistent inhibitory, but not bactericidal, activity against *E. faecalis*. *E. faecium* are less susceptible to β-lactam antibiotics than *E. faecalis* because the penicillin-binding proteins of the former have markedly lower affinities for the antibiotics (Huycke et al. 1998).

Enterococci often acquire antibiotic resistance through exchange of resistance-encoding genes carried on conjugative transposons, pheromone-responsive plasmids and other broad host range plasmids (Rice et al. 1995). Among several phenotypes for vancomycin resistant enterococci (VRE), *van*A (resistance to vancomycin and teicoplanin) and *van*B (resistance to vancomycin alone) are most common (Arthur et al. 1993). *van*A and *van*B types of resistance are primarily found among enterococci isolated from clinical, veterinary and food specimens (Klare et al. 1995) but not other common intestinal or environmental bacteria.

Growth and survival

Enterococci are facultative anaerobes, with chemoorganotrophic fermentative metabolism which requires nutritionally rich medium for growth. They use a wide range of carbohydrates, the main product of fermentation of L (+)–lactic acid but there is no gas formation. They are catalase negative, often showing (α-hemolysis) on blood agar. They can grow in a wide temperature range from 10°C to 45°C. Some strains can grow at temperatures as low as 1°C. The maximum reported growth is 50°C, but optimum is 37°C. They can also grow in presence of 10% NaCl or 40% of bile in the medium and at pH till 9.6 (Sedlácek 2007). Minimum water activity for growth is generally dependent on the presence of solute; *E. faecalis* is reported to grow at 0.93.

Heat resistance

The enterococci have many times been associated with spoilage of some cooked meats and pasteurized milk. The reason for their presence is that they are relatively heat resistant and able to survive many mild pasteurization processes.

The Patient

Clinical picture

Enterococci used to be known only as causes of endocarditis and rare cases of meningitis. This picture has changed considerably in the last decade: these bacteria have become one of the leading causes of nosocomial (hospital-acquired) bacteremia, and of surgical and urinary tract infections (Vogel et al. 2010). In the past, the presence of safer strains within food-borne enterococci was emphasized (Giraffa and Sisto 1997). This suggestion has been recently contradicted after finding a marked strain-specific incidence of virulence factors associated with enterococci (Franz et al. 2001). The most unfortunate is the high efficiency of enterococci to transfer different genes via conjugative plasmid transfer, transposons and insertion sequences (Abriouel et al. 2008), e.g., antibiotic resistance genes, to pathogenic bacteria is a worrying trait which weakens any selection criteria (Giraffa 2003). At least 20 *Enterococcus* species have been described, but the most common species associated with foods and human diseases are *E. faecalis* and *E. faecium*.

Risk groups

All individuals are thought to be susceptible to food poisoning caused by enterococci. Immunocompromised and long term or intensive care hospitalized patients in particular have a greater risk to acquire such infections (Vogel et al. 2010).

Incidence and outbreaks, food and water

There is a very little data available on incidence and outbreaks of food borne enterococcal infections. It is the presence of high numbers of *Enterococcus* species and the absence of other food borne pathogens that has caused some outbreaks of food borne disease to be linked with the enterococci (Richard et al. 2008). Food borne outbreaks have been associated with sausages, ham, milk, cheeses and chocolate pudding.

Legislation

EU directive 98/83/EC of 3 November 1998 on the quality of water intended for human consumption has set regulations for level of enterococci in drinking water and water used in food industry. These requirements are a level of 0/100 ml. For water on sale in bottles or containers there is stricter requirement of 0/250 ml.

Enterococcal infection

The infectious dose for food borne outbreaks is found high (> 10^7 cells), and the incubation time is reported to vary widely between 2–60 h. Symptoms normally include abdominal cramps diarrhea, nausea, vomiting and dizziness. The disease is thought to be typically of short duration and self limiting (Richard et al. 2008).

Risk Food

Enterococci can become an important part of the fermented food microflora (especially fermented meats, cheeses and other dairy products, olives), because they can withstand normal conditions of food processing and can also contaminate finished products during food production. *E. faecium* and *E. faecalis* are the two species mostly found in food and to a less extent *E. durans* is detected. *E. hirae* and *E. casseliflavus* are encountered occasionally (Giraffa 2003). The presence of enterococci in the gastrointestinal tract of animals leads to a high possibility of contamination of meat at the time of slaughter. Enterococci may enter the milk either directly from human or animal faeces or indirectly from contaminated water sources, outside of the animals and/or from the milking equipment and bulk storage tank (Giraffa 2003). Enterococci occur in a variety of cheeses, especially artisanal cheeses produced in southern Europe from raw or pasteurized milk, and in natural milk starters.

Route of infection/transmission

Poor hygiene, under-processed food and cross-contamination, when cooked material comes into contact with raw produce or contaminated materials (cutting boards), are the main causes of infection.

Control and prevention

Proper cooking and hygienic food handling can prevent *Enterococcus* infections to a large extend. If strict adherence to heat processing regimes and the control of chilling chains is maintained, it minimizes the numbers of any enterococci present in pasteurized foods (Richard et al. 2008). Also, strict adherence to cleaning regimes and the use of proper sanitizers can control the organisms in food processing establishments (Richard et al. 2008).

Concluding Remarks

Enterococci are commonly found as part of the natural microflora of the gastrointestinal tract in humans and animals. They are used in production of some meat and dairy products, due to their important role in flavor, development and fermentation. There has been a history of safe use for the organisms. But, still the risk prevails. Fermented foods of both animal and plant origin and enterococci strains used as probiotics or starter cultures should be put under detailed control.

References

Abriouel, H., N. Ben Omar, A.C. Molinos, R.L. Lo´pez, M.J. Grande, P. Martinez Viedma, E. Ortega, M.M. Cañamero and A. Galvez. 2008. Comparative analysis of genetic diversity and incidence of virulence factors and antibiotic resistance among enterococcal populations from raw fruit and vegetable foods, water and soil, and clinical samples. Int. J. Food Microbiol. 123: 38–49.

Arthur, M. and P. Courvalin. 1993. Genetics and mechanisms of glycopeptide resistance in enterococci. Antimicrob. Agents Ch. 37: 1563–1571.

Bover-Cid, S. and W.H. Holzapfel. 1999. Improved screening procedure for biogenic amine production by lactic acid bacteria. Int. J. Food Microbiol. 53: 33–41.

Chow, J.W., L.A. Thal, M.B. Perri, J. Vazquez, A.S.M. Donabedian, D.B. Clewell and M.J. Zervous. 1993. Plasmid encoded hemolysin and aggregation substance production contribute to virulence in experimental enterococcal endocarditis. Antimicob. Agents Ch. 37: 2474–2477.

Clewell, D.B. 1990. Movable genetic elements and antibiotic resistance in enterococci. Eur. J. Clin. Microbiol. 9: 90–102.

Commission Regulation (EC) No 2073/2005 of 15 November 2005 on microbiological criteria for foodstuffs available at http://eur-lex.europa.eu/LexUriServ/LexUriServ.do?uri=CELEX:32005R2073:en:NOT.

Commission Regulation (EC) No 98/83 of 3 November 1998 on the quality of water intended for human consumption available at http://eurlex.europa.eu/LexUriServ/LexUriServ.do?uri=CELEX:31998L0083:en:NOT.

Devriese, L.A., K. Ceyssens, U.M. Rodrigues and M.D. Collins. 1993. *Enterococcus columbae*, a species from pigeon intestines. Int. J. Syst. Bacteriol. 43: 188–189.

Deveriese, L.A. and B. Pot. 1995. The genus *Enterococcus In:* B.J.B. Wood and W.H. Holzapfel. (eds.). The genera of lactic acid bacteria. Blackie Academie and Professional, London. pp. 327–367

Domig, K.J., H.K. Mayer and W. Kneifel. 2003. Methods used for the isolation, enumeration, characterization and identification of *Enterococcus* spp.: 1. Media for isolation and enumeration. Int. J. Food Microbiol. 88: 147–164.

Dupré, I., S. Zanetti, A.M. Schito, G. Fadda and L.A. Sechi. 2003. Incidence of virulence determinants in clinical *Enterococcus faecium* and *Enterococcus faecalis* isolates collected in Sardinia (Italy). J. Med. Microbiol. 52: 491–498.

Eaton, T.J. and M.J. Gasson. 2001. Molecular screening of *Enterococcus* virulence determinants and potential for genetic exchange between food and medical isolates. Appl. Environ. Microb. 67: 1628–1635.

Franz, C.M.A.P., A.B. Muscholl-Silberhorn, M.K. Yousif, M. Vancanneyt, J. Swings and W.H. Holzapfel. 2001. Incidence of virulence factors and antibiotic resistance among enterococci isolated from food. Appl. Environ. Microbiol. 67: 4385–4389.

Franz, C.M.A.P., M.E. Stiles, K.H. Schleifer and W.H. Holzapfel. 2003. Enterococci in foods—a conundrum for food safety. Int. J. Food Microbiol. 88: 105–122.

Galli, D., F. Lottspeich and R. Wirth. 1990. Sequence analysis of *Enterococcus faecalis* aggregation substance encoded by the sex pheromone plasmid pAD1. Mol. Microbiol. 4: 895–904.

Giraffa, G. and F. Sisto. 1997. Susceptibility to vancomycin of enterococci isolated from dairy products. Lett. Appl. Microbiol. 25: 335–338.

Giraffa, G. 2002. Enterococci from foods. FEMS Microbiol. Rev. 26: 163–171.

Giraffa, G. 2003. Functionality of enterococci in dairy products. Int. J. Food Microbiol. 88: 215–222.

Halász, A., Á Baráth, L.S. Sarkadi and W.H. Holzapfel. 1994. Biogenic amines and their production by microorganisms in food. Trends Food Sci. Tech. 5: 42–49.

Huycke, M.H., D.F. Sahm and M.S. Gilmore. 1998. Multiple-drug resistant enterococci: the nature of the problem and an agenda for the future. Emerg. Infect. Dis. 4: 239–249.

Jett, B.D., M.M. Huycke and M.S. Gilmore. 1994. Virulence of enterococci. Clin. Microbiol. Rev. 7: 462–478.

Jett, B.D., R.V. Atkuri and M.S. Gilmore. 1998. *Enterococcus faecalis* localization in experimental endophthalamitis: Role of plasmid-encoded aggregation substance. Infect. Immun. 66: 843–848.

Kalač, P., J. Šavel, M. Křížek, T. Pelikánová and M. Prokopová. 2002. Biogenic amine formation in bottled beer. Food Chem. 79: 431–434.

Karovičová, J. and Z. Kohajdová. 2003. Biogenic amines in food. Chemical paper. 59: 70–79.

Klare, I., H. Heier, H. Claus, R. Reissbrodt and W. Witte. 1995. *van*A mediated high level glycopeptide resistance in *Enterococcus faecium* from animal husbandry. FEMS Microbiol. Letters. 125: 165–172.

Klein, G. 2003. Taxonomy, ecology and antibiotic resistance of enterococci from food and the gastro-intestinal tract. Int. J. Food Microbiol. 88: 123–131.

Komprda, T., R Burdychová, V. Dohnal, O. Cwiková, P. Sládková and H. Dovráčková. 2008. Tyramine production in Dutch-type semi-hard cheese from two different producers. Food Microbiol. 25: 219–227.

Ladero, V., M. Fernández, I. Cuesta and M.A. Alvarez. 2010. Quantitative detection and identification of tyramine producing enterococci and lactobacilli in cheese by multiplex qPCR. Food Microbiol. 27: 933–939.

Landete, J.M., B. De las Rivas, A. Marcobal and R. Munoz. 2007. Molecular methods for the detection of biogenic amine producing bacteria on foods. Int. J. Food Microbiol. 117: 258–269.

Lowe, A.M., P.A. Lambert and A.W. Smith. 1995. Cloning of and *Enterococcus faecalis* endocarditis antigen: homology with adhesions from some oral streptococci. Infect. Immun. 63: 5218–5224.

Nallapareddy, S.R., X. Qin, G.M. Weinstock, M. Hook and B.E. Murray. 2000. *Enterococcus faecalis* adhesion, *ace*, mediates attachment to extracellular matrix proteins collagen type IV and laminin as well as collagen type I. Infect. Immun. 68: 5218–5224.

Pircher, A., F. Bauer and P. Paulsen. 2007. Formation of cadaverine, histamine, putrescine and tyramine by bacteria isolated from meat, fermented sausages and cheeses. Eur. Food Res. Technol. 226: 225–231.

Rice, E.W., J.W. Messer, C.H. Johnson and D.J. Reasoner. 1995. Occurrence of high level aminoglycoside resistance in environmental isolates of enterococci. Appl. Environ. Microbiol. 61: 374–376.

Rice, L.B., I. Carias, S. Rudin, C. Vael, H. Goossens, C. Konstabel, I. Klare, S.R. Nallapareddy, W. Huang and B.E. Murray. 2003. A potential virulence genes hyl_{efm} predominates in *Enterococcus faecium* of clinical origin. J. Infect. Dis. 187: 508–512.

Rich, R.L., B. Kreikemeyer, R.T. Owens, S. LaBrenz, S.V. Narayana, G.M. Weinstock, B.E. Murray and M. Hook. 1999. Ace is a collagen-binding MSCRAMM from *Enterococcus faecalis*. J. Biol. Chem. 274: 26939–26945.

Richard, L., L. Curtis and J. Davis. 2008. The Food Safety Hazard: Guidebook. The Royal Society of Chemistry, UK. pp. 42–44.

Russo, P., G. Spano, M.P. Arena, V. Capozzi, D. Fiocco, F. Grieco and L. Beneduce. 2010. Are consumers aware of the risks related to biogenic amines in food? *In:* A. Méndez-Vilas (eds.). Current Research, Technology and Education Topics in Applied Microbiology and Microbial Biotechnology. Formatex Research Center, Spain. pp. 1087–1093.

Schleifer, K.H. and R. Kilpper-Balz. 1984. Transfer of *Streptococcus faecalis* and *Streptococcus faecium* to the genus *Enterococcus* nom.rev.as *Enterococcus faecalis* comb. nov. and *Enterococcus faecium* comb.nov. Int. J. Syst. Bacteriol. 34: 31–34.

Sedláček, I. 2007. Taxonomie prokaryot (Taxonomy of Prokaryotes). 1. edi. Brno: Masarykova Univerzita, pp. 248–249.

Singh, K.V., T.M. Coque, G.M. Weinstock and B.E. Murray. 1998. *In vivo* testing of *Enterococcus faecalis efa*A mutant and use of *efa*A homologs for species identification. Fems Immunol. Med. Mic. 21: 323–331.

Silla Santos, M.H. 1996. Biogenic amines: their importance in foods. Int. J. Food Microbiol. 29: 213–231.

Špička, J., P. Kalač, S. Bover-Cid and M. Křížek. 2002. Application of lactic acid bacteria starter cultures for decreasing the biogenic amines levels in sauerkraut. Eur. Food Res. Technol. 215: 509–514.

Stratton, J.E., R.W. Hutkins and S.L. Taylor. 1991. Biogenic amines in cheese and other fermented foods a review. J. Food Protect. 54: 460–470.

Su, Y.A., M.C. Sulavik, P. He, K.K. Makinen, P.L. Makinen, S. Fiedler, R.Wirth and D.B. Clewell. 1991. Nucleotide sequence of the gelatinase gene (*gel*E) from *Enterococcus faecalis* subsp. *liquefaciens*. Infect. Immun. 59: 415–420.

ten Brink, B., C. Damink, H.M.L.J. Joosten and J.H.J. Huisint Veld. 1990. Occurrence and formation of biologically active amines in foods. Int. J. Food Microbiol. 11: 73–84.

Trivedi, K., I. Borkovcová and R. Karpíšková. 2009. Tyramine production by enterococci from various foodstuffs: A threat to the consumers. Czech J. Food Sci. 27: 357–360.

Trivedi, K., Š. Cupakova and R. Kárpíšková. 2011. Virulence factors and antibiotic resistance in enterococci isolated from foodstuffs. Vet. Med. Czech. 56: 352–357.

Vankerckhoven, V., T. van Autgaerden, C. Vael, C. Lammens, S. Chapelle, R. Rossi, D. Jabes and H. Goossens. 2004. Development of multiplex PCR for the detection of *asa*1, *gel*E, *cyl*A, *esp*, and *hyl* genes in enterococci and survey for virulence determinants among European hospital isolates of *Enterococcus faecium*. J. Clin. Microbiol. 42: 4473–4479.

Vogel, R.F., C.M.H. Frestl and C.A. Lenz. 2010. Sub-lethal stress effect on virulence gene expression in *Enterococcus faecalis*. Food Microbiol. 27: 317–326.

Willems, R.J., W. Homan, J. Top, M. van Santen-Verheuvel, D. Tribe, X. Manzioros, C. Gaillard, C.M. Vandenbroucke-Grauls, E.M. Mascini, E. van Kregten, J.D. van Embden and M.J. Bonten. 2001. Variant esp gene as a marker of a distinct genetic lineage of vancomycin –resistant *Enterococcus faecium* spreading in hospitals. Lancet. 357: 853–855.

Wirth, R. 1994. The sex pheromone system of *Enterococcus faecalis* more than just a plasmid collection mechanism. Eur. J. Biochem. 222: 235–246.

Woodford, N., M. Soltani and K.J. Hardy. 2001. Frequency of *esp* in *Enterococcus faecium* isolates. Lancet. 358: 584.

Zaman, M.Z., F.A. Bakar, J. Selmat and J. Bakar. 2010. Occurrence of biogenic amines and amines degrading bacteria in fish sauce. Czech J. Food Sci. 28: 440–449.

CHAPTER 10

Listeria monocytogenes—Very Food-borne Bacteria

Wilhelm Tham[#,*] and
Marie-Louise Danielsson-Tham[#]

Introduction

In both man and animals *Listeria monocytogenes* is the most important cause of the disease listeriosis. The main manifestations of listeriosis (septicemia, abortion, meningoencephalitis/encephalitis) are similar in both animals and humans. The most common mode of transmission is through consumption of food and feed.

Classification

The genus *Listeria* comprises catalase positive, non-spore-forming, non-encapsulated, facultative anaerobic Gram-positive rods. The genus includes both pathogenic and nonpathogenic species. The species are *Listeria monocytogenes, L. innocua, L. seeligeri, L. welshimeri, L. grayi, L. ivanovii* (subspecies *ivanovii* and subspecies *londoniensis*), *L. marthii* and *L. rocourtiae* (Seeliger and Jones 1986, Boerlin et al. 1992, Leclercq et al. 2010, Graves et al. 2010). First described by Gustav Hülphers in 1911 (Hülphers 1911), the bacterium was given its final name in 1940. The species name

School of Hospitality, Culinary Arts and Meal Science, Örebro University, Box 1, S-712 02 Grythyttan, Sweden.
Emails: Listeria@Telia.com; marie-louise.danielsson-tham@oru.se
* Corresponding author

"*L. monocytogenes*" is related to its ability to survive in monocytes and to cause monocytocis in animals, but not in man. After 36h incubation on blood agar, *L. monocytogenes* presents a narrow, clear hemolysis around the colony. *L. monocytogenes* ferments L-rhamnose, but not D-xylose, and cell tumbling motility at room temperature, but not at 37°C, is characteristic (Seeliger and Jones 1986).

L. monocytogenes is commonly divided through serotyping into 13 serovars. However only four serovars 1/2a, 1/2b, 1/2c and 4b are reported to be responsible for 93% of 1363 listeriosis patients in the United Kingdom between 1967 and 1988 (McLauchlin 1990). *L. monocytogenes* is suggested to have a clonal genetic structure, i.e., it consists of different clonal lineages and the horizontal gene flow between the different serovar-associated clonal lineages is considered low (Kathariou 2002). The species is divided into four lineages where lineage I includes serovars 1/2b, 3b, 4b, 4d and 4e, and lineage II includes serovars 1/2a, 3a, 1/2c, and 3c. Lineages III and IV include serovars 4a, 4c and an atypical 4b, which is less seen in human cases (Wiedmann et al. 2002, Nightingale et al. 2007, O'Connor et al. 2010, Orsi et al. 2011, Tsai et al. 2011). While Lineage I is highly clonal, lineage II shows greater diversity and evidence of horizontal gene transfer (Nightingale et al. 2005).

Distribution in Humans and Animals (healthy carriers)

Husu (1990) isolated *L. monocytyogenes* in feces from 6.7% of 3878 randomly selected dairy cows, representing 240 farms in Finland during two years. The shedding of *L. monocytogenes* from clinically healthy cows means that cattle feces could contaminate milk, dairy products and carcasses with *L. monocytogenes*, and thus contribute to food-borne listeriosis. Around 5 to 10% of humans could be healthy carriers of *L. monocytogenes* (Farber and Peterkin 1991).

Characterization of Strains by Phenotyping and Genotyping Method

The ability to distinguish accurately between different strains within a bacterial species is a fundamental requirement for epidemiological surveillance and microevolution studies (Cooper and Feil 2004). Beyond the species or subspecies level, isolates can be further differentiated by subtyping (Wiedmann 2002). The most widely used phenotypical method is serotyping, either with classical methods (Seeliger and Höhne 1979) or through PCR. A good agreement was obtained between molecular and classical serotyping methods using two PCR assays, a multiplex PCR and a *fla*A PCR (Kérouanton et al. 2010) or a multiplex PCR-based assay, in combination with an antibody-based serogrouping assay (Burall et al. 2011).

Further subtyping is usually performed by genotyping, i.e., molecular methods. These methods are either band-based methods with DNA fragment pattern data (DNA fingerprints) or methods that generate DNA sequence data.

Fragment-based/Band-based/Gel-based methods with DNA fragment pattern data (DNA fingerprints)

Pulsed-field gelelectrophoresis (PFGE)

PFGE involves the analysis of DNA fragment patterns, and cluster isolates according to the similarity in banding pattern, which can be illustrated through a dendrogram. Although a dendrogram constructed from a PFGE procedure may appear similar to and sometimes correlates well with a phylogenetic tree, dendrograms are not based on DNA sequences and should not be used to provide expression of the evolutionary history of certain isolates. However, PFGE can be used for assigning lineages (Nightingale et al. 2007, Orsi et al. 2011).

PFGE has the advantage that it can be performed in the absence of detailed information about the genome (Cooper and Feil 2004). PFGE probes the entire genome, whereas, e.g., *multi-locus sequence typing* (MLST) only analyses nucleotides within specific genes (Revazishvili et al. 2004). However, PFGE is labor-intensive and often relies on computer-based band marking, which can be inaccurate and requires manual interpretation by skilful personal; thus, the results are difficult to compare between laboratories.

Methods that generate DNA sequence data

L. monocytogenes is one of the most highly sequenced human pathogens (Nightingale 2010). As more DNA sequencing data becomes available, further sequence-based subtyping methods for *L. monocytogenes* are introduced. These methods will probably supersede band-based methods as they are more objective and the sequence data can be used to infer the evolutionary history of a group of isolates. Methods that generate DNA sequence data can be used to construct phylogenetic trees, reveal genetic lineages and clonal groups, and for studying horizontal gene transfer (Nightingale 2010).

Multiple-locus variable-number tandem repeat analysis (MLVA)

MLVA is based on multiple *Variable-Number of Tandem Repeat* (VNTR) loci. Tandem repeat sequences are among the most variable regions within a bacterial genome. A full genome sequence is usually needed to identify VNTR as targets for an MLVA typing scheme. MLVA could be used to study the short-term evolution of food-borne pathogens due to the rapid accumulation of changes in VNTR regions (Nightingale 2010). However, as the mechanisms behind mutations in repeated DNA are unknown, it is uncertain whether MLVA can be used for studying the evolution of *L. monocytogenes*. There is a correlation between PFGE, MLST and MLVA for *L. monocytogenes* (Miya et al. 2008), which indicates that all methods follow *Listeria* evolution, but more studies comparing results from the different methods are needed to confirm this and it is too early to state whether MLVA or PFGE can be used for evolution studies (Lindstedt et al. 2008).

Multi-locus sequence typing (MLST) or multi-virulence-locus sequence typing (MVLST)

Direct sequencing of DNA is the most accurate method for evaluating the genetic relationship of organisms and for conducting long-term studies on the evolution of bacterial pathogens, i.e., predicting the evolution of the genome (Kersting 2008). Housekeeping genes (necessary for bacterial cells to survive) evolve slowly and are useful for providing reliable information about evolutionary relationships/reconstruction of phylogenies. However, genes with the potential for positive selection, e.g., bacterial virulence genes or virulence-associated genes, evolve quickly and can assist in understanding the evolution of virulence-associated characteristics in bacterial pathogens. Traditional MLST targets housekeeping genes and MVLST targets virulence genes. A combination can be used for phylogenetic studies and identify epidemic clones (Nightingale et al. 2005, Nightingale 2010, den Bakker et al. 2010).

Whole-genome sequencing

Besides total chromosome sequencing, all available molecular approaches to genomic comparison represent incomplete datasets, where some information is lacking (Goering 2010). When sequencing technologies become cheaper, whole genome sequencing may be an appropriate subtyping tool for replacing all other subtyping methods. Next-generation sequencing has greatly improved the speed with which whole-genome sequences can be generated. One application for a generation of whole-genome sequences is in producing epidemiologically relevant data quickly. This information can be used for informing infection control through tracking transmission routes in a clonal outbreak or for identifying virulence genes in an epidemic that is still unfolding. This method is expected to become increasingly popular as a tool in public health scenarios (Baldry 2010).

Virulence/Virulence Factors

The main infectious route for both humans and animals is gastrointestinal, where, after oral intake of contaminated food/feed, *L. monocytogenes* induces its own uptake in the intestines by invading enterocytes. From there, *L. monocytogenes* can spread to neighboring cells and into the blood stream, reaching the liver and spleen haematogenously. If the system is normal, hepatic and splenic macrophages kill the invading bacteria. However, in an immuno-compromised individual, *L. monocytogenes* causes bacteremia and/or spreads to other organs. *L. monocytogenes* has a predilection for the placenta and the central nervous system, leading to meningitis, encephalitis, and maternofoetal infection (Oevermann et al. 2010).

The most important virulence factors involved in infection are internalin A (InlA), internalin B (InlB), listeriolysin (LLO), phospholipase C, and ActA. Internalin A is a surface protein essential for the passage of intestinal epithelial cells, as it recognizes and binds to (the human) receptor protein E-cadherin present on the host cell membrane. E-cadherin is expressed on cells of three host barriers: intestinal, blood-brain and placental barriers (Lecuit 2005): this attachment stimulates host cell phagocytosis of *L. monocytogenes*. Different types of host cells, such as fibroblasts, hepatocytes, epithelial and endothelial cells, can be invaded by *L. monocytogenes* through the surface protein Internalin B (InlB), which interacts with the hepatocyte growth factor "Met". There are similarities between the molecular mechanism behind InlB-mediated invasion into host cells and classical phagocytosis. On phagocytosis by the host cells, *L. monocytogenes* is enveloped in a small single-layer membrane phagocytic vacuole in the cell (phagosome). Listeriolysin O (LLO), which is responsible for the hemolytic properties in *L. monocytogenes*, creates pores in the phagosomal membrane: this blocks acidification of this compartment, and prevents its fusion with lysosomes (Shaughnessy et al. 2006), thus, allowing the listeria bacteria to escape into the cytoplasm of the host cell. Two different phospholipase C are involved in *L. monocytogenes* invasion and spread to host cells: phosphatidyl-inositol phospholipase C (PI-PLC) and phosphatidyl-choline phospholipase C (PC-PLC). Together with listeriolysin (LLO), these phospholipases facilitate the escape of *L. monocytogenes* from the first phagosome, the cell-to-cell-spread, and the creation of the second phagosome. As soon as *L. monocytogenes* bacteria have escaped from the first phagosome into the cytoplasm of the host cell, they start to multiply rapidly. To be able to move to another cell, a bacterial surface protein, Actin A (ActA), is needed to induce polymerization of the actin cytoskeletal molecules of the host cell onto the surface of the bacteria. Within three to four hours from the start of the infection, long actin-filaments are created at one bacterial pole, which rapidly propel *L. monocytogenes* bacteria through the cytoplasm towards the neighboring cell: part of the host cell membrane bulges and the membrane of the neighboring cell curves inwards. Thus, the double-membrane vacuole harboring listeria bacteria is swallowed by the neighboring cell and a second phagosome is created. This phagosome is lysed by LLO and PC-PLC and a new infection begins. Once phagocytosed, the listeria bacteria are never extracellular again and become an intracytoplasmic parasite, such as *Shigella flexneri* and *Rickettsia rickettsii*. This process allows the bacteria to spread from one cell to another without leaving the intracellular room, and thus, do not have to confront with the immune system.

The patient

Listeriosis affects primarily pregnant women, unborn children, newborns, the elderly and adults with weakened immune systems. The overall case fatality rate from invasive listeriosis is from 20 to over 50% (Schlech 2000). Invasive listeriosis in humans can be divided into two categories: perinatal acquired listeriosis and listeriosis acquired after the neonatal period.

Perinatal acquired listeriosis

There are two different forms of perinatal acquired listeriosis, early and late onset, depending on when the disease makes its début. In early onset, the foetus will be infected in utero through transplacental transmission from the mother. The foetus may die intrauterine with or without spontaneous abortion (miscarriage, stillbirth, premature delivery). Children born alive are septic when born; alternatively, symptoms present during the first hours or days post partus: the mortality rate for live infants is about 20% (Schlech 2000). In 40% of cases, the pregnant women have flu-like symptoms such as fever, headache, and myalgia due to listeria bacteremia, and it is probably during this stage, the foetus/child becomes infected.

Late onset perinatal acquired listeriosis affects children contracting meningitis during the second to fourth week of life. These children are born healthy, but are probably infected during the passage through the birth canal or from the environment. With adequate treatment, the prognosis is good and the fatality rate is circa 10% (Schlech 2000).

Listeriosis acquired after the neonatal period

Immunosuppressed individuals, e.g., AIDS and chemotherapy patients, patients taking immunosuppressive agents, alcoholics, diabetics, organ transplacental recipients, and older people are over represented in this group. The most common manifestations are acute meningitis or meningoencephalitis, usually associated with bacteremia. In contrast to meningitis, encephalitis targeting the brainstem (rhombencephalitis) appears to occur, predominantly in previously healthy people (Oevermann et al. 2010). Healthy individuals can suffer a non-invasive form of listeriosis with gastroenteritis and fever as the main symptoms: a less common form is cutaneous listeriosis, which presents as a rash on the arms and a light fever, and is seen among veterinarians and farmers assisting during animal delivery.

Infectious dose

In the 1980s in the USA, Switzerland and UK, analyses of *L. monocytogenes* levels from unopened packages from retailers revealed infectious doses ranging from $< 10^2$ to 10^7 cfu/g (Mclauchlin 1996). During shelf life, the European legal limit is ≤ 100 cfu/g (colony-forming units per gram) in ready-to-eat food. With the recent rise of listeriosis among people > 60 years within the EU, Gillespie et al. (2009) question this limit and recommend it be revaluated in order to protect the immuno-suppressed population. The Center for Disease Control and Prevention in US (CDC 2012) emphasizes that people over 50, especially those over 65, should avoid hot dogs, lunchmeats, cold cuts and other delicatessen meats, unless they are "steaming hot" or reheated to 73.89°C (165°F).

Incubation time

A longer incubation period is observed in pregnancy-associated cases (median: 28 d, range 14–88 d) than for cases with CNS involvement (median: 10 d, range 2–19 d), which has implications for the investigation of outbreaks (Goulet et al. 2013). The incubation period is 24–48 h for the gastrointestinal form (Schlech 2000) and 1–4 d for cutaneous listeriosis.

Treatment

In the treatment of listeriosis, ampicillin and gentamycin are still the drugs of choice; however, aminoglycosides, such as gentamycin, may be harmful to patients with renal failure (Mitjà et al. 2009). Vancomycin in combination with an aminoglycoside has been successfully used as treatment for penicillin-allergic patients with listeriosis (Schlech 2000). The length of treatment is three weeks for sepsis and meningitis and six weeks for brain abscess or endocarditis (Lorber 1997). Listeria is not sensitive to cephalosporins, however, human isolates have not yet acquired clinically relevant antibiotic resistance (Morvan et al. 2010, dos Reis et al. 2011).

Incidence

In developed countries where listeriosis is documented, the incidence rates is around 0.3 per 100,000 population (Todd and Notermans 2011). Scandinavian countries have consistently higher rates than other countries possibly due to the consumption of smoked fish (Todd and Notermans 2011). Since 2005 in Sweden, there has been an increase in listeriosis. In 2009, there were 0.78 cases per 100,000, which is the largest number of cases reported in Sweden (SMI 2012). The PFGE types (*Asc* I) identified in listeriosis patients and in vacuum-packaged cold-smoked and gravad salmons in Sweden are often identical and *L. monocytogenes* is frequently isolated from these products bought in retail stores. The bacterium has been isolated from 11 (circa 20%) of 56 products of gravad or cold-smoked salmons sold in Sweden and the level of *L. monocytogenes* contamination is between < 100 to 1500 cfu/g (Peiris et al. 2009).

Since 2001, the incidence of listeriosis in patients aged over 60 years in England has doubled and patients report the consumption of cooked meats (beef and ham/pork), cooked fish (specifically smoked salmon) and shellfish (prawns), dairy products and mixed salads (Gillespie et al. 2010). The increased incidence of listeriosis among patients over 60 years old in both England and Wales between 2001 and 2007 was higher in both cancer patients and patients treated with acid-suppressing medication to reduce gastric acid production. Thus, this vulnerable patient group may need specific dietary advice about avoiding listeriosis (Gillespie et al. 2009).

An increased number of bacteremic listeriosis among the elderly is also reported from Scotland and Northern Ireland, Germany, Belgium, France, Lithuania, the Netherlands and Spain and as more "elderly patients" receive immunosuppressive therapies for chronic conditions, this can increase the risk of listeriosis (ACSMF 2009). The elderly and older people have difficulties reading labels with 'use by' dates and purchase items closer to use by dates because they are sometimes cheaper (Hudson and Hartwell 2002): these items are sometimes kept in a refrigerator with incorrect temperatures for up to a month before consumption.

Outbreaks

The first outbreak which finally identified *L. monocytogenes* as a food-borne agent involved 41 patients (17 deaths) and occurred in Nova Scotia, Canada, in 1981. The vehicle of infection was coleslaw made from cabbage fertilized with sheep manure: among the sheep providing the manure, two sheep had died from listeriosis (Schlech et al. 1983).

During the 1980s, two large cheese-borne outbreaks provided an insight into the problems of listeriosis and "*raised listeriosis to higher level of concern among food manufacturers and regulatory agencies"* (Farber and Peterkin 1991). In Switzerland during 1983–1987, a Swiss soft cheese, Vacherin Mont d'Or, caused 122 cases of listeriosis, and in California, USA, in 1985, a Mexican-style fresh cheese caused 142 cases of listeriosis. The cheese-borne epidemic strains from the two outbreaks shared serovar 4b and the same PFGE pattern and was classed as Epidemic Clone I (ECI) (Buchrieser et al. 1993, Kathariou 2002). The epidemic strain (with identical PFGE profiles) from Switzerland and California has also been identified among at least 80 human cases in Denmark and Sweden during 1976–1990 (Buchrieser et al. 1993, Jensen et al. 1994, Ericsson et al. 1996), with cheese as a probable common source. Other large outbreaks involved epidemic strains also belonging to ECI, however, not with an identical PFGE pattern: 41 cases from coleslaw in Canada in 1981; 49 cases from milk in Massachusetts, USA, in 1983; and, 247 cases from pig tongue in aspic in France in 1992 (Schlech et al. 1983, Fleming et al. 1985, Jacquet et al. 1995, Kathariou 2002). A new clonal group, serovar 4b, ECII, was identified during the outbreaks in US 1998–99 (hot dogs, 101 cases) and 2002 (turkey delicatessen meats, 54 cases) (Kathariou et al. 2006, Gottlieb et al. 2006).

In the 1980s and 1990s, outbreaks were mainly caused by serovars 4b and 1/2b, although there were outbreaks with serovar 1/2a combined with 4b and a pure 1/2a outbreak due to salmon (Miettinen et al. 1999). Since 2000, several 1/2a outbreaks have been reported: fresh cheese (Danielsson-Tham 2004), sliced processed delicatessen turkey meat (Olsen et al. 2005), flat whipping cream (Pagotto et al. 2006), tomme cheese (Bille et al. 2006), Quargel, a sour-milk curd-cheese (acid curd cheese) (Fretz et al. 2010), delicatessen meats (Garrido et al. 2010, Farber et al. 2011), soft washed-rind cheese (Gaulin et al. 2012). Sequence typing confirms that a predominant *Listeria monocytogenes* serovar 1/2a clone caused human listeriosis cases and outbreaks in Canada from 1988–2010 (Knabel et al. 2012). Todd & Notermans (2011) list recent outbreaks and emphasize that soft cheeses and delicatessen meats are the products most often implicated in outbreaks of listeriosis.

Route of Infection

L. monocytogenes causes listeriosis in both man and a wide range of animal species. One of the most common routes of infection to humans is through food; a less common route is through skin lesions that can cause dermatitis. Occasionally animals and humans can harbor *L. monocytogenes* in the intestines without displaying any symptoms of disease. In nature, listeria bacteria spread through feces and can settle in soil, on grass, and in water. Grazing animals then ingest the bacteria in the pasture or through silage. The environment surrounding milk-producing animals is contaminated by feces and milk can be easily contaminated with *L. monocytogenes* during milking. *L. monocytogenes* can be introduced from livestock into the home via shoes and gloves, and *L. monocytogenes*-contaminated shoes are identified more frequently from ruminant farmhouses than rural households not raising ruminants on site (Kersting et al. 2010). During slaughter, carcasses and meat can become contaminated through intestinal content. Thus, raw milk and raw meat can harbor *L. monocytogenes*. In 2056 samples (fecal, feed, and environment) collected from ruminant farms, Nightingale et al. (2004) found 20.1% of samples were positive for *L. monocytogenes*. In 1805 samples (soil, water and other environmental samples) from urban and pristine environments (locations with little direct animal and human contact, such as state parks, national forests), the percentage of positive samples were 7.3% from urban areas and 1.3 pristine environments (Nightingale et al. 2004). As *L. monocytogenes* is psychrotrophic and can withstand unfavorable living conditions, they can easily become established in food production plants. Several cases of listeriosis have been caused by ready-to-eat foods harboring *L. monocytogenes* that are consumed without any further control step, such as cooking, and the consumer has limited opportunities for destroying the pathogen before the food is consumed (Todd & Noterman 2011). The food types with the highest prevalence of *L. monocytogenes*, in the United Kingdom are pre-packed sandwiches and pre-packed mixed salad vegetables, that is, products consumed without any further treatment. These products sometimes exceed the limit of 100 cfu/g, especially if they contain salad ingredients, and serovar 1/2a is the most common serovar in those products (Little et al. 2009, Meldrum et al. 2010). In the United Kingdom, 236 (17.4%) of 1344 samples of cold smoked fish contained *L. monocytogenes*, but this was present below the 100 cfu/g legal limit (Food Standards Agency 2008).

Microbiological Criteria for *L. monocytogenes*

One debated aspect of the microbiological criteria for *Listeria* is the issue of whether the target should be the absence of *L. monocytogenes* or whether low levels should be allowed, i.e., the limit 100 cfu/g being permissible to the end of shelf life. Some countries require absence in ready-to-eat foods (in a 25 g sample) intended for infants and other vulnerable groups and ready-to-eat foods supporting the growth of *L. monocytogenes*, e.g., refrigerated ready-to-eat foods with extended shelf life, such as pre-packed sandwiches, sliced meat/delicatessen meat, soft/semi-soft cheeses, pâté and gravad and cold-smoked fish, marinated smoked mussels, undercooked hot dogs and chicken are at most risk. Up to 100 cfu/g, or even higher levels, are permitted where growth would be very slow or unlikely.

Control and Prevention

Food-borne listeriosis can be prevented in two ways—through controlling the organism in the food-processing environment and through careful attention to the preparation and choice of foods in the home. Although cooking and pasteurizing all foods eliminates the risk of food-borne listeriosis entirely, modern food preferences emphasize "wholesomeness" of raw and minimally processed foods as part of the normal diet (Schlech 2000).

L. monocytogenes grow in high salt concentrations (10% NaCl) and over both wide pH (pH 4.5–9) and temperature (0–45°C) ranges (Velge & Marie-Roche 2010). Foods unable to support the growth of *L. monocytogenes* have a pH $\leq$ 4.4 or $a_w \leq$ 0.92 (water activity). If a product has pH of $\leq$ 5.0, the a_w must be $\leq$ 0.94 or *vice versa* (EC No. 2073/2005). Although cooking procedures kill the bacterium, it can be potentially present in raw food and certain ready-to-eat foods. Even though several industry guidelines are available, the food industry should consider whether products are ready-to-eat foods where *Listeria* growth is possible or where growth does not occur in order to prevent contamination and/or growth of *L. monocytogenes* in ready-to-eat foods. The determination of an adequate shelf life for ready-to-eat foods is a key factor for controlling *L. monocytogenes*.

Canada's general approach to *L. monocytogenes* control includes a combination of industry controls, environmental and end-product testing. The purpose of environmental testing of food contact surfaces is to monitor the effectiveness of sanitation and good manufacturing practice (GMP) in preventing the contamination of ready-to-eat processing environments and products with *L. monocytogenes*. Establishments are strongly recommended to perform environmental testing on both food contact and non-food contact surfaces, as this could help locate potential niches and harbor sites for *Listeria* spp. within food processing plants, and this needs to be controlled by the manufacturer (Farber et al. 2011).

Slicing equipment could be one source of pathogen dissemination (cross contamination). Raw products entering the production line may already be contaminated with *L. monocytogenes*; with proper treatment (cooking, hot smoking), the pathogen could be eliminated to provide a safe product. Without treatment the pathogen may become a domestic flora, thus, establishing itself and surviving for long periods.

Detection

L. monocytogenes can be readily cultured from clinical specimens, as it is easy to identify in normally sterile sites such as blood and cerebrospinal fluid. *L. monocytogenes* will usually grow on unselective media such as blood agar plates incubated at 35–37°C for 24–48 h. However, the organism is more difficult to find in fecal or food samples due to competing microflora and a complex matrix. Normally detection involves an enrichment procedure, with an enrichment broth containing ingredients that inhibit competing organisms and stimulate *Listeria*. The broth (half-Fraser) is incubated for 24–48 h (24h for qPCR) at 30°C, followed by subcultivation of a smaller portion of the broth in another selective broth (Fraser). After an additional incubation for 24–48 h (4h for qPCR) at 37°C, a small fraction of the broth is plated on a selective media, and *Listeria* present will grow after 24–48 h incubation at 30–37°C. More new optimal liquid and solid selective media are constantly invented.

Molecular methods are increasingly applied in the detection, quantification and study of microbial populations in food or during food processing. Among these methods, polymerase chain reaction (PCR)-based techniques has received considerable focus as it uses primers that target virulence and non-virulence factors, such as hemolysin (*hly*), invasion associated protein (*iap*), and 16S rRNA genes. Real-time PCR, also called quantitative real time PCR (qPCR) is a laboratory technique based on PCR and used to amplify and simultaneously quantify a targeted DNA molecule as the reaction progresses in real time and each new DNA fragment created is fluoresced by a special fluorescent molecule and the amount of certain DNA or RNA in a group of bacterial cells is measured. The advantage of qPCR is the shorter time required to obtain the results. With qPCR, including the enrichment step, the detection of *L. monocytogenes* takes two working days instead of seven days with the conventional methods. However, the disadvantage of the enrichment step is that it is impossible to quantify the initial contamination level (Postollec et al. 2011).

A capture enzyme-linked immunosorbent assay (ELISA) has been standardized and validated for identifying *L. monocytogenes* in food and refined through analyzing samples of meat, seafood, dairy products, pasta and flour. A comparison of *L. monocytogenes* capture ELISA with the ISO 11290-1:1996 for isolating and identifying *L. monocytogenes* in food matrices produced a significant concordance index (Portanti et al. 2011).

Enumeration of Agent

A way of enumerating *L. monocytogenes* in food is to dilute the food sample. A small portion (usually 0.1 ml) of the different dilutions is plated onto both unselective and selective media. Typical colonies are counted on the most appropriate plate and the number is transformed into cfu/g or cfu/ml. The *L. monocytogenes* colonies may, however, be difficult to identify due to competitive micro flora. Problems due to a detection limit of 10 cfu/g are solved with the use of three plates instead of one, or the most probable number (MPN) methods.

Concluding Remarks

The world's population is ageing and has a greater chance of developing debilitating chronic conditions. As people live longer into the 80s and 90s, they lose their immunity and are more vulnerable to infections. The demographic shift and the widespread use of immunosuppressive medications have increased the immunocompromised population that have an increased risk for listeriosis (Allerberger and Wagner 2010, Todd and Notermans 2011, Luber et al. 2011). In Canada, people aged 65–69 years have a 4-fold increased risk of contracting listeriosis than healthy individuals aged 40-59 years, and those aged 75–79 years of age have nearly a 9-fold increased risk (Luber et al. 2011).

Consumer lifestyle has changed and less time is available for food preparation. Modern lifestyle has markedly changed eating habits worldwide, with a consequent increased demand for ready-to-eat foods (Aparecida et al. 2010), therefore, more ready-to-eat foods and take away foods are consumed. Within the food industry, extended shelf life ready-to-eat foods and new ready-to-eat food types are an area of expansion, and as these foods pose the most concern for listeriosis, this area deserves much attention (Allerberger and Wagner 2010). To control *L. monocytogenes* in ready-to-eat products, ready-to-eat products at risk should be properly labeled with regard to time and temperature of storage, and consumers should be educated about food storage practices. In addition, the shelf life for some products with a known risk of *L. monocytogenes* contamination and growth (e.g., vacuum-packed gravad and cold smoked salmon) should be restricted.

References

ACMSF (Advisory Committee on the Microbiological Safety on Food). 2009. *Ad Hoc* Group on Vulnerable Groups. Report on the increased incidence of listeriosis in the UK. Crown copyright. Published by Food Standards Agency July 2009. FSA/1439/0709. www.food.gov.uk/multimedia/pdfs/committee/acmsflisteria.pdf.

Allerberger, F. and M. Wagner. 2010. Listeriosis: a resurgent foodborne infection. Clin. Microbiol. Infect. 16: 16–23.Aparecida de Oliveira, M., E.G. Abeid Ribeiro, A.M. Morato Bergamini and E.C. Pereira De Martinis. 2010. Quantification of *Listeria monocytogenes* in minimally processed leafy vegetables using a combined method based on enrichment and 16S rRNA real-time PCR. Food Microbiol. 27: 19–23.

Baldry, S. 2010. Attack of the clones. Nature Rev. Microbiol. 8: 390.

Bille, J., D.S. Blanc, H. Schmid, K. Boubaker, A. Baumgartner, H.H. Siegrist, M.L. Tritten, R. Lienhard, D. Berner, R. Anderau, M. Treboux, J.M. Ducommun, R. Malinverni,

D. Genné, P.H. Erard and U. Waespi. 2006. Outbreak of human listeriosis associated with tomme cheese in northwest Switzerland, 2005. Eurosurveill. 11: 91–93.

Boerlin, P., J. Rocourt, F. Grimont, P.A.D. Grimont, C. Jacquet and J.C. Piffaretti. 1992. *Listeria ivanovii subsp. londoniensis* subsp nov. Int. J. Syst. Bacteriol. 42: 69–73.

Buchrieser, C., R. Brosch, B. Catimel and J. Rocourt. 1993. Pulsed-field gel electrophoresis applied for comparing *Listeria monocytogenes* strains involved in outbreaks. Can. J. Microbiol. 390: 395–401.

Burall, L.S., A.C. Simpson and A.R. Datta. 2011. Evaluation of a serotyping scheme using a combination of an antibody-based serogrouping method and a multiplex PCR assay for identifying the major serotypes of *Listeria monocytogenes*. Food Protect. 74: 403–409.

(CDC) Center for Disease Control and Prevention in US. dec 2012. http://www.cdc.gov/listeria/prevention.html.

Cooper, J.E. and E.J. Feil. 2004. Multilocus sequence typing—what is resolved? Trends Microbiol. 12: 373–377.

Danielsson-Tham, M.L., E. Eriksson, S. Helmersson, M. Leffler, L. Lüdtke, M. Steen, S. Sørgjerd and W. Tham. 2004. Causes behind a human cheese-borne outbreak of gastrointestinal listeriosis. Foodborne Pathog. Dis. 1: 153–159.

den Bakker, H.C., E.D. Fortes and M. Wiedmann. 2010. Multilocus sequence typing of outbreak-associated *Listeria monocytogenes* isolates to identify epidemic clones. Foodborne Pathog. Dis. 7: 257–265.

dos Reis, C.M.F., A.V. Barbosa, L.A. Rusak, D.C. Vallim and E. Hofer. 2011. Antimicrobial susceptibilities of *Listeria monocytogenes* human strains isolated from 1970 to 2008 in Brazil. Rev. Soc. Bras. Med. Trop. 44: 173–176.

EC No 2073/2005. Commission Regulation of 15 November 2005 on microbiological criteria for foodstuffs (http://eur-lex.europa.eu/LexUriServ/LexUriServ.do?uri=CELEX:32005R2073:en:NOT).

Ericsson, H., E. Bannerman, J. Bille, M.L. Danielsson-Tham, Ch. Jacquet, J. Rocourt, P. Stålhandske, J. Ursing and W Tham. 1996. Division into five groups by REA of the most frequently isolated phagovar of *Listeria monocytogenes* in Sweden 1976–1985. Med. Microbiol. Lett. 5: 145–155.

Farber, J.M. and P.I. Peterkin. 1991. *Listeria monocytogenes*, a food-borne pathogen. Microbiol. Rev. 55: 476–511.

Farber, J.M., G.K. Kozak and S. Duquette. 2011. Changing regulation: Canada's new thinking on Listeria. Food Control. 22: 1506–1509.

Fleming, D.W., S.L. Cochi, K.L. MacDonald, J. Brondum, P.S. Hayes, B.D. Plitkaytis, M.B. Holmes, A. Audurier, C.V. Broome and A.L. Reingold. 1985. Pasteurised milk as a vehicle of infection in an outbreak of listeriosis. N. Engl. J. Med. 312: 404–407.

Food Standards Agency in United Kingdom. 2008. Retail survey of *Listeria monocytogenes* in smoked fish (http://www.food.gov.uk/science/surveillance/fsisbranch2008/fsis0508).

Fretz, R., J. Pichler, U. Sagel, P. Much, W. Ruppitsch, A.T. Pietzka, A. Stöger, S. Huhulescu, S. Heuberger, G. Appl, D. Werber, K. Stark, R. Prager, A. Flieger, R. Karpíšková, G. Pfaff and F. Allerberger. 2010. Update: Multinational listeriosis outbreak due to 'Quargel', a sour milk curd cheese, caused by two different *L. monocytogenes* serotype 1/2a strains, 2009-2010. Euro Surveill. 2010;15(16):pii=19543. Available online: http://www.eurosurveillance.org/ViewArticleaspx?ArticleId=19543.

Garrido, V., A.I. Vitas and I. Garcia-Jalon. 2010. The problem of listeriosis and ready-to-eat products: prevalence and persistence. *In:* A. Méndez-Vilas (ed.). Current Research, Technology and Education Topics in Applied Microbiology and Microbial Biotechnology. Microbiology Book Series, no 2, Vol 2. Formatex Research Center. FORMATEX Badajoz, Spain ISBN-13: 978-84-614-6195-0. pp. 1182–1189.

Gaulin, C., D. Ramsay and S. Bekal. Widespread listeriosis outbreak attributable to pasteurized cheese, which led to extensive cross-contamination affecting cheese retailers, Quebec, Canada, 2008. 2012. J. Food Protect. 75: 71–78.

Gillespie, I.A., J. McLauchlin, C.L. Little, C. Penman, P. Mook, K. Grant and S.J. O'Brien. 2009. Disease presentation in relation to infection foci for non-pregnancy-associated human listeriosis in England and Wales, 2001 to 2007. J. Clin. Microbiol. 47: 3301–3307.

Gillespie, I.A., P. Mook, C.L. Little, K. Grant and G.K. Adak. 2010. *Listeria monocytogenes* infection in the over-60s in England between 2005 and 2008: A retrospective case-control study utilizing market research panel data. Foodborne Pathog. Dis. 7: 1373–1379.

Goering, R.V. 2010. Pulsed field gel electrophoresis. A review of application and interpretation in the molecular epidemiology of infectious disease. Inf. Gen. Evol. 10: 866–875.

Gottlieb, S.L., E.C. Newbern, P.M. Griffin, L.M. Graves, R.M. Hoekstra, N.L. Baker, S.B. Hunter, K.G. Holt, F. Ramsey, M. Head, P. Levine, G. Johnson, D. Schoonmaker-Bopp, V. Reddy, L. Kornstein, M. Gerwel, J. Nsubuga, L. Edwards, S. Stonecipher, S. Hurd, D. Austin, M.A. Jefferson, S.D. Young, K. Hise, E.D. Chernak, J. Sobel and the Listeriosis Outbreak Working Group. 2006. Multistate outbreak of listeriosis linked to Turkey deli meat and subsequent changes in US regulatory policy. Clin. Infect. Dis. 42: 29–36.

Goulet, V., L. King, V. Vaillant and H. De Valk. 2013. What is the incubation periods for listeriosis? BMC Infect Dis. 13: 11. doi: 10.1186/1471-2334-13-11. http://www.biomedcentral.com/1471-2334/13/11.

Graves, L.M., L.O. Helsel, A.G. Steigerwalt, R.E Morey, M.I. Daneshvar, S.E. Roof, R.H. Orsi, E.D. Fortes, S.R. Milillo, H.C. den Bakker, M. Wiedmann, B. Swaminathan and B.D. Sauders. 2010. *Listeria marthii* sp. nov., isolated from the natural environment, Finger Lakes National Forest. International J. Syst. Evol. Microbiol. 60: 1280–1288.

Hudson, P.K. and H.J. Hartwell. 2002. Food safety awareness of older people at home: a pilot study. J. Royal Soc. Prom. Health. 122: 165–169.

Husu, J.R. 1990. Epidemiological studies on the occurrence of *Listeria monocytogenes* in the feces of dairy cattle. J. Vet. Med. B37: 276–282.

Hülphers, G. 1911. Lefvernekros hos kanin orsakad av en ej förut beskriven bakterie. Svens Veterinärtidskrift 15: 2: 265–273 (In Swedish). Liver necrosis in rabbit caused by a hitherto unknown bacterium, ISOPOL XV. XV International Symposium on Problems of Listeriosis, Uppsala, Sweden, September 12–15, 2004. Abstract 147A (In English) plus McLauchlin J. Was Hülphers the first to describe *Listeria monocytogenes*. Abstract 147b (In English).

Jacquet, C., B. Catimel, R. Brosch, C. Buchrieser, P. Dehaumont, V. Goulet, A. Lepoutre, P. Veit and J. Rocourt. 1995. Investigations related to the epidemic strain involved in the french listeriosis outbreak in 1992. Appl. Environ. Microbiol. 61: 2242–2246.

Jensen, A., W. Frederiksen and P. Gerner-Smidt. 1994. Risk factors for listeriosis in Denmark 1989–1990. Scand. J. Infect. Dis. 26: 171–178.

Kathariou, S. 2002. *Listeria monocytogenes* virulence and pathogenicity, a food safety perspective. J. Food Protect. 65: 1811–1829.

Kathariou, S., L. Graves, C. Buchrieser, P. Glaser, R.M. Siletzky and B. Swaminathan. 2006. Involvement of closely related strains of a new clonal group of *Listeria monocytogenes* in the 1998-99 and 2002 multistate outbreaks of foodborne listeriosis in the United States. Foodborne Pathog. Dis. 3: 292–302.

Kérouanton, A., M. Marault, L. Petit, J. Grout, T.T. Dao and A. Brisabois. 2010. Evaluation of a multiplex PCR assay as an alternative method for *Listeriamonocytogenes* serotyping. J. Microbiol. Meth. 80: 134–137.

Kersting, A.L. 2008. *Listeria monocytogenes*, zoonotic exposure, rural residency, and prevention. Dissertation (Doctor of Philosophy). The Ohio State University, Veterinary Preventive Medicine, 164 pages.

Kersting, A.L., L.C. Medeiros and J.T. LeJeune. 2010. Differences in *Listeria monocytogenes* contamination of rural Ohio Residents with and without Livestock. Foodborne Pathog. Dis. 7: 57–62.

Knabel, S.J., A. Reimer, B. Verghese, M. Lok, J. Ziegler, J. Farber, F. Pagotto, M. Graham, C.A. Nadon and M.W. Gilmour. 2012. Sequence typing confirms that a predominant *Listeria monocytogenes* clone caused human listeriosis cases and outbreaks in Canada from 1988-2010. J. Clin. Microbiol. 50: 1748–1751doi:10.1128/JCM.06185-11.

Leclercq, A., D. Clermont, C. Bizet. P.A. Grimont, A. Le Flèche-Matéos, S.M. Roche, C. Buchrieser, V. Cadet-Daniel, A. Le Monnier, M. Lecuit and F. Allerberger. 2010. *Listeria rocoutiae* sp. no. Int. J. Syst. Evol. Microbiol. 60: 2210–2214.

Lecuit, M. 2005. Understanding how *Listeria monocytogenes* targets and crosses host barriers. Clin. Microbiol. Infect. 11: 430–436.

Lindstedt, B.A., W. Tham, M.L. Danielsson-Tham, T. Vardund, S. Helmersson and G. Kapperud. 2008. Multi-locus variable-number tandem-repeat analysis of *Listeria monocytogenes* using multicolour capillary electrophoresis and comparison with pulsed-field gel electrophoresis typing. J. Microbiol. Meth. 72: 141–148.

Little, C.L., S.K. Sagoo, I.A. Gillespie, K. Grant and J. McLauchlin. 2009. Prevalence and level of *Listeria monocytogenes* and other *Listeria* species in selected retail ready-to-eat foods in the United Kingdom. J. Food Prot. 72: 1869–1877.

Lorber, B. 1997. Listeriosis. Clin. Infect. Dis. 24: 1–9.

Luber, P., S. Crerar, C. Dufour, J. Farber, A. Datta and E.C.D. Todd. 2011. Controlling *Listeria monocytogenes* in ready-to-eat foods: Working towards global scientific consensus and harmonization–Recommendations for improved prevention and control. Food Control 22: 1535–1549.

McLauchlin, J. 1990. Distribution of serovars of *Listeria monocytogenes* isolated from different categories of patients. Eur. J. Clin. Microbiol. Infect. Dis. 9: 210–213.

Mclauchlin, J. 1996. The relationship beween *Listeria* and listeriosis. Food Control 7: 187–193.

Meldrum, R.J., P.W. Ellis, P.T. Mannion, D. Halstead and J. Garside on behalf of the Welsh Food Microbiological Forum. 2010. Prevalence of *Listeria monocytogenes* in ready-to-eat foods sampled from the point of sales in Wales, United Kingdom. J. Food Protect. 73: 1515–1518.

Miettinen, M.K., A. Siitonen, P. Heiskanen, H. Haajanen, K.J. Björkroth and H.J. Korkeala. 1999. Molecular epidemiology of an outbreak of febrile gastroenteritidis caused by *Listeria monocytogenes* in cold-smoked rainbow trout. J Clin. Microbiol. 37: 2358–2360.

Mitjà, O., C. Pigrau, I. Ruiz, X. Vidal, B. Almirante, A.M. Planes, I. Molina, D. Rodríguez and A. Pahissa. 2009. Predictors of mortality and impact of aminoglycosides on outcome in listeriosis in a retrospective cohort study. J. Antimicrob. Chemother. 64: 415–423.

Miya, S., B. Kimura, M. Sato, H. Takahashi, T. Ishikawa, T. Suda, C. Takakura, T. Fujii and M. Wiedmann. 2008. Development of a multilocus variable-number of tandem repeat typing method for *Listeria monocytogenes* serotype 4b strains. Int. J. Food Microbiol. 124: 239–249.

Morvan, A., C. Moubareck, A. Leclercq, M. Herve-Bazin, S. Bremont, M. Lecuit, P. Courvalin and A. Le Monnier. 2010. Antimicrobial Resistance of *Listeria monocytogenes* strains isolated from humans in France. Antimicrob. Agents Chemother. 54: 2728–2731.

Nightingale, K.K., Y.E. Schukken, C.R. Nightingale, E.D. Fortes, A.J. Ho, Z. Her, Y.T. Grohn, P.L. McDonough and M. Wiedmann. 2004. Ecology and transmission of *Listeria monocytogenes* infecting ruminants and in the farm environment. Appl. Environ. Microbiol. 70: 4458–4467.

Nightingale, K.K., K. Windham and M. Wiedmann. 2005. Evolution and molecular phylogeny of *Listeria monocytogenes* isolated from human and animal listeriosis cases and foods. J. Bacteriol. 187: 5537–5551.

Nightingale, K., L. Bovell, A. Grajczyk and M. Wiedmann. 2007. Combined *sigB* allelic typing and multiplex PCR provide improved discriminatory power and reliability for *Listeria monocytogenes* molecular serotyping. J. Microbiol. Meth. 68: 52–59.

Nightingale, K. 2010. *Listeria monocytogenes*: Knowledge gained through DNA sequence-based subtyping, implications, and future considerations. J AOAC Int. 93: 1278–1286.

O'Connor, L., M. O'Leary, N. Leonard, M. Godinho, C. O'Reilly, L. Coffey, J. Egan and R. O'Mahony. 2010. The characterization of *Listeria* spp. isolates from food products and the food-processing environment. Lett. Appl. Microbiol. 51: 490–498.

Oevermann, A., A. Zurbriggen and M. Vandevelde. 2010. Review article–Rhombencephalitis caused på *Listeria monocytogenes* in humans and ruminans: A zoonosis on the rise? Interdis Perspect. Infect. Dis. Article ID 632513, doi:10.1155/2010/632513, 22 pages.

Olsen S.J., M. Patrick, S.B. Hunter, V. Reddy, L. Kornstein, W.R. MacKenzie, K. Lane, S. Bidol, G.A. Stoltman, D.M. Frye, I. Lee, S. Hurd, T.F. Jones, T.N. LaPorte, W. Dewitt, L. Graves, M. Wiedmann, D.J. Schoonmaker-Bopp, A.J. Huang, C. Vincent, A. Bugenhagen, J. Corby, E.R. Carloni, M.E. Holcomb, R.F. Woron, S.M. Zansky, G. Dowdle, F. Smith, S. Ahrabi-Fard, A.R. Ong, N. Tucker, N.A. Hynes and P. Mead. 2005. Multistate outbreak of *Listeria monocytogenes* infection linked to delicatessen turkey meat. Clin. Infect. Dis. 40: 962–967.

Orsi, R.H., H.C. den Bakker and M. Wiedmann. 2011. *Listeria monocytogenes* lineages: Genomics, evolution, ecology, and phenotypic characteristics. Int. J. Med. Microbiol. 301: 79–96.

Pagotto, F., L.K. Ng, C. Clark and J. Farber. 2006. Canadian listeriosis reference service. Foodborne Pathog. Dis. 3: 132–137.

Peiris, W.I.P., G. Lopez-Valladares, V.S. Parihar, S. Helmersson, S.B. Barbuddhe, W. Tham and M.L. Danielsson-Tham. 2009. Gravad (Gravlax) and cold-smoked salmon (*salmo salar*), a potential source of listeriosis. J. Foodservice. 20: 15–20.

Portanti, O., T. Di Febo, M. Luciani, C. Pompilii, R. Lelli and P. Semprini. 2011. Development and validation of an antigen capture ELISA based on monoclonal antibodies specific for *Listeria monocytogenes* in food. Vet. Ital. 47: 281–290.

Postollec, F., H. Falentin, S. Pavan, J. Combrisson and D. Sohier. 2011. Recent advances in quantitative PCR (qPCR) applications in food microbiology. Food Microbiol. 28: 848–861.

Revazishvili, T., M. Kotetishvili, O. Colin Stine, A.S. Kreger, J. Glenn Morris and A. Sulakvelidze. 2004. Comparative analyis of multilocus sequence typing and pulsed-field gel electrophoresis for characterizing *Listeria monocytogenes* strains isolated from environmental and clinical sources. J. Clin. Microbiol. 42: 276–285.

Schlech, W.F. III., P.M. Lavigne, R.A. Bortolussi, A.C. Allen, E.V. Haldane, A.J. Wort, A.W. Hightower, S.E. Johnson, S.H. King, E.S. Nicholls and C.V. Broome. 1983. Epidemie listeriosis. Evidence for transmission by food. New Engl. J. Med. 308: 203–206.

Schlech III, W.F. 2000. Foodborne listeriosis. Clin. Infect. Dis. 31: 770–775.

Seeliger, H.P.R. and K. Höhne. 1979. Serotyping of *Listeria monocytogenes* and related species. *In*: T. Bergan and J. Norris [eds.]. Methods in Microbiology, Vol. 13. Academic Press, New York, USA. pp. 31–49.

Seeliger, H.P.R. and D. Jones. 1986. Genus *Listeria*. *In*: P.H.A. Sneath, N.S. Mair, M.E. Sharpe and J.G. Holt (eds.). Bergey's Manual of Systematic Bacteriology, Vol. 2. Williams & Wilkins, Baltimore, USA. pp. 1235–1245.

Shaughnessy, L.M., A.D. Hoppe, K.A. Christensen and J.A. Swanson. 2006. Membrane perforations inhibit lysosome fusion by altering pH and calcium in *Listeria monocytogenes* vacuoles. Cell. Microbiol. 8: 781–792.

SMI (Smittskyddsinstitutet) 2012. http://www.smittskyddsinstitutet.se/statistik/listeriainfektion/.

Todd, E.C.D. and S. Notermans. 2011. Surveillance of listeriosis and its causative pathogen, *Listeria monocytogenes*. Food Control. 22: 1484–1490.

Tsai, Y.H.L., S.B. Maron, P. McGann, K.K. Nightingale and M. Wiedmann. 2011. Recombination and positive selection contributed to the evolution of *Listeria monocytogenes* lineages III and IV, two distinct and well supported uncommon *L. monocytogenes* lineages. Inf. Gen. Evol. 11: 1881–1890.

Velge, P. and S. Marie-Roche. 2010. Variabiliy of *Listeria monocytogenes* virulence: a result of the evolution between saprophytism and virulence? Future Microbiol. 5: 1799–1821.

Wiedmann, M. 2002. Molecular subtyping methods for *Listeria monocytogenes*. J. AOAC Int. 85: 524–531.

CHAPTER 11

Moulds as a Threat to Food Safety

Dereje T. Asefa[1,*] and *Ida Skaar*[2]

Introduction

Moulds

Moulds are part of the natural environment and grow practically everywhere. They are microscopic fungi that are found indoors and/or outdoors throughout the year (Carlile et al. 2001, Deacon 1997, Samson et al. 2004). Moulds grow best in warm, damp, humid conditions and spread with the help of the spores they are producing. On moist, nutrient-rich substrates the spores can germinate into tiny filaments called hyphae that form mycelia upon further extension (Deacon 2006, Pitt and Hocking 2009, Webster and Weber 2007). Mould spores can survive harsh environments, such as, dry conditions that are not suitable for microbial growth (Pitt and Hocking 2009). Moulds are heterotrophic that break down organic materials with the help of extra cellular enzymes and absorb carbon for energy (Pitt and Hocking 2009, Webster and Weber 2007).

[1] Reeco-san International, Gamle Åsvei 25, 1400, Ski.
Email: d.asefa@reecosan.no
[2] Section of Mycology, The Norwegian Veterinary Institute, Ullevålsveien 68, Pb 750 Sentrum, Oslo, Norway.
Email: ida.skaar@vetinst.no
* Corresponding author

Species concept

Moulds are multi-cellular eukaryotic organisms belonging to the fungal kingdom (Pitt and Hocking 2009). Species is the unit of measure that biologists use to categorize and differentiate diversity. Moulds were classically identified at a species level based on their differences in macro (colony characters) and micro morphological characters (reproductive structures) in a specific environmental niche (Deacon 1997, Pitt and Hocking 2009, Webster and Weber 2007). The method is good to identify the most common fungi but cannot be employed on the majority of fungi as they are uncultivable and lack the important reproductive structures useful for identification (Pitt and Hocking 2009, Samson et al. 2004). Recent works on fungal systematic advocate the use of the combination of morphological, chemical and molecular characters to identify mould species (Frisvad 1991, Frisvad 2004, Pitt and Hocking 2009, Samson et al. 2004). It is known as the polyphasic approach and defines species as "fungi that show evidence for evolutionary divergence from other taxa, unique DNA characters at multiple loci, in addition to any distinctive chemical and morphological characters" (Frisvad 2004).

Notification/Surveillance of Agent

Any surveillance activity for mycological agents on food should take and analyse samples from raw materials, the production environment and the final products. Hence, the sampling method, isolation procedures, media and incubation conditions have to be selected carefully to obtain a reliable result (Frisvad 1991, Frisvad 2004, Pitt and Hocking 2009). The following standard mycological techniques are applicable to analyse mycological agents associated with food production activity.

Sampling and isolation procedures

Techniques like direct transferring, direct plating, dilution plating and surface sampling are important to take samples from food raw materials and products whereas surface and air samples are important to study mould contaminants in food production facilities.

Direct transferring is a simple transfer of fungal spores and/or fragments from their natural habitat to culture media using a sterile inoculating needle (Malloch 1982, Pitt and Hocking 2009). In comparison direct plating involves the placement of small pieces of food samples on a nutrient rich agar medium for fungal isolation (Malloch 1982, Pitt and

Hocking 2009). Moulds associated with raw materials, intermediate and final products can be recovered by these method. A direct transferring technique is practical to study the types of visible fungal colonies (Malloch 1982, Pitt and Hocking 2009) and provide data as to the type of the mould dominating the food. Additional information can be obtained in the case of direct plating by estimating the percentage of contaminated samples (Pitt and Hocking 2009). Dilution plating involves a repeated decimal dilution of a known amount of food sample or culture to estimate the number of viable fungal contaminants (Malloch 1982, Pitt and Hocking 2009). When fixed volumes of the decimal dilution series are spread on agar media and incubated, different numbers of colonies will be obtained. By noting the number of colonies for each dilution, the volume of inoculant added and the mass or volume of sample diluted, moulds present in the original sample can be quantified (Malloch 1982, Pitt and Hocking 2009). This technique has been widely used to quantify the level of fungal contamination in liquid, grain and powdered food samples (Pitt and Hocking 2009).

Surface sampling is a useful and convenient method to isolate fungal contaminants from the surfaces of the different food raw materials, products, processing machines and room installations of food processing facilities (Pitt and Hocking 2009, Samson et al. 2004). It can be performed using a clear adhesive tape, contact agar plate, Petri films or sterile swabs (Pitt and Hocking 2009). A clear adhesive tape helps to get the prints of fungal spores on the surface of a given sample. Contact plate is a plastic dish filled with convex surfaced agar that can gently be rolled on the area to be tested. Petri film is a layer of medium enclosed in a plastic film (Pitt and Hocking 2009) to perform microbial testing on flat surfaces. The swab method has been applied widely for surface sampling and involves the rubbing of a moistened sterile cotton swab over the test surface (Malloch 1982, Pitt and Hocking 2009, Samson et al. 2004).

Assessment of the mycological air quality in food production facilities is an important aspect of identifying, evaluating and controlling potential fungal food safety hazards (Andon 2006, Asefa et al. 2009, Pitt and Hocking 2009, Samson et al. 2004). Air sampling can be performed either by passive (gravitational) or active (volumetric) sampling techniques (Andon 2006, Asefa et al. 2009, Portnoy et al. 2004, Samson et al. 2004). In passive air sampling (settle plate) Petri dishes containing nutrient rich agar medium have to be exposed to the air for a certain period of time so that spores of different fungal species fall by the help of gravitational force (Andon 2006). In contrast, active air sampling uses devices that draw a fixed volume of air at a specific speed over a specific period of time (Andon 2006). Spores and/or fungal fragments are captured by mounting Petri dishes containing agar medium on the sampler.

Types of isolation media

The physio-chemical need of moulds, such as a_w, temperature and pH affects the choice and composition of the culturing media (Dijksterhuis and Samson 2007, Frisvad 1991, Pitt and Hocking 2009, Samson et al. 2004). A number of widely accepted culture media exist for fungal isolation as a single medium cannot provide all the requirements. The a_w of the culture medium is exceptionally important and can be determined by taking the a_w of the food on which the moulds are growing in to consideration (Pitt and Hocking 2009, Samson et al. 2004). Most food spoilage moulds, like species of *Penicillium,* require a_w of 0.9–0.85, while fast growing moulds, such as *Mucor* spp. need a a_w higher than 0.9 (Pitt and Hocking 2009, Samson et al. 2004). In extreme dry products (a_w = 0.75–0.65) it is only slow growing xerophilic moulds that can survive (Pitt and Hocking 2009, Samson et al. 2004). In addition the pH and incubation temperature have to be adjusted to the range at which the moulds can grow comfortably (Filtenborg et al. 1996, Pitt and Hocking 2009, Samson et al. 2004).

Incubation

Incubation temperature can vary depending upon the origin of the food to be analysed, even if the International Commission on Food Mycology (ICFM) has set 25°C for 5 days as a standard (Pitt and Hocking 2009). For foods of tropical origin, an incubation temperature of 30°C is recommended while 22–25°C would be ideal for moulds associated with foods of temperate regions (Pitt and Hocking 2009). Petri dishes should be stored upright during incubation to minimize the accumulation of spores on the lid and reduce the chance of contamination during inspection (Pitt and Hocking 2009). After a few days of incubation, growing mould colonies can be identified morphologically at a genus level. They should be re-cultivated into their pure culture form using selective media to perform species identification.

Identification and Characterization of Moulds

Morphological characterization

Macroscopic and/or microscopic features of moulds are valuable for species identification (Frisvad 2004, Pitt and Hocking 2009, Samson et al. 2004). Colony size on nutrient rich agar media, appearance on both forward and reverse side of the growth media, production of diffusible pigments in the growth media, occurrence of exudates on the colony, mycelium type and colony topography, are important macro-morphological for identification (Frisvad 2004, Pitt and Hocking 2009, Samson et al. 2004). The microscopic features useful for identification and classification of moulds are characteristics of their reproductive structures (Frisvad 2004, Pitt and Hocking 2009, Samson et al. 2004). The way sexual and asexual spores are formed and the spore types, the nature and arrangement of fruiting structures (conidiophores, sporangiophores, basidioma, ascoma) are some of the most important features (Frisvad 2004, Pitt and Hocking 2009, Samson et al. 2004). The use of selective culture media can facilitate the morphological identification processes by revealing the unique macro- and micro-morphological features of moulds (Frisvad 2004, Pitt and Hocking 2009, Samson et al. 2004) (Fig. 1 and 2). Identification is then performed by comparing the observed morphological and physiological data with keys and descriptions published in several text books (Frisvad 2004, Pitt 1979, Pitt and Hocking 2009, Samson et al. 2004).

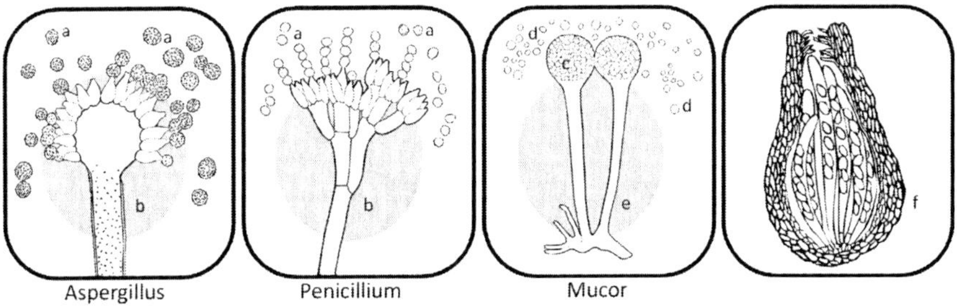

Figure 1. Micro-morphological pictures of moulds important for identification **a**-conidia, **b**-conidiophores, **c**-sporangia, **d**-sporangiospores, **e**-Sporangiophore, **f**- Ascoma.

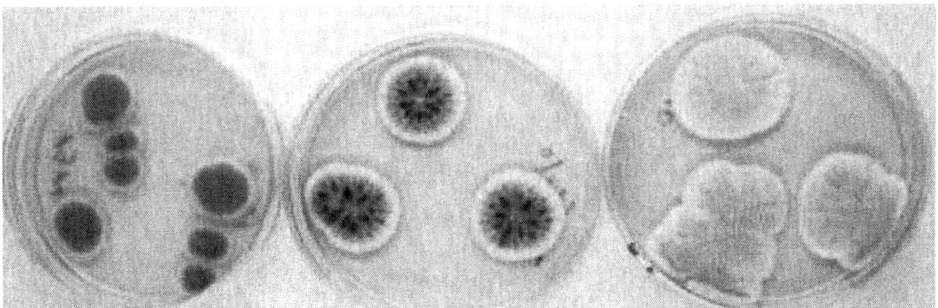

Figure 2. *Penicillium nalgiovense* (from the left side) on Malt Extract Agar (MEA), Czapek Yeast Extract Agar (CYA), and Yeast Extract Sucrose (YES) agar medium (Photo: Dereje T. Asefa 2007).

Color image of this figure appears in the color plate section at the end of the book.

Chemotaxonomy

The use of biochemical substances produced by moulds for species identification and characterization is helping a lot to increase the reliability of the morphological identification processes (De Joungh and Nilsen 2007, Frisvad 2004, Frisvad et al. 2008). Secondary metabolites are especially important with this regard, and some have a discriminatory power useful for identification and classification (Frisvad 2004, Frisvad et al. 2008). Some are characteristic of certain taxonomic groups and can be species specific. Secondary metabolites have been used extensively in the identification and classification of species of *Alternaria, Aspergillus, Fusarium* and *Penicillium* (Frisvad 2004, Frisvad et al. 2008).

Chemotaxonomic information can be generated using different analytical methods and are important also for the surveillance of foods contaminated with mycotoxins. One can use simple laboratory tests like bioassay analysis, Erlich test, thin layer chromatography (TLC) in addition to a more sophisticated chromatographic techniques such as high performance liquid chromatography (HPLC), gas chromatography (GC), mass spectrometry (MS) and their variants (Frisvad et al. 2008, Toscani et al. 2007).

Molecular characterization

Molecular tools are becoming important as a part of a polyphasic mould species identification approach. It involves the extraction and analysis of genetic material for taxonomic purpose. Sequence-based strategies are considered as the "golden standard" for fungal species identification (Guarro et al. 1999, Shenoy et al. 2007, Weber 2009). One has to select the target genes that are easy to amplify and sequence using standardized "universal" primer sets. Ribosomal genes of internal transcribed spacer (ITS) regions within the fungal genome are generally described as good for fungal species identification (Guarro et al. 1999, Shenoy et al. 2007, Weber 2009, White et al. 1990). The ITS region is the most widely sequenced DNA region of most fungi and has been used for molecular systematics at the species level (Glass and Donaldson 1995, Bellemain et al. 2010, Nilsson et al. 2009). Recently, mycologists have designated ITS as a DNA bar-coding region for species identification of fungi (Seiffer 2009). On the other hand, DNA fingerprinting involves the generation of a set of distinct DNA profiles containing taxonomically important genotypes. They could be culture dependent, at which the genetic material is obtained after culturing the microbe or culture independent where culturing is not necessary to analyse the genetic material (Giraffa and Neviani 2001). The most common culture dependent fingerprinting methods are restriction fragment length polymorphism (RFLP), randomly amplified polymorphic DNA (RAPD) analysis, amplified fragment length polymorphism (AFLP) and PCR amplification of repetitive DNA elements (rep-PCR) (Cocolin et al. 2011, Olive and Bean 1999). Among the culture-independent techniques, denaturing gradient gel electrophoresis (DGGE) is the most common (Giraffa and Neviani 2001) even if sequence-based analyses are also expanding to provide snapshots of the microbes present in foods and their significance (Lee et al. 2007, Weng et al. 2006). Culture independent methods are well suited for analyzing microbial communities over time and may provide the possibility of exploring details of microbial dynamics (Weng et al. 2006).

Polyphasic identification and characterization

Recent works on fungal systematics advocate the use of a polyphasic approach by which combinations of morphological, chemical and molecular techniques are used for a reliable identification of moulds (Frisvad 2004, Frisvad et al. 2008, Pitt and Hocking 2009, Samson et al. 2004). In this case species is defined as "an organism that show evidence for evolutionary divergence from other taxa, unique DNA characters at multiple loci, in addition to any distinctive extrolites and morphological characters" (Frisvad 2004).

After a reliable identification of the moulds associated with foods, the potential mycotoxins that can be produced on the food and the level can be determined using varieties of biochemical analyses techniques such as HPLC, TLC, Bioassyas and others (Frisvad et al. 2008, Petzinger and Weidenbach 2002).

Virulence: Moulds as Threat to Food Safety and their Route to the Human Food Chain

Food spoilage moulds and their toxic secondary metabolites are among the common contaminants that reduce nutritional value, quality and safety of food products (Filtenborg et al. 1996, Frisvad et al. 2007, Moss 1996). Species of *Aspergillus, Alternaria, Cladosporium, Fusarium* and *Pencillium* are the prominent filamentous fungi with this regard (Filtenborg et al. 1996, Moss 1996, Peraica et al. 2006). Species of *Alternaria, Cladosporium and Fusarium* contaminate food mostly at pre-harvest stage and are referred to as field fungi (Pitt and Hocking 2009). Species of *Aspergillus* and *Penicillium* on the other hand prevail at the post-harvest food production conditions and are known by the name storage fungi (Pitt and Hocking 2009, Samson et al. 2004).

It is estimated that moulds cause 5–10% direct productivity loss and defect in quality by contaminating food raw materials and products (Food and Agriculture Organization of the United Nations (FAO) 2003a).

In addition they might affect food safety by producing toxic secondary metabolites, such as mycotoxins as their metabolic by-products. Mycotoxins are fungal secondary metabolites that are toxic to vertebrates when introduced via a natural route (Food and Agriculture Organization of the United Nations (FAO) 2003a, Moss 1996). Over 300 identified mycotoxins are known to induce signs of toxicity in a very small concentrations and their number is increasing (Food and Agriculture Organization of the United Nations (FAO) 2003a, Frisvad 1991, Lopez-Garcia 2010, Moss 1996, Petzinger and Weidenbach 2002). They might be produced at the pre-harvest and post-harvest food production conditions by toxigenic moulds (Food and Agriculture Organization of the United Nations (FAO) 2003a, Frisvad 1991, Lopez-Garcia 2010). Mycotoxins are reported to contaminate 25% of the world's annual crop production (Food and Agriculture Organization of the United Nations (FAO) 2003a).

Toxins of field fungi can enter human food chain via plant based food products (Food and Agriculture Organization of the United Nations (FAO) 2003a). Toxins of field fungi could also be introduced indirectly through the consumption of animal food products containing residues of mycotoxins (Food and Agriculture Organization of the United Nations (FAO) 2003a, Lopez-Garcia 2010, Moss 1996). Some mycotoxins can be stored in meat, milk and eggs if the animals have been exposed to contaminated feed and end up into the human food chain (Food and Agriculture Organization of the United Nations (FAO) 2003a, Lopez-Garcia 2010, Moss 1996). Many of them are stable and difficult to destroy by food treatment activities such as cooking and freezing.

The important sources of mycotoxins on food products at the post-harvest production conditions are toxigenic moulds associated with storage and post-harvest food processing activities (Frisvad 1991, Lopez-Garcia 2010, Moss 1996, Pitt and Hocking 2009, Samson et al. 2004). A variety of moulds can contaminate food, but it is only the associated mycobiota that affect food quality and safety (Filtenborg et al. 1996, Frisvad 1991, Frisvad et al. 2007). Associated mycobiota are some specific numbers of moulds causing spoilage by growing comfortably on specific food products (Filtenborg et al. 1996). The associated mycobiota of a particular food product usually prevails inside the production facility where the food is processed and stored (Filtenborg et al. 1996, Frisvad et al. 2007, Petzinger and Weidenbach 2002, Samson et al. 2004). Species of *Aspergillus* and *Penicillium* prevail under such conditions and have frequently been isolated from post-harvest food processing facilities. Many of them are known for their ability to produce mycotoxins (Filtenborg et al. 1996, Food and Agriculture Organization of the United Nations (FAO) 2003a, Frisvad 1991, Frisvad 2004, Frisvad et al. 2007, Iacumin et al. 2009, Lopez-Garcia 2010, Moss 1996, Petzinger and Weidenbach 2002, Pitt and Hocking 2009, Samson et al. 2004) (Table 1).

Table 1. Common mycotoxins, their producers, effect on health and the foods they are associated with (Food and Agriculture Organization of the United Nations (FAO) 2003a, Frisvad 1991, Frisvad 2004, Frisvad et al. 2007, Moss 1996, Muture and Ogana 2005, Peraica et al. 1999, Peraica et al. 2006, Pitt and Hocking 2009, Samson et al. 2004).

Important mycotoxins	Producers	Effect on health	Associated with
Prevail at a pre-harvest food production conditions			
Fumonisins B1, B2, B3	*Fusarium verticillioides, F. proliferatum, Aspergillus niger*	Hepatotoxicity Genotoxicity Immunomodulation Neurotoxicity	Cereals and their products
Trichothecenes (T-2 toxins, DON)	*Fusarium graminearum, F. culmorum, F. poae, F. sporotrichioides, F. tricinctum*	Hematotoxicity Immunomodulation Skin toxicity	Cereals and their products
Zearalenone	*F. graminearum, F. culmorum*	Affect fertility and reproduction	Cereals and their products
Ergot Alkaloids	*Claviceps purpurea, C. paspali*	Ergotism	Cereals (oat)
Citrinine	*Aspergillus terreus, A. carneus, Penicillium verrucosum, P. citrinum, P. expansum*	Nephrotoxic	Rice, wheat, flour, barley, maize, rye, oats, peanuts and fruit
Alternaria toxins (alternariol, alternariol methyl ether)	*Alternaria alternata* *Alternaria solani*	Oesophageal cancer Mutagenic	Cereals
Prevail at a post-harvest food production conditions			
Cyclopiazonic acid	*Aspergillus flavus, A. versicolor, A. tamarii, Penicillium commune, P. palitans, P. camemberti*	Organ damages	Cereals, nuts, pulses, cheese, ham, sausage
Aflatoxins B1, B2, G1, G2, M	*Aspergillus flavus, A. parasiticus, A. nomius*	Hepatotoxicity Genotoxicity Oncogenicity Immunomodulation	Cereals, nuts, pulses, cheese, ham, fermented sausage, milk
Ochratoxin A	*Penicillium verrucosum, P. nordicum, Aspergillus ochraceus, A. carbonarius*	Nephrotoxicity Genotoxicity Immunomodulation	Cereals pulses, dried vine coffee, spices, groundnuts, dried meat, cheese
Patulin	*Penicillium expansum, Aspergillus clavatus, Byssochlamys nivea*	Neurotoxicity Mutagenesis	Mouldy fruit, vegetables, cereals, other foods
Sterigmatocystin	*Aspergillus nidulans, A. versicolor, A. flavus*		
Tremorgenic toxins	*Penicillium roquefortii, P. crustosum, P. puberulum, Aspergillus clavatus, A. fumigatus*	Tremorgenic Cardiotoxic Neurotoxic	

The growth of toxigenic moulds on food products does not necessarily mean that they produce mycotoxins all the time (Food and Agriculture Organization of the United Nations (FAO) 2003a, Frisvad 1991, Iacumin et al. 2009, Pitt and Hocking 2009, Samson et al. 2004). It is crucial that the complex interactions of moisture, temperature, substrate, aeration, microbial interactions and a_w are optimal (Fig. 3). The interaction between a_w and temperature is especially critical for fungal growth and mycotoxin production (Frisvad 1991, Pitt and Hocking 2009, Samson et al. 2004). The species of *Penicillium* in general need lower temperature compared to *Aspergillus,* to produce mycotoxins at a relatively higher a_w. Conditions for mycotoxin production are generally more restrictive than those for growth (Frisvad 1991, Pitt and Hocking 2009, Samson et al. 2004). Different mycotoxins might be produced by the same species at different environmental conditions. The type of mycotoxins contaminating food products also depend on the types of the moulds dominating on the raw materials, in the production facility and the environmental conditions in the processing facility (Frisvad 1991, Pitt and Hocking 2009, Samson et al. 2004, Neus 2006). Some fungal species are capable of producing more than one mycotoxin and a single mycotoxin can be produced by more than one fungus (Pitt and Hocking 2009, Samson et al. 2004). The coproduction of mycotoxins may result in additive or synergistic effects on consumers and may thus increase the toxigenic potential (Moss 1996, Pitt and Hocking 2009).

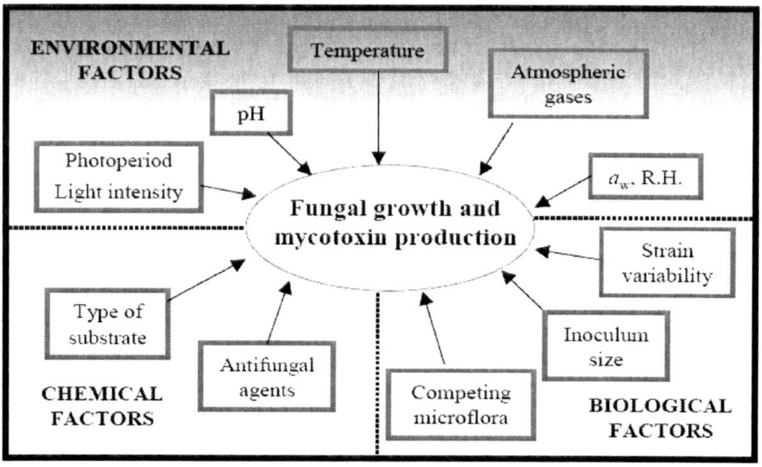

Figure 3. Factors affecting mycotoxin production by toxigenic moulds (modified from Neus 2006).

Mycotoxins are usually produced on the surfaces of food products where fungal contaminants are proliferating. The toxins might also diffuse to the central parts of the foods depending on chemical properties of the mycotoxin and physio-chemical properties of the food such as water content (Moss 1996, Neus 2006).

Historical Outbreaks

Mycotoxicosis is a poisoning caused by mycotoxins ingested with foods (Filtenborg et al. 1996, Frisvad et al. 2007, Moss 1996, Pitt and Hocking 2009). Several pathological syndromes can be observed in human and animals at a very small concentration. The problem of ergotism resulting from the infestation of rye by *Claviceps* has been recognized as an important cause of human mortality in medieval Europe (Peraica et al. 1999). Since then incidences of ergotism have been reported in 1954 and 1978 in France and Ethiopia respectively (Peraica et al. 1999, Peraica et al. 2006). At the beginning of the 1900s, the possibile occurence of human diseases as a result of the consumption of mould-damaged rice was reported in Japan. In the USSR, the consumption of overwintered millet had been linked to mycotoxicosis in 1940s (Peraica et al. 1999). The serious worldwide concern of mycotoxins began in the early 1960s after the outbreak of the Turkey "X" disease by aflatoxins (Peraica et al. 2006). Acute toxicity and fatality could happen in humans at relatively high concentration. Such incidents are usually restricted to the less developed parts of the world where resources for control are limited (Lopez-Garcia 2010, Moss 1996, Peraica et al. 1999, Peraica et al. 2006). An outbreak of acute aflatoxicosis in Kenya in 2004 with 125 fatalities is the most recent case reported (Muture and Ogana 2005). On the other hand, frequent exposure to low doses of mycotoxins (often below detection level) is responsible for chronic effects and increased susceptibility to infectious disease (Filtenborg et al. 1996, Frisvad 1991, Moss 1996, Muture and Ogana 2005, Peraica et al. 1999, Peraica et al. 2006). Some of the most common mycotoxins are known for their carcinogenicity, genotoxicity, teratogenicity, nephrotoxicity, hepatotoxicity and immunosuppressive effect (Filtenborg et al. 1996, Frisvad 1991, Moss 1996, Muture and Ogana 2005, Peraica et al. 1999, Peraica et al. 2006). Aflatoxins were placed on the list of known human carcinogens leading to primary heptocarcinoma by the International Agency for Research on Cancer (Frisvad 1991, Moss 1996, Peraica et al. 1999, Peraica et al. 2006). Ochratoxin A (OTA) is an important hepatotoxic, immunosuppressive, teratogenic and nephrotoxic compound (Food and Agriculture Organization of the United Nations (FAO) 2003a, Frisvad 1991, Moss 1996, Peraica et al. 1999, Peraica et al. 2006) and has been linked with Balkan Endemic Nephropathy (Food and Agriculture Organization of the United Nations (FAO) 2003a, Moss 1996, Peraica et al.

1999, Peraica et al. 2006). Some act as neurotoxins while others interfere with the immune system and other important metabolic activities (Filtenborg et al. 1996, Food and Agriculture Organization of the United Nations (FAO) 2003a, Frisvad 2004, Moss 1996, Peraica et al. 1999, Peraica et al. 2006). The challenge with mycotoxicosis is that it is very difficult to clearly diagnose as the toxic effect of mycotoxins depends on a number of factors such as the intake levels, duration of exposure, the types of toxin present on the food, their mode of action, metabolism and defense mechanism.

Control and Prevention

Effective control requires a combination of good agricultural practice, carefully controlled storage conditions and surveillance at every stage from farm to fork. However, moulds and mycotoxins are still a challenge in the developing countries where the effective control systems are not well established and this can affect the developed world too via a globalized food market. Many countries have recognized mycotoxins as an unavoidable public health risks. As a result many national and international standards are developed as effective measure to keep the level of mycotoxins at an acceptable level (Food and Agriculture Organization of the United Nations (FAO) 2003a, Moss 1996, Peraica et al. 1999, Peraica et al. 2006) (Table 2). This is important to minimise the exposure of the public to mycotoxins.

Table 2. Maximum allowed levels of mycotoxins in on selected foods in the EU (Food and Agriculture Organization of the United Nations (FAO) 2003a, Moss 1996, Peraica et al. 1999, Peraica et al. 2006).

Toxins	Europe (ppb)	Food types
Total aflatoxin (B1, B2, G1, G2, M)	4	• Processed cereals, groundnuts
Citrinin	100	• Cereals
Deoxynivalenol	1750 200	• Unprocessed cereals • Processed cereals for infants and young children
Total Fumonisins	800 200	• Maize-based breakfast • Processed maize based foods for infants and young children
Ochratoxin A	3 0.5	• Processed cereals and cereal products • Baby food
Zearalenone	100 50 20	• Maize intended for direct human consumption • Bread and pastries • Cereal-based food for children

Concluding Remarks and Future Trends

Mycotoxins are often produced only under certain environmental conditions such as in warm, damp and humid conditions. In addition toxigenic moulds require high level of a_w, usually > 0.9 to produce mycotoxins. The prevention of these suitable environmental conditions will play an important role in minimizing their adverse effect on health. However, it is difficult to control many of these environmental conditions. As a result many countries have surveillance programs for toxigenic moulds and their toxins on foods to make sure that their concentrations do not exceed the maximum recommended levels. The threat from mycotoxins to public health and global economy is expected to increase with the increasing agricultural production and global warming. To alleviate the challenge, food industries should be proactive towards establishing a system that minimises the association of toxigenic moulds with food production activities. The implementation of a HACCP based food safety and quality assurance system at every segment of the food chain and in every sector of the food industry is recommended internationally to reduce such food safety hazards on human (Food and Agriculture Organization of the United Nations (FAO) 2003b, Food and Agriculture Organization of the United Nations (FAO) 2008, Mortimore 2001, Oriss and Whitehead 2000, Food Products Association (FPA) 2006, Asefa et al. 2011).

References

Andon, B.M. 2006. Active air and Passive air (settle plate) monitoring in routine environmental programs. J. Pharm. Sci. Technol. 60: 350–355.

Asefa, D.T., C.F. Kure, R.O. Gjerde, S. Langsrud, M.K. Omer, T. Nesbakken and I. Skaar. 2011. A HACCP plan for mycotoxigenic hazards associated with dry-cured meat production processes. Food Contr. 22: 831–837.

Asefa, D.T., S. Langsrud, R.O. Gjerde, C.F. Kure, M.S. Sidhu, T. Nesbakken and I. Skaar. 2009. The performance of SAS-super-180 air sampler and settle plates for assessing viable fungal particles in the air of dry-cured meat production facility. Food Contr. 20: 997–1001.

Bellemain, E., T. Carlsen, C. Brochmann, E. Coissac, P. Taberlet and H. Kauserud. 2010. ITS as an environmental DNA barcode for fungi: an *in silico* approach reveals potential PCR biases. BMC Microbiol. 10: 189.

Carlile, M.J., S.C. Watkinson and G.W. Gooday. 2001. The Fungi. 2nd ed. Elsevier, Acadamic Press. San Diego, CA.

Cocolin, L., P. Dolci and K. Rantsiou. 2011. Biodiversity and dynamics of meat fermentations: The contribution of molecular methods for a better comprehension of a complex ecosystem. Meat Sci. 89: 296–302.

De Joungh,W.A. and J. Nilsen. 2007. Filamentous fungi as cell factories for metabolite production. *In:* J. Dijksterhuis and R.A. Samson (ed.). Food mycology: a multifaceted approach to fungi and food. CRC Press. Boca Raton, Florida.

Deacon, J.W. 1997. Introduction to Modern Mycology. 3rd ed. Blackwell Science Ltd., Carlton. Australia.

Deacon, J.W. 2006. Fungal Biology. 4th ed. Blackwell publishing Ltd., Carlton. Australia.

Dijksterhuis, J. and R.A. Samson. 2007. Food Mycology: a multifaceted approach to fungi and food. CRC Press. Boca Raton, Florida.

Filtenborg, O., J.C. Frisvad and U. Thrane. 1996. Moulds in food spoilage. Int. J. Food Microbiol. 33: 85–102.

Food and Agriculture Organization of the United Nations (FAO). 2003a. Code of practice for the prevention and reduction of mycotoxin contamination in Cereals, including annexes on Ochratoxin A, Zearalenone, Fumonisins and trichothecenes: CAC/RCP 51–2003.

Food and Agriculture Organization of the United Nations (FAO). 2003b. Codex Alimentarius: Food hygine-Basic text. Rome.

Food and Agriculture Organization of the United Nations (FAO). 2008. Risk-based food inspection manual. Rome.

Food Products Association (FPA). 2006. V.N. Scott and K.E. Stevenson (eds.). HACCP, a systemic approach to food safety. Cornell University, USA. 219 pp.

Frisvad, J.C. 1991. Filamentous fungi in foods and feeds: ecology, spoilage and mycotoxin production Vol. 3. Foods and Feeds New York.

Frisvad, J.C. 2004. Polyphasic taxonomy of *Penicillium* subgenus *Penicillium*: A guide to identification of food and air-borne terverticillate Penicillia and their mycotoxins. Stud. Mycol. 49: 1–174.

Frisvad, J.C., B. Andersen and R.A. Samson. 2007. Association of moulds to food. *In:* J. Dijksterhuis and R.A. Samson (ed.). Food mycology: a multifaceted approach to fungi and food. CRC Press. Boca Raton, Florida.

Frisvad, J.C., B. Andersen and U. Thrane. 2008. The use of secondary metabolite profiling in chemotaxonomy of filamentous fungi. Mycol. Res. 112: 231–240.

Giraffa, G. and E. Neviani. 2001. DNA-based, culture-independent strategies for evaluating microbial communities in food-associated ecosystems. Int. J. Food Microbiol. 67: 19–34.

Glass, N.L. and G.C. Donaldson. 1995. Development of primer sets designed for use with the PCR to amplify conserved genes from filamentous ascomycetes. Appl. Environ. Microbiol. 61: 1323–1330.

Guarro, J., J. Gene and A.M. Stchigel. 1999. Developments in Fungal Taxonomy. Clin. Microbiol. Rev. 12: 454–500.

Iacumin, L., L. Chiesa, D. Boscolo, M. Manzano, C. Cantoni, S. Orlic and G. Comi. 2009. Moulds and ochratoxin A on surfaces of artisanal and industrial dry sausages. Food Microbiol. 26: 65–70.

Lee, L., S. Tin and S. Kelley. 2007. Culture-independent analysis of bacterial diversity in a child-care facility. BMC Microbiol. 7: 27.

Lopez-Garcia, R. 2010. Mycotoxin Management: An International Challenge. *In:* Christine E. Boisrobert, S. Aleksandra, O. Sangsuk and L.M.L. Huub (ed.). Ensuring Global Food Safety. Academic Press, San Diego. pp. 209–216.

Malloch, D. 1982. Moulds: Their Isolation, Cultivation and Identification. University of Toronto Press.

Mortimore, S. 2001. How to make HACCP really work in practice. Food Contr. 12: 209–215.

Moss, M.O. 1996. Mycotoxins. Mycol. Res. 100: 513–523.

Muture, B.N. and G. Ogana 2005. Aflatoxin levels in maize and maize products during the 2004 food poisoning outbreak in Eastern Province of Kenya. East Afr. Med. J. 82: 275–279.

Neus, B.M. 2006. Ochratoxin A and ochratoxigenic moulds in grapes, must and wine. Ecophysiological studies: Ecophysiological studies. University of Lleida, Departament of food Technology.

Nilsson, R.H., M. Ryberg, K. Abarenkov, E. Sjokvist and E. Kristiansson. 2009. The ITS region as a target for characterization of fungal communities using emerging sequencing technologies. FEMS Microbiol. Lett. 296: 97–101.

Olive, D.M. and P. Bean. 1999. Principles and Applications of Methods for DNA-Based Typing of Microbial Organisms. J. Clin. Microbiol. 37: 1661–1669.

Oriss, G.D. and A.J. Whitehead. 2000. Hazard analysis and critical control point (HACCP) as a part of an overall quality assurance system in international food trade. Food Contr. 11: 345–351.

Peraica, M., S. Hendrichs, C. Landgern and C.M. Bryant. 2006. Food Mycotoxins: An Update. J. Food Sci. 71: 51–65.

Peraica, M., B. Radic, A. Lucic and M. Pavlovic. 1999. Toxic effects of mycotoxins in humans. Bulletin of the World Health Organization. 77: 754–766.

Petzinger, E. and A. Weidenbach. 2002. Mycotoxins in the food chain: the role of ochratoxins. Livestock Prod. Sci. 76: 245–250.

Pitt, J.I. 1979. The Genus *Penicillium* and its teleomorphic states *Eupenicillium* and *Talaromyces*. Academic Press, London, New York. 632 pp.

Pitt, J.I. and A.D. Hocking. 2009. Fungi and food spoilage. 3rd ed. Springer Dordrechi Heidelberg, Newyork.

Portnoy, J.M., C.S. Barnes and K. Kennedy. 2004. Sampling for indoor fungi. J. Allergy Clin. Immunol. 113: 189–198.

Samson, R.A., J.C. Frisvad and E.S. Hoekstra. 2004. Introduction to Food- and Airborne fungi. Centraalbureau voor Schimmelcultures, Utrecht.

Seiffer, K. 2009. Progress toward DNA barcoding of fungi. Mol. Ecol. Resources 9: 83–89.

Shenoy, B.D., R. Jeewon and K.D. Hyde. 2007. Impact of DNA sequence-data on the taxonomy of anamorphic fungi. Fungal Diversity. 26: 1–54.

Toscani, T., A. Moseriti, A. Dossena, C. Dall'Asta, N. Simoncini and R. Virgili. 2007. Determination of ochratoxin A in dry-cured meat products by a HPLC-FLD quantitative method. J. Chemother. 855: 242–248.

Weber, R.W.S. 2009. Recent Developments in the Molecular Taxonomy of Fungi. *In:* K. Esser (ed.). The Mycota. Physiology and Genetics: selected basics and Applied aspects. Springer, Bochum, Germany. pp. 1–15.

Webster, J. and R.W.S. Weber. 2007. Introduction to Fungi. 3rd ed. Cambridge University Press. Cambridge.

Weng, L., W.M. Rubin and J. Bristow. 2006. Application of sequence-based methods in human microbial ecology. Genome Res. 16: 316–322.

White, T.J., T. Burns, S. Lee and J. Tayler. 1990. Amplification and direct sequencing of fungal ribosomal RNA genes for phylogenetics. Academic Press, New York.

CHAPTER 12

The Role of Mycobacteria in Food Safety

Charles O. Thoen[1,*] and *John B. Kaneene*[2]

Introduction: The Mycobacteria

Bacteria of the genus *Mycobacterium*, including *Mycobacterium tuberculosis*, the causative agent of pulmonary tuberculosis in humans, were first identified by Robert Koch in 1882. At this time tuberculosis was not uncommon or novel: the disease was known to the ancient Greeks, and was known by several names over the centuries (consumption, phthisis, white plague). Pulmonary tuberculosis kills more people today than at any time in history. It kills more people than any other infectious disease. It is estimated that one of three people alive are infected with the tubercle bacillus, and more than 9 million new cases were reported in 2008 (WHO 2008).

Other forms of mycobacterial infection have been important public health issues in the past. The disease scrofula (tuberculous cervical adenitis) (Wolinsky 1995) has a long history: it was believed that the touch of a king could cure sufferers (Grzybowski and Allen 1995). After discovery of mycobacteria in the late 19th and early 20th centuries, evidence that linked

[1] Department of Veterinary Microbiology and Preventive Medicine, College of Veterinary Medicine, Iowa State University, Ames, IA 50011, USA.
Email: cthoen@iastate.edu

[2] Center of Comparative Medicine, College of Veterinary Medicine, Michigan State University, East Lansing, MI 48824, USA.
Email: kaneene@cvm.msu.edu

* Corresponding author

scrofula infants and children that consumed raw milk from cattle infected with *M. bovis*, a member of the *M. tuberculosis* complex was found (Grange 2001). Accurate information is not available on the incidence of *M. bovis* in humans (Thoen et al. 2006). This is especially true in developing countries where the diagnosis of TB is based on microscopic examination of specimens for presence of acid-fast bacilli and techniques for differentiating *M. bovis* from *M. tuberculosis* are not widely accessible (de Kantor et al. 2010).

Hansen's disease, or leprosy, is a chronic disease caused by *M. leprae* has a history extending over 4,000 years (Sasaki et al. 2001). Diseases caused by other mycobacterial species, including members of the *M. avium* complex and saprophytic mycobacteria, have been responsible for respiratory, lymphatic, dermatologic, systemic, and gastrointestinal infection in humans (Woods and Washington 1987). The advent of HIV-AIDS has seen an increase in all forms of mycobacterial disease in co-infected patients, making previously rare atypical forms of mycobacteria more common (Chin et al. 1994, Thoen et al. 2006, Woods and Washington 1987).

The genus *Mycobacterium* includes *M. tuberculosis* complex, *M. avium* complex, *M. chelonae* group, and other pathogenic and nonpathogenic mycobacteria present in water and soil. The *Mycobacterium tuberculosis* complex includes: *M. tuberculosis, M. africanum, M. bovis, M. cannnetii, M. caprae* and *M. microti* (Brosch et al. 2002), and the *Mycobacterium avium* complex includes: *M. avium* ssp. *avium, M. avium* ssp. *hominisuis, M. avium* ssp. *paratuberculosis*), and *M. intracellulare* (Mijs et al. 2002). Other mycobacteria of clinical significance include *M. fortuitum, M. chelonae, M. marinum, M. kansassi* and *M. scrofulaceum* (Olsen et al. 2010). Mycobacteria not in the *M. tuberculosis*-complex are often referred to as MOTT (Mycobacteria Other than Tubercule Bacilli).

Mycobacterial pathogenicity and host response

Development of mycobacterial disease in a host depends on the ability of the mycobacteria to survive and multiply within the macrophages of the host (Olsen et al. 2010). Pathogenic mycobacteria produce granulomatous lesions in a wide range of species, including mammals (Thoen et al. 2009), birds (Tell et al. 2001), fishes (Decostere et al. 2004), reptiles (Soldati et al. 2004) and amphibians (Densmore and Green 2007).

The pathogenicity of mycobacteria is a multifactorial phenomenon, requiring the participation and cumulative effects of several cell components (Cole et al. 1998). Mycobacteria survive in non-activated macrophages by inhibition of fusion of phagosome and lysosomes, by avoiding the generation of reactive oxygen and nitrogen intermediates or may escape into the cytoplasm (Olsen et al. 2010). The mycobacterial cell wall is composed of peptidoglycan, arabinogalactan, and mycolic acids, such as cord factor (trehalose 6, 6-dimycolate) or other mycolic acid-containing glycolipids, have been implicated in the pathogenesis of mycobacterial disease. Glycolipids and sulfur-containing glycolipids reportedly promote the survival of virulent tubercle bacilli within macrophages by inhibiting phagolysosome.

Numerous host mechanisms have been associated with susceptibility and development of mycobacterial disease. The macrophage act as the principal effector cells in regulating mycobacterial infection in the host. Cell-mediated immune (CMI) responses including, specific T lymphocytes and activated mononuclear macrophages, are important in host resistance to virulent acid-fast bacilli (Russell et al. 2010). CD4 T-cells secrete gamma interferon, limiting bacterial growth and may enhance clearance of the bacillus. The cytolytic activity of CD8 T-cells directly destroys intracellular bacteria by releasing the antimicrobial protein granulysin. Various cytokines have been shown to activate tuberculostatic macrophage functions and limit the replication of mycobacteria. Reactive nitrogen intermediates (i.e., nitric oxide) and oxygen intermediates singlet oxygen and superoxide anions produced by peritoneal macrophages have been proposed to be important in killing virulent *M. tuberculosis* (Olsen et al. 2010).

Although a great deal of new information and understanding has been acquired in recent years on the pathogenesis and immunogenesis of mycobacterial infections, definitive information on the mechanisms of virulence remains obscure (Olsen et al. 2010, Russell et al. 2010). Recently, the genome of *M. tuberculosis, M. bovis*, and *M. avium* ssp. *paratuberculosis*, have been sequenced; therefore, it is anticipated that information on the role of factors associated with virulence will soon be available (Cole et al. 1998, Garnier et al. 2003, Li et al. 2005).

Routes of transmission

The great majority of tuberculosis cases in humans, caused by *M. tuberculosis*, are usually spread by aerosol transmission of droplet nuclei. The initial infection usually resolves in most people, but tuberculosis can reactivate several years following primary infection. In addition to person-to-person aerosol spread, mechanically generated aerosols from the use of pressure washers to clean animal housing (Dalovisio et al. 1992, Eisenberg et al. 2011), or aerosols generated in abattoirs (Thoen et al. 2006) have been shown to contain mycobacteria.

Ingestion is another important route of infection for mycobacteria. Mycobacteria known to infect livestock (*M. tuberculosis*—complex organisms, including *M. bovis* and *M. caprae*; *M. avium* ssp. *paratuberculosis*) have been found in milk associated with mycobacterial disease in humans (Eltholth et al. 2009, Pavlik et al. 2002, Thoen et al. 2009) and water has been recognized as an important vehicle for environmental MOTT (Falkinham 1996). In addition to inhalation and ingestion, many mycobacteria can also be transmitted through contact with contaminated animals or animal products (de la Rua-Domenech 2006) and with soil, water, and other environmental substrates (Decostere et al. 2004, Falkinham 1996, Whan et al. 2005, Woods and Washington 1987).

Diagnosis

Detection and diagnosis of pulmonary tuberculosis is well-established (Dunlap et al. 2000), and several techniques (Mantoux skin test, radiography smear microscopy, culture) have been in use for decades, and considerable efforts are being made to develop more rapid and sensitive molecular diagnostic tests for tuberculosis (Pai et al. 2010). Diagnosis of extrapulmonary tuberculosis in the past was dependent upon clinical signs and detection of mycobacteria in biopsy specimens; this has been supplemented by new diagnostic approaches using molecular techniques (e.g., nucleic acid amplification, PCR, Southern blot) (Fontanilla et al. 2011). Efforts have been made to develop alternative tests, particularly in resource-poor regions, including serological tests for antibody detection and urine tests for antigen detection. Urine antigen detection is in its infancy and requires further development (Lawn 2012), and performance of serum antibody tests have been highly variable, and the WHO has issued policy statements against the use of these tests (Steingart et al. 2011).

The classic gold standard for diagnosis of mycobacteria has been culture; but may take weeks to months to complete: to overcome this problem, several rapid culture methods have been developed (Nahid et al. 2012). Once identified, reliable biochemical tests are available for differentiating bacteria of the *M. tuberculosis* complex; however, molecular techniques have been described and are now widely used (Harris 2006, Collins 2011). For *M. avium* complex bacteria, a characteristic that is useful in differentiating MAP from other *M. avium* complex organisms is dependency on mycobactin, an iron-chelating agent required for *in vitro* growth. Since mycobactin-dependence has also been reported for certain strains of *M. avium* ssp. *avium*, definitive molecular methods such as PCR and restriction endonuclease analysis that have been developed for identifying *M. avium* ssp. *paratuberculosis* (Harris 2006, Pavlik et al. 2000).

Treatment

The primary course of treatment for all mycobacterioses is treatment with antimicrobials, including isoniazid, rifampin, pyrazinamide, and ethambutol (CDC 2003). Unfortunately, drug resistance is of increased concern in the treatment of tuberculosis. Multidrug resistant and extensively drug resistant strains of *M. tuberculosis* have been identified in countries around the globe (WHO 2008). Another complication has been the rise of HIV-tuberculosis co-infection, which has required new regimens of treatment (Lienhardt et al. 2012).

The primary means of preventing mycobacterial disease has been to reduce exposure to viable bacteria. An outstanding success story of this approach has been pasteurization or heat-treatment of milk, which has nearly eliminated foodborne *M. bovis* infections in countries with rigorous milk quality standards (Thoen et al. 2006). Recently, electron beam irradiation at 4 kGy has been used to successfully kill *M. avium* complex organisms in ground pork (Thoen and Beran 2012). A secondary approach to reducing the public health risks of mycobacterial disease has been to reduce levels of infection in the reservoir hosts, such as cattle and goats. Unfortunately, outbreaks of *M. bovis* in wildlife have emerged as an impediment to eradicating *M. bovis* from cattle (Gilsdorf et al. 2006, Kaneene et al. 2002, O'Brien et al. 2006, Thoen et al. 2009).

Foodborne Mycobacterioses

Disease associated with *M. bovis*

Mycobacterium bovis, a slowly-growing non-photochromogenic organism is the etiologic agent of bovine tuberculosis and causes disease in other domestic and wild animals and humans (Karlson and Carr 1970, Thoen 2010). *Mycobacterium bovis* has been isolated from cattle, sheep, swine and other food producing animals with tuberculosis in many non-industrialized and developed countries around the globe (Thoen et al. 2009).

Milk is considered the main source for transmission of *M. bovis* to humans, as the organism has been isolated from cheese and other food prepared from unpasteurized milk (CDC 2005, Harris et al. 2007). Infection in humans related to ingestion of contaminated milk and cheese has been documented (LoBue et al. 2010). In San Diego, California it was found that 7% of human TB was caused by *M. bovis (*LoBue et al. 2003*)*. Ninety percent of these cases occurred in Mexican-born immigrants or U.S.-born Hispanic children. More recently in New York City *M. bovis* was isolated from thirty five TB patients (CDC 2005). The ethnic composition of the patients was very similar to that found in San Diego, and *M. bovis* infection appeared to be associated with consumption of unpasteurized dairy products from Mexico, particularly fresh cheese.

The possibility of meat serving as a vehicle for *M. bovis* infection cannot be ignored, if contamination occurs at slaughter and the meat is not heated to temperatures sufficient to inactivate the bacteria. Tuberculous lesions are most often associated with the lung and lymph nodes in food producing animals, which are not often included in food. The isolation of *M. bovis* or *M. tuberculosis* from skeletal muscle of bovidae or other species is very rare (Ashford et al. 2006).

The primary hosts of *M. bovis* are cattle; whereas goats are considered the primary hosts of *M. bovis* ssp. *caprae*. Aerosol exposure to *M. bovis* generally leads to involvement of pulmonary lymph nodes and lungs of cattle and other susceptible species, while animals exposed by ingestion of contaminated food and water usually develop primary foci in lymph nodes of the head and other lymph tissues associated with the gastrointestinal tract (Thoen et al. 2009). Mycobacteria multiply within macrophages, and after about 10–14 days, cell-mediated immune (CMI) responses develop and host macrophages acquire an increased capacity to kill the intracellular bacilli. The CMI responses are mediated by lymphocytes, which release lymphokines (γ-IFN) that attract, immobilize, and activate additional blood-borne mononuclear cells at the site where virulent mycobacteria or their products exist (Olsen et al. 2010). The cellular hypersensitivity that develops contributes to cell death and tissue destruction (caseous necrosis). In some instances, liquefaction and cavity formation occur due to enzymatic action on proteins and lipids. Rupture of these cavities into the bronchi allows aerosol spread of bacilli.

Lymph nodes are more commonly infected than other tissues because fluids in an animal eventually pass through the nodes where the meshwork of trabeculae entraps organisms. The enlargement and presence of macrophages in impenetrable passageways between reticular cell fibers of the lymph node provide an environment for mycobacterial growth and development of the granulomatous lesion in the node. Primary lesions often become localized in a node(s) and may become large and firm. Fibrous connective tissue development probably contributes to localization of the granulomatous lesions.

Disease associated with *M. tuberculosis*

M. tuberculosis, the human tubercle bacillus, produces progressive generalized disease in nonhuman primates, dogs, swine, and certain exotic animals; cats, rabbits and cattle are quite resistant. Recently, tuberculosis due to *Mycobacterium tuberculosis* in grazing cattle has been reported in central Ethiopia (Ameni et al. 2011) and in cattle in Slovenia (Ocepek et al. 2005). However, the organism fails to cause progressive disease in bovidae (Thoen and Karlson 1984) and transmission of *M. tuberculosis* from animal to animal in food producing species is considered rare. Lesions are most often observed in lymph nodes of the head and of the intestinal tract. There has been no direct evidence of *M. tuberculosis* infection associated with consumption of milk or meat from infected animals, but the potential for contamination of meat at slaughter in unhygienic conditions is possible.

Disease associated with *M. avium* complex organisms

The MOTT includes organisms that have the widest host range among all mycobacteria (Olsen et al. 2010, Thoen et al. 1981). *M. avium* complex bacteria have been isolated from tissues of swine and other food producing animals (Komijn et al. 1999, Thoen 2010, Olsen et al. 2010, Wellenberg et al. 2010). These organisms have also been isolated from water, sphagnum bogs, wood shavings and other environmental specimens (Matlova et al. 2004, Kazda 1979). It has been reported that *M. avium* could survive in carcasses buried in soil for more than four years in North Dakota (Schalk et al. 1935).

The *M. avium* complex organism of greatest public health concern is *M. avium* ssp. *paratuberculosis* (*MAP*). MAP has been isolated from powdered milk and cheese (Grant et al. 2002, Ayele et al. 2005, Slana et al. 2008) and from pasteurized milk (Grant et al. 2002); recently it has been isolated from cheek muscle of cows at slaughter (Okura et al. 2011). Evidence also exists that MAP may be resistant to high-temperature short-time pasteurization, but this finding is controversial (Vaerewijck et al. 2005). Some researchers have found that MAP has been isolated from patients with Crohn's disease; however, recent reports fail to provide epidemiologic evidence that supports MAP as the etiologic agents of Crohn's disease (Qual et al. 2010, Van Kruningen 2011).

One important consideration is that MAP is highly resistant to acid and bile salts, which increases its survivability in the human intestine after ingestion (Primm et al. 2004). In addition to MAP and other *M. avium* complex organisms have been isolated from humans with HIV infections and other immune comprised patients (Primm et al. 2004).

MAP is the cause of a transmissible intestinal disorder of ruminants commonly known as Johne's disease (Thoen 2007). Infection with MAP occurs primarily in the young (< 30 days of age) through the faecal-oral route; other routes that have been suggested include placental and intrauterine. After an incubation period of 2–5 years (range 6 months to 8 years), animals develop a general unthriftiness, rough hair coat, chronic weight loss, and intermittent diarrhea. The clinical course usually lasts only a few months, terminating in severe diarrhea, emaciation, ventral edema, debilitation, and death. Vaccines are available which alter the clinical course of disease and limit the shedding of bacteria in feces.

Paratuberculosis is characterized by a granulomatous inflammation of the terminal region of the small and large bowel and regional lymph nodes. Clinical Johne's disease usually involves impaired intestinal function associated with chronic inflammatory responses. Lesions in cattle are primarily in the intestinal wall and characterized by diffuse granulomatous changes with little or no evidence of necrosis. Granulomas containing numerous acid-fast bacilli are often present in lymph nodes associated with the intestinal tract in cattle. Caseous necrosis has been observed in such lesions in sheep and goats.

Other members of the *M. avium* complex are known to infect other animals. *M. avium* ssp. *hominissuis* is often isolated from swine (Dvorska et al. 2002, Wellenberg et al. 2010, Thoen 2012), and *M. avium* ssp. *intracellulare* are often associated with granulomatous lesions in cold-blooded animals (Thoen and Schliesser 1984, Thoen and Williams 1994). In birds, disease due to *M. avium* ssp. *avium* is usually progressive, with lesions in the liver and spleen; in nonhuman primates, cattle and swine, lesions are usually confined to lymph nodes associated with the intestinal tract. Rabbits are highly susceptible to experimental infection with *M. avium* serovars 1 and 2, but are relatively resistant to other serovars of *M. avium complex* (Thoen 2010). Lesions of lymph nodes associated with the gastrointestinal tract, as with *M. avium* ssp. *avium* in cattle and swine, suggest infection by ingestion. The medial retropharyngeal lymph nodes are a frequent site of infection and are the most commonly infected site in the head. Other lymph nodes of the head (mandibular, parotid, and lateral retropharyngeal) are occasionally involved.

Disease associated with other Mycobacteria

There are several species of mycobacteria that have been found in fishes, including *M. chelonae*, *M. marinum*, and *M. fortuitum* (Decostere et al. 2004). The disease results in systemic infections in affected animals, which then shed the bacteria into the water where it can be acquired by humans that come into contact with the contaminated water. MOTT has been identified in a variety of wild and farmed fish (Brocklebank et al. 2003, Decostere et al. 2004, Ghittino et al. 2003, Mediel et al. 2000, Rhodes et al. 2004), and have been isolated from frozen fish at market (Mediel et al. 2000). While there have currently been no documented cases of mycobacteriosis from consumption of infected fish, the possibility of infection has been assessed by the USDA and the Canadian Food Inspection Agency (CFIA) and deemed remote (Brocklebank et al. 2003).

Several species of mycobacteria are saprophytes, and are considered generally non-pathogenic except in unusual cases (Primm et al. 2004, Vaerewijck et al. 2005). Environmental mycobacteria, including *M. avium* ssp. *avium*, *M. chelonae*, *M. fortuitum*, *M. gordonae*, *M. intracellulare*, *M. kansasii*, *M. xenopi*, and *M. mucogenicum*, have been found in public drinking water, hospital water distribution systems, and home tap water (Le Dantec et al. 2002, Fernandez-Rendon et al. 2012, Vaerewijck et al. 2005). There have also been instances where environmental mycobacteria have been found in prepared foods (Cerna-Cortéz et al. 2009), which have been attributed to the use of contaminated water in food preparation. Given the widespread presence of environmental mycobacteria, it has been suggested that most humans have been exposed to these mycobacteria and have experienced transient, self-curing colonizations (Primm et al. 2004, Vaerewijck et al. 2005). While no documented cases of mycobacterial infection from drinking water have been reported to date, the presence of mycobacteria in drinking water is a potential public health issue, particularly for the severely ill or immunocompromised (Primm et al. 2004, Vaerewijck et al. 2005).

Summary

The genus *Mycobacterium* has a documented history as a food-borne pathogen, particularly through the consumption of dairy products infected with *M. bovis*. Recent research has provided evidence for the presence of other mycobacteria in food and water available for human consumption; however, it should be emphasized that there is little or no evidence for the occurrence of foodborne outbreaks by these species. The risk of foodborne infection from these sources is considered to be very low for the general population, especially if proper food preparation is used. Risks may still exist, though, for some individuals: the very young and old, the very ill, and patients suffering with compromised immune systems. In these instances, extra precautions should be taken to reduce the risk of mycobacterial infection from any high-risk foods consumed by these individuals.

References

Ashford, D.A., L. Volker and J.H. Steele. 2006. Bovine tuberculosis: environmental public health preparedness considerations for the future. *In:* C.O. Thoen, J.H. Steele and M.J. Gilsdorf (eds.). *Mycobacterium bovis* Infection in Animals and Humans. Blackwell Publ., Ames, USA. pp. 305–315.

Ameni, G., M. Vordemeier, R. Firdessa, A. Aseffa, G. Hewinson, S.V. Gordon and S. Berg. 2011. *Mycobacterium tuberculosis* infection in grazing cattle in central Ethiopia. Vet. J. 188: 359–361.

Ayele, W.Y., P. Svastova, P. Roubal, M. Bartos and I. Pavlik. 2005. *Mycobacterium avium* subspecies *paratuberculosis* cultured from locally and commercially pasteurized cow's milk in the Czech Republic. Appl. Environ. Microbiol. 71: 1210–1214.

Brocklebank, S., S. Raverty and J. Robinson. 2003. Mycobacteriosis in Atlantic salmon farmed in British Columbia. Can. Vet. J. 44: 486–489.

Brosch, R., S.V. Gordon, M. Marmiesse, P. Brodin, C. Buchrieser, K. Eiglmeier, T. Garnier, C. Gutierrez, G. Hewinson, K. Kremer, L.M. Parsons, A.S. Pym, S. Samper, D. van Soolingen and S.T. Cole. 2002. A new evolutionary scenario for the *Mycobacterium tuberculosis* complex. Proc. Natl. Acad. Sci. USA. 99: 3684–3689.

Centers for Disease Control and Prevention (CDC). 2003. Treatment of Tuberculosis, American Thoracic Society, CDC, and Infectious Diseases Society of America. *MMWR* 52(RR11): 1–77.

Centers for Disease Control and Prevention (CDC). 2005. Human tuberculosis caused by *Mycobacterium bovis*—New York City, 2001–2004. *MMWR*. 54: 605–608.

Cerna-Cortéz, J.F., T. Estrada-García and J.A. González-y-Merchand. 2009. Isolation of *Mycobacterium mucogenicum* from street-vended chili sauces: A potential source of human infection. J. Food Prot. 72: 182–184.

Chin, D.P., P.C. Hopwell, D.M. Yajko, E. Vittinghoff, C.R. Horsburgh, W.K. Hadley, E.N. Stone, P.S. Nassos, S.M. Ostrogg, M.A. Jacobson, C.C. Matkin and A.L. Reingold. 1994. *Mycobacterium avium* complex in the respiratory or gastrointestinal tract and the risk of *M. avium* complex bacterium in patients with human immunodeficiency virus infection. J. Infect. Dis. 169: 289–295.

Cole, S.T., R. Brosch, J. Parkhill, T. Garnier, C. Churcher, D. Harris, S.V. Gordon, K. Eiglmeier, S. Gas, C.E. Barry III, F. Tekaia, K. Badcock, D. Basham, D. Brown, T. Chillingworth, R. Connor, R. Davies, K. Devlin, T. Feltwell, S. Gentles, N. Hamlin, S. Holroyd, T. Hornsby, K. Jagels, A. Krogh, J. McLean, S. Moule, L. Murphy, K. Oliver, J. Osborne, M.A. Quail, M.A. Rajandream, J. Rogers, S. Rutter, K. Seeger, J. Skelton, R. Squares, S. Squares, J.E. Sulston, K. Taylor, S. Whitehead and B.G. Barrell. 1998. Deciphering the biology of *Mycobacterium tuberculosis* from the complete genome sequence. Nature. 393: 537–544.

Collins, D.M. 2011. Advances in molecular diagnostics for *Mycobacterium bovis*. Vet. Microbiol. 151: 2–7.

Dalovisio, J.R., M. Stetter and S. Mikota-Wells. 1992. Rhinoceros' rhinorrhea: cause of an outbreak of infection due to airborne *Mycobacterium bovis* in zookeepers. Clin. Infect. Dis. 15: 598–600.

Decostere, A., K. Hermans and F. Haesbrouck. 2004. Piscine mycobacteriosis: a literature review covering the agent and the disease it causes in fish and humans. Vet. Microbiol. 99: 159–166.

de Kantor, I.N., P.A. LoBue and C.O. Thoen. 2010. Human tuberculosis caused by *Mycobacterium bovis* in the United States, Latin America and the Caribbean. Intern. J. Tuberc. Lung Dis. 14: 1369–1373. 84: 200–06.

de la Rua-Domenech, R. 2006. Human *Mycobacterium bovis* infection in the United Kingdom: Incidence, risks, control measures and review of the zoonotic aspects of bovine tuberculosis. Tuberculosis. 86: 77–109.

Densmore, C.L. and D.E. Green. 2007. Diseases of Amphibians. ILAR J. 48: 235–254.

Dvorska, L., M. Bartos, O. Ostadal, J. Kaustova, L. Matlova and I. Pavlik. 2002. IS1311 and IS1245 restriction fragment length polymorphism analyses, serotypes, and drug susceptibilities of *Mycobacterium avium* complex isolates obtained from a human immunodeficiency-virus negative patient. J. Clin. Microbiol. 40: 3712–3719.

Dunlap, N.E., J. Bass, P. Fujiwara, P. Hopewell, C.R. Horsburgh Jr., M. Salfinger and P.M. Simone. 2000. American Thoracic Society: Diagnostic Standards and Classification of Tuberculosis in Adults and Children. Am. J. Respir. Crit. Care Med. 161: 1376–1395.

Eisenberg, S., M. Nielen, J. Hoeboer, M. Bouman, D. Heederik and A. Koets. 2011. *Mycobacterium avium* subspecies *paratuberculosis* in bioaerosols after depopulation and cleaning of two cattle barns. Vet. Rec. 168: 587.

Eltholth, M.M., V.R. Marsh, S. Van Winden and F.J. Guitian. 2009. Contamination of food products with *Mycobacterium avium paratuberculosis*: a systematic review. J. Appl. Microbiol. 107: 1061–1071.

Falkinham, J.O. 1996. Epidemiology of infection by nontuberculous mycobacteria. Clin. Microb. Rev. 9: 177–215.

Fernandez-Rendon, E., J.F. Cerna-Cortéz, M.A. Ramirez-Medina, A.C. Helguera-Repetto, S. Rivera-Gutierrez, T. Estrada-García and J.A. González-y-Merchand. 2012. *Mycobacterium mucogenicum* and other non-tuberculous mycobacteria in potable water of a trauma hospital: a potential source for human infection. J. Hosp. Infect. 80: 74–76.

Fontanilla, J.M., A. Barnes and C. Fordham von Reyn. 2011. Current diagnosis and management of peripheral tuberculous lymphadenitis. Clin. Infect. Dis. 53: 555–562.

Garnier, T., K. Eiglmeier, J.C. Camus, N. Medina, H. Mansoor, M. Pryor, S. Duthoy, S. Grondin, C. Lacroix, C. Monsempe, S. Simon, B. Harris, R. Atkin, J. Doggett, R. Mayes, L. Keating, P.R. Wheeler, J. Parkhill, B.G. Barrell, S.T. Cole, S.V. Gordon and R.G. Hewinson. 2003. The complete genome of *Mycobacterium bovis*. Proc. Natl. Acad. Sci. USA. 100: 7877–7882.

Ghittino, C., M. Latini, F. Agnetti, C. Panzieri, L. Lauro, R. Ciappelloni and G. Petracca. 2003. Emerging pathologies in aquaculture: effects on production and food safety. Vet. Res. Comm. 27(Suppl. 1): 471–479.

Grant, I.R., H.J. Ball and. M.T. Rowe. 2002. Incidence of *Mycobacterium paratuberculosis* in bulk raw and commercially pasteurized cows' milk from approved dairy processing establishments in the United Kingdom. Appl. Environ. Microbiol. 68: 2428–2435.

Gilsdorf, M.J., L. Judge and E.D. Ebel. 2006. Current challenges to and impacts on the U.S. National bovine tuberculosis eradication program: *Mycobacterium bovis* outbreaks in alternative species and surveillance performance. *In:* Thoen, J.H. Steele and M.J. Gilsdorf (eds.). *Mycobacterium bovis* infection in animals and humans. 2nd ed. C.O. Ames: Blackwell Publ., Ames. pp. 211–225.

Grange, J.M. 2001. *Mycobacterium bovis* infection in human beings. Tuberculosis. 81: 71–77

Grzybowski, S. and E.A. Allen. 1995. History and importance of scrofula. Lancet. 346: 1472–1474.

Harris, N.B. 2006. Molecular techniques: application in epidemiologic investigations. *In:* Thoen, J.H. Steele and M.J. Gilsdorf (eds.). *Mycobacterium bovis* infection in animals and humans. 2nd ed. C.O. Ames: Blackwell Publ. pp. 54–62.

Harris, N.B., J. Payeur, D. Bravo, R. Osorio, T. Stuber, D. Farrell, D. Paulson, S. Trevisio, S. Mikolon, A. Rodriquez-Lainz, Cernek-Hoskins, R. Rast, M. Ginsberg and H. Kinde. 2007. Recovery of *Mycobacterium bovis* from soft fresh cheese originating in Mexico. Appl. Environ. Microbiol. 73: 1025–1028.

Kaneene, J.B., C.S. Bruning-Fann, L.M. Granger, R. Miller and B.A. Porter-Spalding. 2002. Environmental and farm management factors associated with tuberculosis on cattle farms in northeastern Michigan. JAVMA. 221: 837–842.

Karlson, A.G. and D.T. Carr. 1970. Tuberculosis caused by *Mycobacterium bovis*; report of six cases. Ann. Intern. Med. 73: 979–983.

Kazada, J. 1979. Sphagnum moss vegetation and other sources of mycobacteria. *In:* J. Viallier (ed.). Colloque Intern, Sur Les Mycobacterterie, Oct. 1–3. Lyon, France Vol II. pp. 318–327.

Komijn, R.E., P.E.W de Haas, M.M.E. Schneider, T. Egar, J.H.M. Nieuwenhuijs, R.J. van den Hoek, D. Bakker, F.G. van Zijd Erveld and D. van Soolingen. 1999. Prevalence of *Mycobacterium avium* in slaughter pigs in the Netherlands and comparison of IS1245 restriction fragment length polymorphism patterns of porcine and human isolates. J. Clin. Microbial. 37: 1254–1259.

Lawn, S.D. 2012. Point-of-care detection of lipoarabinomannan (LAM) in urine for diagnosis of HIV-associated tuberculosis: a state of the art review. BMC Infect. Dis. 12: 103.

Le Dantec, C., J.P. Duguet, A. Montiel, N. Dumoutier, S. Dubrou and V. Vincent. 2002. Occurrence of Mycobacteria in water treatment lines and in water distribution systems. Appl. Environ. Microbiol. 68: 5318–5325.

Li, L., J.P. Bannantine, Q. Zhang, A. Amonsin, B.J. May, D. Alt, N. Banerji, S. Kanjilal and V. Kapur. 2005. The complete genome sequence of *Mycobacterium avium* subspecies *paratuberculosis*. Proc. Natl. Acad. Sci. USA. 102: 12344–12349.

Lienhardt, C., M. Raviglione, M. Spigelman, R. Hafner, E. Jaramillo, M. Hoelscher, A. Zumla and J. Gheuens. 2012. New drugs for the treatment of tuberculosis: needs, challenges, promise, and prospects for the future. JID 205, Suppl. 2: S241–S249.

LoBue, P.A., W. Betancourt, C. Peter and K. Moser. 2003. Epidemiology of *Mycobacterium bovis* disease in San Diego County, 1994–2000. Intern. J. Tuber. Lung Dis. 7: 180–185.

LoBue, P.A., D.A. Enarson and T.C. Thoen. 2010. Tuberculosis in humans and its epidemiology, diagnosis and treatment in the United States. Intern. J. Tuberc. Lung Dis. 14: 2226–2232.

Matlova, L., L. Dvorska, K. Palecek, L. Maurenc, M. Bartos and I. Pavlik. 2004. Impact of sawdust and wood shavings in bedding on pig tuberculous lesions in lymph nodes, and IS*1245* RFLP analysis of *Mycobacterium avium* subsp. *hominissuis* of serotypes 6 and 8 isolated from pigs and environment. Vet. Microbiol. 102: 227–236.

Mediel, M.J., V. Rodriquez, G. Codina and N. Martin-Casabona. 2000. Isolation of mycobacteria from frozen fish destined for human consumption. Appl. Environ. Microbiol. 66: 3637–3638.

Mijs, W., P. de Haas, R. Rossau, T. van Der Laan, L. Rigouts, F. Portaels and D. van Soolingen. 2002. Molecular evidence to support a proposal to reserve the designation *Mycobacterium avium* subsp. *avium* for bird-type isolates and *M. avium* subsp. *hominissuis* for the human/porcine type of *M. avium*. Int. J. Syst. Evol. Microbiol. 52: 1505–1518.

Nahid, P., P.S. Kim, C.A. Evans, D. Alland, M. Barer, J. Diefenbach, J. Ellner, R. Hafner, C.D. Hamilton, M.F. Iademarco, G. Ireton, M.E. Kimerling, C. Lienhardt, W.R. MacKenzie, M. Murray, M.D. Perkins, J.E. Posey, T. Roberts, C. Sizemore, W.S. Stevens, L. Via, S.D. Williams, W.W. Yew and S. Swindells. 2012. Clinical research and development of tuberculosis diagnostics: moving from silos to synergy. JID 205, Suppl. 2: S159–S168.

O'Brien, D.J., S.M. Schmitt, S.D. Fitzgerald and D.E. Berry. 2006. Managing the wildlife reservoir of *Mycobacterium bovis*: the Michigan, USA, experience. Vet. Microbiol. 112: 313–323.

Ocepek, M., M. Pate, M. Zolnir-Dovc and M. Poljak. 2005. Transmission of *Mycobacterium tuberculosis* from human to cattle. J. Clin. Microbiol. 43: 3555–3557.

Okura, H., N. Toft, N. Pozazato, A. Tondo and S.S. Nielsen. 2011. Apparent prevalence of beef carcasses contaminated with *Mycobacterium avium* subsp. *paratuberculosis* sampled from Danish slaughter cattle. Vet. Med. Intern. 152867.

Olsen, I., R.G. Barletta and C.O. Thoen. 2010. *Mycobacterium. In:* C.L. Gyles, J.F. Prescott, J.G. Songer and C.O. Thoen (eds.). Pathogenesis of Bacterial Infections in Animals. Wiley/Blackwell, Ames, USA. pp. 113–132.

Pai, M., J. Minion, K. Steingart and A. Ramsay. 2010. New and improved tuberculosis diagnostics: evidence, policy, practice, and impact. Curr. Opin. Pulmon. Med. 16: 271–284.

Pavlik, I., P. Svastova, J. Bartl, L. Dvorska and I. Rychlik. 2000. Relationship between IS901 in the *Mycobacterium avium* complex strains isolated from birds, animals, humans, and the environment and virulence for poultry. Clin. Diag. Lab. Immunol. 7: 212–217.

Pavlik, I., F. Bures, P. Janovsky, P. Pecinka, M. Bartos, L. Dvorska, L. Matlova, K. Kremer and D. Van Soolingen. 2002. The last outbreak of bovine tuberculosis in cattle in the Czech Republic in 1995 was caused by *Mycobacterium bovis* subspecies *caprae*. Vet. Med.—Czech. 47: 251–263.

Primm, T.P., C.A. Lucero and J.O. Falkinham III. 2004. Health impacts of environmental mycobacteria. Clin. Microbiol. Rev. 17: 98–106.

Qual, D.A., J.B. Kaneene, T.J. Varty, R. Miller and C.O. Thoen. 2010. Lack of association between the occurrence of Crohn's disease and occupational exposure to dairy and beef cattle herds infected with *Mycobacterium avium* subspecies *paratuberculosis*. J. Dairy Sci. 93: 2371–2376.

Rhodes, M.W., H. Kator, I. Kaattari, D. Gauthier, W. Vogelbein and C.A. Ottinger. 2004. Isolation and characterization of mycobacteria from striped bass *Morone saxatilis* from the Chesapeake Bay. Dis. Aquat. Org. 61: 41–51.

Russell, D.G., C.E. Barry and J.L. Flynn. 2010. Tuberculosis: what we don't know can, and does, hurt us. Science 328: 852–856.

Sasaki, S., F. Takeshita, K. Okuda and N. Ishii. 2001. *Mycobacterium leprae* and leprosy: a compendium. Microbiol. Immunol. 45: 729–736.

Schalk, A.F., L.M. Roderick, H.F. Foust and G.S. Harshfield. 1935. Avian tuberculosis; collected studies. North Dakota Agricultural Experiment Station Technical Bulletin. No. 279.

Slana, I., F. Paolicchi, B. Janstova, P. Natratilova and I. Pavlik. 2008. Detection methods for *Mycobacterium avium* subsp. *paratuberculosis* in milk and milk products: a review. Vet. Med. 53: 283–306.

Soldati, G., Z.H. Lu, L. Vaughan, A. Polkinghome, D.R. Zimmermann, J.B. Huder and A. Pospichil. 2004. Detection of mycobacteria and chlamydiae in granulomatous inflammation of reptiles: a retrospective study. Vet. Pathol. 41: 388–397.

Steingart, K.R., L.L. Flores, N. Dendukuri, I. Schiller, S. Laal, A. Ramsay, P.C. Hopewell and M. Pai. 2011. Commercial serological tests for the diagnosis of active pulmonary and extrapulmonary tuberculosis: An updated systematic review and meta-analysis. PLoS Med. 8:e1001062.

Tell, L.A., L. Woods and R.L. Cromie. 2001. Mycobacteriosis in birds. Rev. Sci. Tech. 20: 180–203.

Thoen, C.O. 2007. Johne's disease. *In:* L. Geller (ed.). Mc Graw-Hill Encyclopedia of Science & Technology. 10th edition. McGraw Hill Publishing. New York. USA. 9: 515–516.

Thoen, C.O. 2010. Tuberculosis and other mycobacterial Infections. *In:* C.M. Kahn and S. Line (eds.). The Merck Veterinary Manual. 10th ed., Merck & Co., Inc. Whitehouse Station. N.J. pp. 621–63.

Thoen, C.O. 2012. Tuberculosis. *In*: J.J. Zimmerman, L.A. Karrriker, A. Ramierz, K.J. Swartz and G.W. Stevensen (eds.). Diseases of Swine. Tenth Edition Wiley/Blackwell Publishing, Ames, USA. Chapter 63. pp. 856–865.

Thoen, C.O. and G.W. Beran. 2012. Effect of electron beam irradiation on *Mycobacterium avium* complex in ground pork. Intern. J. Tuberc. Lung Dis. 16: 1134–1135.

Thoen, C.O. and A.G. Karlson. 1984. Experimental tuberculosis in rabbits. *In:* G.P. Kubica and L.G. Wayne (eds.). The Mycobacteria: a Sourcebook. Marcel Dekker, Inc. New York and Basel. pp. 978–89.

Thoen, C.O., A.G. Karlson and E.M. Himes. 1981. Mycobacterial infections in animals. Rev. Infect. Dis. 3: 960–972.

Thoen, C.O. and P.A. LoBue. 2007. *Mycobacterium bovis* tuberculosis: forgotten, but not gone. Lancet 369(9569): 1236–1238.

Thoen, C.O., P.A. LoBue, D.A. Enarson, J.B. Kaneene and I.N. de Kantor. 2009. Tuberculosis: a re-emerging disease in animals and humans. Vet. Ital. 45: 135–181.

Thoen, C.O., P.A. LoBue and I. de Kantor. 2006. The importance of *Mycobacterium bovis* as a zoonosis. Vet. Microbiol. 112: 339–345.

Thoen, C.O. and T.A. Schliesser. 1984. Mycobacterial Infections in cold-blooded animals. *In:* G.P. Kubica and L.G. Wayne (eds.). The Mycobacteria: A sourcebook. Marcel Dekker, Inc. New York and Basel. pp. 1297–1311.

Thoen, C.O. and D.E. Williams. 1994. Tuberculosis, tuberculoidoses and other mycobacterial infections. *In:* G.W. Beran (ed.). Handbook of Zoonoses. 2nd. CRC Press, Boca Raton, Florida. pp. 41–59.

Vaerewijck, M.J.M., G. Huys, J.C. Palomino, J. Swings and F. Portaels. 2005. Mycobacteria in drinking water distribution systems: ecology and significance for human health. FEMS Microbiol. Rev. 29: 911–934.

Van Kruningen, H.J. 2011. Where are the weapons of mass destruction—the *Mycobacterium paratuberculosis* in Crohn's disease? J. Crohn's and Colitis. 5: 638–644.

Wellenberg, G.J., P.E.W. de Hass, J. van Ingen, D. van Sooloingen and I.J.R. Visser. 2010. Multiple strains of *Mycobacterium avium* subspecies *hominissuis* infections associated with aborted fetuses and wasting pigs. Vet. Rec. 167: 451–454.

Whan, L., H.J. Ball, I.R. Grant and M.T. Rowe. 2005. Occurrence of *Mycobacterium avium* subsp. *paratuberculosis* in untreated water in Northern Ireland. Appl. Environ. Microbiol. 71: 7107–7112.

WHO (World Health Organization) 2008. (www.who.int/tb/publications/global_report/en/index.html, 2008).

Woods, G.L. and J.A. Washington. 1987. Mycobacteria other than *Mycobacterium tuberculosis*: review of microbiologic and clinical aspects. Rev. Infect. Dis. 9: 275–294.

Wolinsky, E. 1995. Mycobacterial lymphadenitis in children: a prospective study of 105 nontuberculous cases with long-term follow-up. Clin. Infect. Dis. 20: 954–963.

CHAPTER 13

Protozoan Parasites: A Plethora of Potentially Foodborne Pathogens

Lucy J. Robertson[#]

Introduction

Although it is now widely accepted that the various species that traditionally are collectively categorized as "protozoa" are not necessarily closely related to each other, it remains a useful term in many situations relating to human and animal health. However, it should be noted that in many scientific contexts, this classification is no longer formally used. Thus, the term 'protozoan parasites' as used within this chapter refers to those eukaryotic, unicellular, organisms that infect another organism, the host, obtaining benefits at the latter's expense. For some infections with protozoan parasites, the host organism is severely weakened or compromised and, in some infections, may be killed.

The human burden of parasitic protozoan infection is enormous. On a global basis, billions of people are infected by protozoan parasites and the DALY (disability-adjusted life year) toll due to protozoan infections is correspondingly huge (WHO 2008). Although vector-borne protozoan infections, such as malaria, are probably most familiar to us, particularly in tropical countries, infections from other protozoan parasites also occur commonly, in temperate regions as well, and add to the burden.

[#] Parasitology Laboratory, Institute for Food Safety and Infection Biology, Norwegian School of Veterinary Science, PO Box 8146 Dep., 0033 Oslo, Norway.
Email: lucy.robertson@nvh.no

Several of the parasitic protozoa that are of importance to human health have an oral infection route. Transmission is often faecal-oral or via ingestion of tissue cysts in meat. For these parasites, the foodborne route of transmission is likely to be highly relevant, or, indeed, for tissue cysts, essential. In Table 1 protozoan parasites that are associated with foodborne transmission are listed, along with their transmission routes and food matrices considered to be of particular importance.

This chapter cannot provide a totally comprehensive documentation for every single protozoan parasite listed in Table 1 and that has the potential to be transmitted by the foodborne route, and is therefore limited to those which are probably of most relevance to the European situation, namely: *Balantidium coli*, *Cryptosporidium* spp., *Cyclospora cayetanensis*, *Entamoeba histolytica*, *Giardia duodenalis*, and *Toxoplasma gondii*. For each of these protozoa, not only is some basic background information provided, but also specific information associated with foodborne transmission. For further information on foodborne transmission of other protozoa that are included in Table 1 but not mentioned further in this chapter, such as *Trypanosoma cruzi* and *Sarcocystis* spp., the reader is referred to the relevant chapters in Robertson and Smith (2012).

Balantidium coli

Introduction

B. coli is a relatively large, ciliated protozoan parasite, in which the lifecycle consists of two stages: the cyst and the trophozoite. Trophozoites are the vegetative stage, residing in the lumen of the large intestine of humans and animals (particularly pigs), where replication occurs by binary fission (during which conjugation may occur). Some trophozoites may invade the colonic wall and multiply. Encystation of the trophozoites results in the production of mature infective cysts (spherical, 40–60 μm diameter, with a thick, cyst wall of one or two layers), which are excreted in the faeces and are immediately infectious.

Between 0.2 and one percent of the world's human population has been estimated to be infected with *B. coli*, with endemic areas in South and Central America, the Philippines, Papua New Guinea and the Middle East (Schuster and Ramirez-Avila 2008). Domestic swine and wild boar are the chief reservoirs for *B. coli*, with reported prevalence rates in domestic pigs ranging from over 55 percent in USA, over 47 percent in China, and over 33 percent in Venezuela down to under 2 percent in Turkey (Theel and Pritt 2012). The highest reported prevalence rate of porcine *B. coli* infection is from Denmark, where in one study nearly all of the adult domestic pigs harboured the ciliate; interestingly human infection with *B. coli* has never

Table 1. Protozoan parasites associated with foodborne transmission.

Protozoan parasite (Phylum)	**Transmission Route**	**Documented foodborne transmission**	**Associated matrices**
Balantidium coli (Ciliophora)	Faecal-oral	Waterborne transmission documented	Water (fresh produce, shellfish)
Cryptosporidium spp. (Apicomplexa)	Faecal-oral	Yes	Water, fresh produce, apple juice
Cyclospora cayetanensis (Apicomplexa)	Faecal-oral	Yes	Fresh produce (especially raspberries), water
Dientamoeba fragilis (Metamonada)	Unknown–robust transmission stage yet to be identified. Postulated to be via helminth eggs.	No	None particularly
Entamoeba histolytica (Amoebozoa)	Faecal-oral	Waterborne transmission documented. Only anecdotal reports from food.	Water (fresh produce)
Giardia duodenalis (Metamonada)	Faecal-oral	Yes	Water, fresh produce, various
Isospora belli (also known as *Cystoisospora belli*) (Apicomplexa)	Faecal-oral	Waterborne transmission documented.	Water
Sarcocystis spp. (Apicomplexa)	For species for which humans are the definitive hosts (*S. bovihominis* and *S. suihominis*), transmission is via ingestion of meat containing sarcocysts. Humans can also act as dead-end hosts for non-human species.	Yes	Beef (*S. bovihominis*) and pork (*S. suihominis*). Contaminated produce and water for other species.

Table 1. contd....

Table 1. contd.

Protozoan parasite (Phylum)	Transmission Route	Documented foodborne transmission	Associated matrices
Toxoplasma gondii (Apicomplexa)	Can either be faecal-oral (from oocysts excreted in felid faeces) or from ingestion of tissue cysts in undercooked meat (Note: the transplacental route of infection is of public health importance)	Yes	Meat (particularly lamb, pork, and game, but meat from other animals as well) for tissue cyst transmission. Water (fresh produce?) for faecal-oral transmission.
Trypanosoma cruzi (Euglenozoa)	Usually vector-borne infection, but may be transmitted via food matrices contaminated with excretions from infected triatomine bugs or from infected reservoir hosts (e.g., opossums).	Yes	Particularly fruit juices

been reported from Denmark, despite the high prevalence in pigs, probably due to standard hygiene measures practiced by Danish pig farmers and good general infrastructure (Theel and Pritt 2012). There is a lack of data regarding prevalence in pigs in endemic countries (Theel and Pritt 2012).

The patient

Human infection with *B. coli* may result in one of four different forms of clinical features: asymptomatic carriage, chronic infection, acute or fulminating balantidiosis, and extra-intestinal dissemination (Theel and Pritt 2012). These are described briefly in Table 2 below.

Diagnosis of *B. coli* infection is traditionally by light microscopic examination of faecal samples (or tissue preparations) and detection of characteristic trophozoites; cysts are less frequently observed in a diagnostic setting.

Current treatment regimens seem to be effective, although therapies have generally not been closely studied or defined. For adults, current therapy options include tetracycline or metronidazole. Following diagnosis and treatment initiation, patient prognosis and recovery are excellent. However, limited access to suitable pharmaceuticals in many developing countries has led to high rates of persistent infection and also death (Theel and Pritt 2012).

Table 2. Clinical features of *B. coli* infection (adapted from information in Theel and Pritt 2012).

Infection type	Clinical features
Asymptomatic carriage	Although this may not be detected by the health care system, such infections serve as important reservoirs and sources of environmental contamination.
Chronic infection	Characterized by intermittent abdominal pain and alternating episodes of constipation and non-bloody diarrhoea, which can last for decades.
Acute or fulminating balantidiosis	Mostly occurs in malnourished or immunocompromised individuals. Characterized by a sudden, severe onset of bloody diarrhoea (indicative of trophozoite invasion of the intestinal epithelium), tenesmus, abdominal pain and weight loss. Complications may include dehydration and haemorrhage, and mortality rates may approach 30 percent, with death occurring within only three to five days after symptoms commence.
Extra-intestinal dissemination	Occurs very rarely; case-by-case descriptions. Has been reported in association with appendicitis, peritonitis, cystitis and vaginitis. Also cases of pulmonary disease (respiratory failure, pulmonary haemorrhage, pulmonary necrosis), which are assumed to be associated with inhalation of aerosolized cysts.

Individuals at highest risk of *B. coli* infection are pig farmers, veterinarians, and those that work with pig products. Direct human-to-human spread of *B. coli* is, unexpectedly, considered to occur infrequently, and volunteer infection studies have apparently not resulted in infection (Theel and Pritt 2012).

Foodborne transmission

Any food matrix, including water, which can be contaminated with faecal matter from a person or animal infected with *B. coli* has the potential to act as a vehicle of infection. Trophozoites of *B. coli* can apparently survive in faeces at room temperature for up to 10 d, while *B. coli* cysts are even more environmentally robust, capable of surviving for several weeks in a humid environment out of direct sunlight (Theel and Pritt 2012). A transmission stage which can survive for prolonged periods is obviously advantageous for foodborne transmission.

Nevertheless, there have been no documented outbreaks of foodborne transmission, perhaps because in endemic areas where this route of infection is likely to be important, it is not possible to associate infection with a particular foodstuff. However, one outbreak of waterborne transmission has been documented from 1971 on the island of Truk (Caroline Islands; at that time US territories) (Walzer et al. 1973). That outbreak, in which 110 people were diagnosed with balantidiosis, was the direct result of typhoon Amy that reached Truk on 1st May. The typhoon devastated homes, animal pens, water collection systems, and sanitation infrastructure, resulting in the local population using rivers and streams as potable water supply. Lack of sanitation, coupled with dispersal of domestic animals, resulted in these water supplies being contaminated, and hence the outbreak of balantidiosis (Walzer et al. 1973).

Although *B. coli* is generally considered an 'emerging pathogen', whilst it continues to be considered of limited importance to public health in industrialized countries, it is unlikely that there will be significant research on it and on its potential for foodborne transmission.

Cryptosporidium spp.

Introduction

Although not identified as a pathogen of particular importance to either public or veterinary health until the mid-1970s, several species of this Apicomplexan parasite are currently recognized to be of significance both to humans and animals. To date, approximately 25 different species of *Cryptosporidium* have been recognized, with different degrees of host specificity. Although several species have been associated with human infection, two are of particular importance: *Cryptosporidium hominis*, which has a predilection for human infection, but has been found infrequently in other animals, including cattle, goats, sheep, and a dugong, and *C. parvum*, which is a host-promiscuous species found in a wide range of vertebrates and is very common in domesticated and wild ruminants. For the purposes of this chapter, just these two species will be considered, but it should be recognized that other species (including particularly *C. cuniculus*, *C. meleagridis*, and *C. ubiquitum*) also are infectious to humans, and that at least one of these species (*C. cuniculus*) has been associated with a community-wide outbreak of waterborne cryptosporidiosis (Chalmers et al. 2009).

The lifecycle of *Cryptosporidium* spp. is direct (no intermediate host) and includes both asexual and sexual stages. Infection is initiated when a susceptible host ingests (or, more rarely, inhales) a viable oocyst (3–5 μm). Each oocyst contains four sporozoites that excyst from the oocyst and invade the epithelial cells of the small intestine, locating epicellularly beneath the host cell membrane, but not inside the cytoplasm nor extracellularly. Repeat cycles of asexual multiplication (merogony) result in destruction of the epithelial cells, and enormous quantities of parasites are produced. Two morphological forms of meronts have been reported, meront I and meront II. Merozoites from the latter eventually undergo gametogony, producing microgamont (male) and macrogamont (female) stages. Following fertilization of the macrogamont by microgametes released from the microgamont, an oocyst is produced that sporulates (forms sporozoites) whilst within the host. Sporozoites in thin-walled oocysts may excyst within the same host, resulting in re-invasion and prolongation of the infection cycle, whereas thick-walled oocysts, which are immediately infectious to a susceptible host, are excreted in the faeces. The repeated asexual cycles can result in billions of oocysts being produced from an infection initiated by very few oocysts (tens).

Cryptosporidium spp. are globally distributed, and, depending on location, it has been estimated that the proportion of any human population that will be infected at some point in life ranges between 20 and 90 percent depending on location (Dillingham et al. 2002). *C. hominis* seems to be most commonly diagnosed species worldwide, although in particular geographic areas (e.g., Middle Eastern countries) or conditions (infections in children in rural areas), *C. paruvm* may predominate (Xiao and Fayer 2008). Among reservoir host populations, *Cryptosporidium* infections may also occur commonly, with pre-weaned calves frequently infected with zoonotic *C. parvum* (Xiao and Feng 2008).

The patient

Although human infection with *Cryptosporidium* spp. can be asymptomatic, a clinical condition, cryptosporidiosis, appears to be more usual. The period between ingestion of oocysts and development of symptoms is variable, but is usually approximately one wk. Infection can be initiated by ingestion of only a small number of viable oocysts; theoretically, a single infectious oocyst should be sufficient to cause infection in a susceptible person (Robertson and Fayer 2012), but a median infectious dose for *C. hominis* of between 10 and 83 oocysts has been demonstrated in volunteer infection studies (Chappell et al. 2006). For *C. parvum*, inter-isolate variation has been shown regarding infective dose (ranging from below 10 to over 1000 oocysts per infective dose in one human volunteer study; Okhuysen et al. 1999).

Cryptosporidiosis is generally characterized by voluminous, watery diarrhoea, which is sometimes mucoid but rarely contains blood, and may be acute (most common) or persistent, lasting several weeks. The diarrhoea may be accompanied by other symptoms, depending on both host factors (age, immunity, nutritional status), and also on the species of *Cryptosporidium*. These symptoms may include nausea, abdominal pains, vomiting, low grade fever, and headache. In immunocompetent patients, in whom cryptosporidiosis tends to resolve within a couple of wk, a relatively high rate of relapse of gastrointestinal symptoms following recovery has been reported (40 percent of patients), while non-intestinal sequelae (joint pain, eye pain, headache) were associated only with *C. hominis* infection and not with *C. parvum* infection (Hunter et al. 2004).

Cryptosporidiosis is usually more severe and persistent in immunocompromised patients, and such cases were prevalent at the beginning of the HIV-epidemic, in the 1980s, before effective anti-retroviral therapies were developed. In immunocompromised patients, daily output of diarrhoea can be as high as 17 L and can contribute to mortality. Weight loss associated with cryptosporidiosis may be significant in all patients, but is particularly marked in immunocompromised patients with chronic, non-resolving cryptosporidiosis.

Also, in severely immunocompromised individuals, such as AIDS patients, cryptosporidiosis may spread beyond the primary site of infection to extra-intestinal sites such as the biliary tract (Warren and Guerrant 2008).

Diagnosis of cryptosporidiosis is traditionally by demonstration of oocysts in stool samples, usually using stains such as Ziehl-Neelsen or Auramine phenol to aid in identification. However, the small size of the oocyst means that identification can be problematic, and immunofluorescent antibody test (IFAT) is considered gold standard. Faecal antigen tests are also commercially available, but low sensitivity and specificity have been reported (e.g., Robertson et al. 2006a), and molecular techniques have also been developed, but are currently not widely used.

Those medications that are used for treating infection with coccidian parasites, amongst which *Cryptosporidium* was initially considered to belong, appear ineffective at combating cryptosporidiosis. One medication (nitazoxanide) has been approved for alleviating symptoms and reducing parasite burden in adults and children. However, treatment of cryptosporidiosis is usually primarily supportive, with oral rehydration, or, less often, intravenous rehydration. In immunocompetent patients, for whom recovery is usually spontaneous and occurs within two weeks, the lack of therapeutic treatment is unlikely to be seriously problematic, but for immunocompromised patients, for whom cryptosporidiosis may be life-threatening, the absence of effective chemotherapy is a considerable concern. Currently, the most useful approach for improving the immune status of immunocompromised patients with HIV-infection has been development of effective anti-retroviral therapies. This has reduced the urgency for developing an effective treatment for cryptosporidiosis. Patients who are immunocompromised due to cancer or organ transplantation, or who suffer from a genetically-based immunodeficiency, lack an effective anti-parasitic therapeutic. Current research on cryptosporidiosis therapy has been directed towards development of novel synthetic isoflavone derivates and molecular-based immunotherapy (Gargala 2008), but effective treatment for cryptosporidiosis in both immunocompromised and immunocompetent patients is unlikely to be available in the immediate future (Robertson and Fayer 2012).

Foodborne transmission

A number of factors in the biology of *Cryptosporidium* favour a foodborne or waterborne route of transmission (Robertson and Fayer 2012). These include the low infectious dose (as described in the previous section), the highly robust transmission stage (the oocyst), which can survive for several months in a moist cool environment, and is also resistant to a number of commonly used disinfectants such as chlorine, the small size of the oocyst (which enables it to penetrate rapid sand filters often used in the water industry), the zoonotic potential, which means that contamination can occur from both animal and human sources, and the high excretion rate from infected humans and animals. Indeed, the number of oocysts excreted during a *Cryptosporidium* infection is variable, but a daily excretion rate of over 10^9 oocysts has been reported from AIDS patients with symptomatic cryptosporidiosis (Goodgame et al. 1993), and similar excretion rates have been reported from calves, with a young calf excreting as many as 4×10^{10}

oocysts during its second wk of life, and around 6 x 10^{11} oocysts during its first mon of life (Uga et al. 2000, Nydam et al. 2001).

A huge number of outbreaks of waterborne cryptosporidiosis have been documented, with *Cryptosporidium* infection accounting for just over 50 percent of all recorded outbreaks of waterborne protozoan disease until 2003, affecting hundreds of thousands of individuals, and considered to pose the greatest threat to the water industry (Karanis et al. 2007). A review of more recent (2004–2010) outbreaks of waterborne protozoan infection (Baldursson and Karanis 2011), identified *Cryptosporidium* as being responsible for more than 60 percent of them, affecting thousands more. That *Cryptosporidium* continues to be a threat to municipal water supplies is probably best exemplified by the outbreak in Östersund, Sweden, during November 2010, in which over 12,000 people were estimated to be infected and considerable disruption and inconvenience was caused to the local population and industries alike (Robertson and Huang 2012). In this outbreak, in which *C. hominis* (sub-genotype IbA10G2) was identified as the species involved, it was suspected that sewage containing oocysts had been discharged to a stream that fed the local water supply. A boil-water notice was implemented as soon as the outbreak was identified, and this was not withdrawn until 18th February 2011; other prevention and control measures instigated included cleaning of the water distribution system, enhanced protection of the water supply area, and incorporation of a UV-treatment at the water treatment works.

Regarding foodborne transmission, food products can be contaminated, and thus serve as vehicles of transmission, at any step of the field-to-fork continuum. Food can be contaminated at production sites (e.g., agricultural produce contaminated from contact with manure, contaminated irrigation water, or from handling by agricultural workers; shellfish filter feeding in contaminated estuaries). Contamination can also occur along the distribution and preparation chain, with food-handlers, wash water, surfaces and equipment, or utensils all of which are possible sources of the contamination.

For many foodborne outbreaks it has been difficult or impossible to pinpoint the actual vehicle of transmission, possibly because more than one item is a vehicle, and this is particularly likely if a food-handler is the source of contamination. A review of more recent (2000–2010) outbreaks of foodborne *Cryptosporidium* infection (Robertson and Fayer 2012), supports this by pointing out the considerable diversity in food products associated with outbreaks in the past 10 years. Although raw produce (fruit, vegetables) are generally regarded as the most likely candidates for transmission vehicles, outbreaks associated with milk, apple cider, raw meat (possibly), and sauce have also been documented. Interestingly, there have been no recorded outbreaks of cryptosporidiosis, to date, that are associated with shellfish, although these are also considered a food product vulnerable to contamination by *Cryptosporidium* (Robertson 2007). Another point of interest from more recent documented outbreaks of foodborne cryptosporidiosis is that whilst the outbreaks apparently occur in diverse areas of the world (with outbreaks recorded from Far East, USA, Australia, and Europe), Nordic countries are strikingly over-represented. Whether over-representation of Nordic countries is a reflection consumer habits, elevated diagnostic proficiency, or that *Cryptosporidium* oocysts thrive best in the moist cool conditions that predominate in these countries has been discussed (Robertson and Fayer 2012, Robertson and Chalmers 2013); it is concluded that foodborne outbreaks of cryptosporidiosis are no more likely to occur in Nordic countries than elsewhere in the world.

The development of validated methods for analysing water samples for *Cryptosporidium* oocysts (e.g., US EPA Method 1622, ISO Method 15553), together with the acknowledgement that different food types are likely candidates for acting as vehicles of transmission for *Cryptosporidium*, has resulted in the development of methods for analysing food for this parasite. Methods that have been developed are generally based upon the four steps used in water analysis; elution, concentration, purification (usually using immunomagnetic separation (IMS)), and detection (usually by IFAT), but may be modified according to particular characteristics of the food matrix being analysed. For example, elution of *Cryptosporidium* from raspberries and other soft red fruits cannot usually be performed using a stomacher as this is likely to crush the fruit, but is effective for analysing lettuce (Cook et al. 2006), whilst digestion of shellfish in acidified pepsin is considered preferable to elution from shellfish tissue homogenates as it provides a higher, more consistent recovery efficiency as well as enabling analysis of a larger representative sample (Robertson and Gjerde 2008).

The development of an ISO method for analysing fresh fruit and vegetables for *Cryptosporidium* oocysts is currently in progress, but is hampered by variable recovery efficiencies during validation trials. For a full discussion of the methodologies being considered in the development of an ISO Method as of 2012, see Robertson and Fayer (2012).

The development of robust analytical methods for different food matrices enables surveys to be conducted. For fresh produce, surveys have been conducted in Asia (Cambodia), Europe (Italy, Norway, Poland, Spain), South America (Costa Rica, Peru), and North America (Canada), and a range of fruits and vegetables have been analysed, including leafy vegetables (herbs, cabbage, lettuce, spinach, water spinach), celery, cauliflowers, broccoli, Brussels sprouts, green onions, leeks, sprouted seeds, mushrooms, and soft fruits (raspberries, strawberries, blackberries). Although some surveys have not identified oocysts on any of the produce analysed, in general a widespread, low level contamination has been observed (Robertson and Fayer 2012), although whether the oocysts are viable or infectious, or whether they are of a species infective to humans has seldom been investigated. It is possible that irrigation water is one of the most likely routes for contamination of fresh produce, and surveys which have analysed irrigation water have generally supported this theory (Robertson and Fayer 2012).

Similarly, surveys have also been conducted on shellfish that have been harvested from different parts of the wold, including Africa (Egypt), Asia (Thailand), Europe (Ireland, Italy, France, Netherlands, Norway, Spain), South America (Brazil), and North America (Canada, USA). A range of different shellfish that are often destined for human consumption have been analysed, including: banded carpet shell clam (*Venerupis rhomboideus*), California mussel (*Mytilus californianus*), carpet shell clam (*Ruditapes decussatus*), common cockle (*Cerastoderma edule*), common (blue) mussel (*Mytilus edulis*), eastern oyster (*Crassostrea virginica*), European flat oyster (*Ostrea edulis*), Gandofli (*Caelatura pruneri*), green-lipped mussel (*Perna canaliculus*), green mussel (*Perna viridis*), horse mussel (*Modiolus modiolus*), mangrove (gaiteira) oyster (*Crassostrea rhizophorae*), Manila clam (Japanese carpet shell) (*Ruditapes philippinarum*), Mediterranean mussel (*Mytilus galloprovincialis*), Pacific cupped oyster (*Crassostrea gigas*), pullet carpet shell (*Venerupis pullastra*), smooth artemis (*Dosinia exoleta*), softshells (*Mya arenaria*), striped venus clam (*Chamelea gallina*), triangular tivela clam (*Tivela mactroides*), warty venus shell (*Venus verrucosa*), and wedgeshell clam (*Donax trunculus limiacus*). As with fresh produce, the occurrence of *Cryptosporidium* in molluscan shellfish seems to be wide-spread, but generally at low concentrations, ranging from < 1 to > 20 oocysts (Robertson 2013).

As *Cryptosporidium* oocysts are so ubiquitous, some research has been directed towards inactivation in food and water. Ultra-violet water treatment is considered an effective barrier against viable oocysts in water supplies, and ozone treatment is also considered to reduce oocyst viability. Various treatments have been investigated for inactivation of oocysts in different food products (see Table 3), particularly as an outbreak of cryptosporidiosis was associated with apple cider that had been ozonated (Blackburn et al. 2006).

The numerous waterborne outbreaks of cryptosporidiosis involving large numbers of people, the extreme robustness of the transmission stage, and the lack of effective pharmacological treatment of cryptosporidiosis mean that *Cryptosporidium* spp. has been the subject of considerable research in recent years.

Table 3. Inactivation of *Cryptosporidium* oocysts on diverse food products (adapted from information in Robertson and Fayer 2012).

Method for inactivation	Food Product	Efficacy of inactivation	Reference
Heat treatment (71.7°C for 15, 10, or 5 s)	Milk	Effective according to mouse infectivity trials	Harp et al. 1996
Heat treatment (70 or 71.7°C for 20, 10, or 5 s)	Apple cider	According to infection of bovine kidney cell cultures, heating for 10 or 20 s at either temperature resulted in at least a 4.9 log reduction in infectivity. For 5 s at either temperature resulted in only slightly less inactivation	Deng and Cliver 2001
Addition of organic acids (malic, citric, and tartaric acids) at concentrations of 1 to 5% (w/v)	Apple cider, orange juice, grape juice	*C. parvum* infectivity for a human ileocecal carcinoma cell-line (HCT-8) inhibited by up to 88%	Kniel et al. 2003
Addition of 0.025% hydrogen peroxide	Apple cider, orange juice, grape juice	*C. parvum* infectivity for HCT-8 cells reduced by > 5 logs	Kniel et al. 2003
High hydrostatic pressure processing (550 megapascal pressure units for 180 s)	Oysters (*Crassostrea virginica*)	Assay in neonatal mice: decrease in numbers of *C. parvum* positive mouse pups by 93.3%	Collins et al. 2005a
E-beam irradiation (2 kGy dose)	Oysters (*Crassostrea virginica*)	Infectivity of *C. parvum* completely eliminated	Collins et al. 2005b
Microwave exposures of 1 to 3 s (43.2 to 62.5°C)	Oysters (*Crassostrea virginica*)	No effect on infectivity	Collins et al. 2005b
Heat treatment (60°C for 45 s or 75°C for 20 s)	Meat (beef)	*C. parvum* oocysts rendered non-infective to HCT-8 cell cultures	Moriarty et al. 2005
Light steaming (as traditionally used for cooking mussels)	Mussels (*Mytilus galloprovincialis*)	Assay in neonatal mice: at least 50% of mice infected, indicating ineffective at inactivating oocysts	Gómez-Couso et al. 2006

Cyclospora cayetanensis

Introduction

Cyclospora cayetanensis, originally referred to as coccidia-like or cyanobacteria-like organism, was first noted in human diarrheic stool in the 1970s, but not fully described and recognized as a protozoan parasite until the early 1990s (Ortega et al. 1994, Shields and Ortega 2012).

To date, around twenty species of *Cyclospora* have been described, of which *C. cayetanensis* is the only one that has been found to infect humans, and seems not to infect any other species, although infection of non-human primates has been discussed (Shields and Ortega 2012).

Infection of a susceptible individual is initiated by ingestion of sporulated *C. cayetanensis* oocysts (8–10 μm in diameter with a 113 nm thick bi-layered wall). Excystation occurs in the intestine, facilitated by gastric acid and bile, and results in the release of two sporocysts (4 x 6 μm, ovoid and each containing two sporozoites). Exposure of a sporocyst to biliary salts results in the release of the sporozoites that infect the intestinal epithelial cells, locating within a supranuclear parasitic vacuole. Sporozoites multiply asexually to form meronts type I, with 8–12 merozoites, and type II meronts, with four merozoites. Sexual multiplication occurs by formation of gametocytes. After fertilization of the macrogametocytes by the microgametocytes, unsporulated oocysts are produced that are then excreted in the faeces and require one to two weeks to differentiate fully and become infectious. The essential environmental conditions for sporulation have not yet been elucidated (Shields and Ortega 2012).

Various studies have investigated the prevalence and incidence of *Cyclospora* in various populations, in both developed and developing countries (Shields and Orgega 2012). It has been estimated that there are over 11,000 cases of *Cyclospora* infection in US annually (Scallan et al. 2011), while in developing countries the prevalence is considered to be higher, particularly in children over 18 mon, and with marked seasonal variability (Shields and Ortega 2012). In various South American countries, particularly Peru and Guatemala, cyclosporiasis is considered endemic, and also in some countries in Asia, particularly Nepal, where it has been particularly associated with diarrhoeal disease in expatriate adults (Shields and Ortega 2012).

The patient

Although infection of immunocompetent adults with *C. cayetanensis* can be asymptomatic, gastrointestinal symptoms, including diarrhoea, flatulence, weight loss, abdominal discomfort, and nausea, are more usual.

The infectious dose is unknown, but is suspected to be low (Shields and Ortega 2012). However, human volunteer studies using *Cyclospora* oocysts recovered from naturally infected individuals have not been successful, with no gastroenteritis and no oocyst excretion detected during the 16-wk period of monitoring (Alfano-Sobsey et al. 2004).

In endemic areas, the infection is typically acquired early in life, usually resulting in diarrhoea, which may be acute or intermittent, and may result in mild to moderate dehydration.

In immunocompromised individuals, cyclosporiasis tends to last longer and be associated with more severe symptoms, including excessive watery diarrhoea and pronounced fatigue. Individual cases of extra-intestinal infection have also been reported from immunocompromised patients (Shields and Ortega 2012).

Currently, the most practical diagnostic method for cyclosporiasis is identification of oocysts in stool specimens by light microscopy. In fresh samples, the oocysts will not be sporulated. Different staining techniques, such as modified acid-fast stain or Safranin stain, may be useful in assisting diagnosis. Also, *Cyclospora* oocyst walls autofluoresce under UV, providing a further useful adjunct for positive identification.

The drug of choice for treatment of cyclosporiasis is a combination drug, co-trimoxazole, containing both trimethoprim (TMP) and sulphamethoxazole (SMX). Ciprofloxacin has also been used for those who cannot tolerate sulpha drugs (Verdier et al. 2000).

Foodborne transmission

Up until the mid-1990s, *C. cayetanensis* was considered as nothing more significant to the US public health services than an occasional cause of travellers' diarrhoea. However, in early summer 1996 over 1400 cases of cyclosporiasis were reported (978 laboratory confirmed) in the US and Canada, of which nearly 50 percent were from clusters, epidemiologically linked to imported fresh raspberries (Herwaldt and Ackers 1997). In 1997, reports of outbreaks of cyclosporiasis began in March and continued through July, followed by further outbreaks in September and December (Herwaldt 2000). *C. cayetanensis* had arrived as a significant protozoan parasite with infection associated particularly with foodborne transmission. Although raspberries predominated as the vehicle of infection, other vehicles have since been recorded including fresh basil, blackberries, various lettuces, and sugar snap peas. The first reported food-associated outbreak of cyclosporiasis in Europe was in 2000 (Döller et al. 2002) and was epidemiologically associated with various lettuces. Some of the foodborne outbreaks and their likely, or confirmed, vehicle of transmission are described in Table 4. It should be noted that for the vast majority of

Table 4. Some foodborne outbreaks of cyclosporiasis (adapted from Shields and Ortega, 2012).

Outbreak location and event	Cases	Suspected food vehicle	Reference
Florida, US. Unspecified event.	6	Cake (with fresh berries and optional berry puree)	Koumans et al. 1998
New York, US. Country club.	32	Fresh raspberries?	Herwaldt 2000
South Carolina, US. Luncheon.	38	Fresh raspberries? Potato salad?	Caceres et al. 1998
Massachusetts, US. Wedding reception.	57	Wine? Dessert containing berries?	Fleming et al. 1998
North York, Canada. Catered luncheon.	35	Strawberry flan (garnished with fresh raspberries & blueberries)	Manuel et al. 1999
Florida, US. Various events.	60	Fresh raspberries	Katz et al. 1999
Ontario & Quebec, Canada. Unspecified events.	160	Fresh raspberries and possibly blackberries	Sewell and Farber 2001
Virginia, DC and Maryland, US. Lunches. 45 clusters.	308	Basil/basil pesto	CDC 1997
Virginia, US. Unspecified event.	21	Fruit plate	Herwaldt 2000
Florida, US. Unspecified event.	12	Mesclun salad	Herwaldt 2000
Ontario, Canada. Restaurant.	31	Fresh raspberries and possibly blackberries	Sewell and Farber 2001
Georgia, US. Unspecified event.	17	Fruit salad	Herwaldt 2000
Ontario, Canada. Various events	192	Berry garnish, fresh raspberries	CDC 1998
Ontario, Canada. Unspecified event.	104	Berry dessert	Herwaldt 2000
Florida, US. Unspecified event.	94	Fruits, berries	Herwaldt 2000
Missouri, US. Birthday party.	32	Chicken pasta salad	Lopez et al. 2001
Missouri, US. Graduation party.	30	Tomato basil salad? Vegetable pasta salad?	Lopez et al. 2001
Philadelphia, US. Wedding reception.	54	Raspberry filling of wedding cake	Ho et al. 2002
Baden-Wuerttemberg, Germany. Four lunches.	34	Side-salad	Döller et al. 2002
Monterrey, Mexico. Wedding reception.	70	Watercress	Ayala-Gaytan et al. 2004
British Columbia, Canada. Various small events.	17	Thai basil (imported through US)	Hoang et al. 2005
Bogor, Indonesia. Scientific meeting.	14	No food identified	Blans et al. 2005

Table 4. contd....

Table 4. contd.

Outbreak location and event	Cases	Suspected food vehicle	Reference
Medellin, Columbia. University.	56	Salad? Juice?	Botero-Garces et al. 2006
Spain via Guatemala. Holiday	7	Raspberry juice.	Puente et al. 2006
Pennsylvania, US. Event.	96	Pasta salad with snow peas	CDC 2004
Lima, Peru. Meals at naval recruitment base.	27	No food identified	Torres-Stimming et al. 2006
Quebec, Canada. Restaurant.	20	Basil	Milord et al. 2011
Stockholm, Sweden. Several small clusters.	12	Sugar snap peas	Insulander et al. 2010
British Columbia, Canada. Cruise ship.	160	No food identified	CDC 2010a
Ontario, Canada. Unspecified event.	210	Basil	Anonymous 2010b

the outbreaks listed in Table 4, a specific food was only implicated using epidemiological tools. In just five outbreaks were laboratory tests of the suspect foods attempted, three times using microscopy and twice using molecular methods. For none of the outbreaks in which microscopy was used to analyse food samples were *Cyclospora* oocysts detected. However, on both occasions when molecular methods were used (the chicken pasta salad outbreak in Missouri, USA (Lopez et al. 2001); and the wedding cake raspberry filling outbreak in Philadelphia, USA (Ho et al. 2002)), positive results were obtained.

Many of the outbreaks listed in Table 4 are associated with imported fresh produce, in particular with produce imported from Guatemala. In May 1997, in consultation with CDC and the US Food and Drug Administration (FDA), the government of Guatemala and the Guatemalan Berries Commission agreed to a voluntary suspension of exports of fresh berries to the US. The following year, exports resumed but only for Guatemalan farms identified as 'safe' after inspection (Herwaldt 2000). Canada did not ban import of berries, and 13 outbreaks were reported in Canada in 1998 (315 cases), but only one in US (Georgia, no food identified) (Herwaldt 2000). Although cyclosporiasis was designated a national notifiable disease in the US in 1999, only 39 states regularly report outbreaks and cases.

Currently samples from all shipments of basil and raspberries imported into US are tested for *Cyclospora*. In 2000, Canada implemented a total ban on the import of wild berries and refuses entry to all blackberries and raspberries from Guatemala between March 14 to August 15 (Anonymous 2010a). In Europe, however, there are currently no import restrictions, policies, or mandatory food testing for *Cyclospora*.

It has been hypothesized that the raspberries associated with outbreaks in the 1990s were contaminated via contaminated water used for dilution of pesticides (Herwaldt 2000). However, it is not possible to prove this either way. Surveillance studies have been conducted on various fresh produce in a number of countries including Cambodia, Costa Rica, Egypt, Nepal, Nigeria, Norway, Peru, and Vietnam, and some have reported low levels of contamination (Shields and Ortega 2012). Some studies have also examined water samples used in irrigation (Nigeria; Alakpa et al. 2003) or for 'refreshment' of produce for sale (Vietnam; Tram et al. 2008), and have also recorded the occurrence of *Cyclospora* oocysts.

Both surveillance studies and analysis of suspected food vehicles in an outbreak situation require reliable, robust methods for analysis. As with *Cryptosporidium*, the steps are basically elution, concentration, and detection, the latter by microscopy or by molecular methods. The methods tend to be based on those that have been more extensively for *Cryptosporidium* and *Giardia*, but antibodies against *Cyclospora* oocysts are not available, so IMS for purification cannot be used, nor IFAT for detection. One of the earlier studies, from Peru (Ortega et al. 1997), used a washing procedure, followed by centrifugation and microscopy, and reported recovery efficiencies from lettuce of around 15 percent. Another study attempted a clarification procedure in which oocysts were separated from the elution fluid using magnetisable beads coated with the lectin wheat germ agglutinin (WGA) (Robertson et al. 2000). Although the recovery efficiency was no better than that reported by Ortega et al. (1997), the resultant pellets were cleaner and thus easier to screen by microscopy. An extensive review of selection of methods for analysis of produce for *Cyclospora* oocysts, including molecular methods for detection, is provided in Shields and Ortega (2012).

The marked seasonality of cyclosporiasis in endemic countries, along with seasonality of crop growth and harvesting, import of crops and occurrence of cyclosporiasis associated with imported fresh produce, suggests that seasonal associations are possible. However, very few studies have looked for seasonal associations and the results are inconsistent (Shields and Ortega 2012).

Survival of *Cyclospora* oocysts exposed to various temperatures and in a variety of food matrices has been studied, and indicates that these transmission stages are highly robust. At least 50 percent of oocysts held in water, milk, or whipped cream at either 4°C or 23°C sporulated after 7 d (Sathyanarayanan and Ortega 2006), and after 1 hour or 24 h at –15°C, approximately 15 percent oocysts sporulated and after 24 hours between 0.4–2.9 percent were still able to sporulate. Of oocysts seeded onto basil, 46.3 percent survived after six d at 4°C, and 0.2–0.3 percent survived after two d at –20°C (Sathyanarayanan and Ortega 2006). These results, which indicate that *Cyclospora* oocysts survive freezing, are supported by the outbreak associated with the raspberry filling of a wedding cake (Ho et al. 2002). In this case, not only had the raspberry filling been frozen briefly before being added to the wedding cake, but the cake itself had also been frozen after assembly but before consumption.

Other studies have examined the effect of pesticides, chlorine-based disinfectants, and microwaves on *Cyclospora*, and none of these have completely abrogated sporulation (Shields and Ortega 2012). Other possibilities for inactivating *Cyclospora* oocysts on food include ultra-violet disinfection, irradiation and high hydrostatic pressure (550 MPa at 40°C for 2 minutes). The latter has been assessed using *Eimeria acervulina* as a surrogate for *Cyclospora* (Kniel et al. 2007) and provided promising results.

Although sporadic cyclosporiasis has been documented in industrialized countries, particularly in North America, it is often associated with foodborne transmission, particularly with fresh produce, especially raspberries, imported from Central America. Nevertheless, there has been relatively little scientific surveillance of produce, and important epidemiological factors, particularly those associated with sporulation, remain unresolved. In addition, analysis of fresh produce for *Cyclospora* is even more problematic than analysis for *Cryptosporidium* or *Giardia*, due largely to a lack of availability of monoclonal antibodies against *Cyclospora* oocysts. It may be anticipated that European outbreaks may become more common in the future, and development of more robust methods may feature more prominently on the European research agenda.

Entamoeba histolytica

Introduction

Entamoeba histolytica is considered the only human pathogen in the *Entamoeba* genus. However, most infections do not produce symptomatic disease; this variability has been associated with antigenic and genetic differences among human infective strains (Theel and Pritt 2012). In one study, 85 unique genotypes were identified from 111 unrelated specimens collected in a single Bangladesh town, with individual disease outcome correlated with subtype (Ali et al. 2007). Thus, there is a remarkable degree of genomic diversity among *E. histolytica* isolates and disease severity is partially a function of the parasite genome (Theel and Pritt 2012). Although *E. histolytica* is not generally considered a zoonotic parasite and humans are the primary hosts, various mammals (including dogs and cats) may also harbour this infection, but are considered to play only a very limited role in transmission to people (Theel and Pritt 2012).

Infection occurs through ingestion of viable cysts. These excyst in the small intestine, releasing a quadrinucleate trophozoite, which then forms eight trophozoites through subsequent divisions. Trophozoites colonize or invade the large intestinal mucosa, localizing to the caecum and the adjacent ascending colon. Trophozoites maintain intestinal infection through binary fission, but the degree of intra-species genomic diversity is higher than would be expected from non-meiotic replication, suggesting that a yet undefined mechanism is employed for the generation of genomic polymorphism (Theel and Pritt 2012). Intestinally-located trophozoites eventually encyst, thought to be triggered by environmental factors such as composition of colonic bacterial flora and intestinal osmotic conditions, and the cysts are excreted in the faeces. Trophozoites can also be excreted in faeces, but are not considered to be capable of further transmission; they are rapidly destroyed once outside the body, and, should they be ingested, would not survive exposure to the gastric environment.

The prevalence of *E. histolytica* varies significantly with geographic location and patient exposure history, with the majority of infections occurring in regions of the world where sanitation infrastructure is limited. Although early studies, based on morphologic identification, predicted that 10 percent of the global population was infected with *E. histolytica*, this estimate is now considered a considerable over-estimate, given the recognition of two prevalent amoebae that are morphologically identical to *E. histolytica* (*E. dispar* and *E. moshkovskii*). It is now estimated that individuals colonized with *E. dispar* out-number those infected with *E. histolytica* by roughly nine to one (Ali et al. 2008). Nevertheless, infection with *E. histolytica* remains a serious health concern in many parts of the world, and it is considered endemic in India, tropical regions of Africa, South and Central America, and the Asian Pacific peninsula. Although most infections are asymptomatic, millions of individuals suffer from symptomatic amoebiasis each year, and complications of invasive disease probably result in tens of thousands of deaths annually (Theel and Pritt 2012).

In industrialized countries, amoebiasis is largely diagnosed in immigrants from endemic countries. A recent immigration study, focusing on Iraqi refugees settling in San Diego Country, California, found that 1.2 percent of 4,520 individuals were colonized with *E. histolytica* (CDC 2010b).

The patient

Although most infections with *E. histolytica* are asymptomatic, approximately 10 percent of infected individuals progress to symptomatic disease. This ranges from mild colitis to severe invasive and/or disseminated infection (Theel and Pritt 2012). The infective dose is considered to be as low as a single viable cyst (Theel and Pritt 2012). Symptomatic amoebiasis can occur with days after initial infection, but may also be up to years. Primary symptomatic *E. histolytica* colitis typically presents with gradual onset of abdominal pain, bloody or mucoid diarrhoea, and tenesmus. Fulminating amoebic colitis is a rare, but serious, form of invasive amoebiasis that occurs in approximately 0.5 percent of infections. Patients typically present with profuse bloody diarrhoea (dysentery), abdominal pain, and fever and there is a significant association with colonic perforation, peritonitis and a 40 percent mortality rate. Another complication of intestinal amoebiasis is the formation of a proliferative, tumour-like inflammatory mass called an "amoeboma" which occur in approximately 1.5 percent of infections and can result in colonic obstruction, and occasionally, misdiagnosis of colon carcinoma or inflammatory bowel disease. It has also been suggested that amoebic colitis can become a chronic condition, with persistent non-dysenteric diarrhoea, abdominal pain and weight loss spanning months to years.

Extra-intestinal spread of *E. histolytica* trophozoites occurs through the portal venous system, and thus the liver is the most common site of extra-intestinal infection. Liver involvement is referred to as amoebic liver abscess (ALA), and typically entails a single abscess, although multiple abscesses are also possible. Patients with ALA most commonly present with right upper quadrant abdominal pain, hepatic tenderness, weight loss and persistent fever. Jaundice is not typically present. Many patients with ALA do not have concurrent amoebic colitis. Progression to ALA has been shown to occur up to 20 times more frequently in adult males between the ages of 18 and 50 yr than in any other cohort population, but the reason is unknown (Stanley 2003).

Spread of *E. histolytica* to sites other than the liver is rare and typically occurs secondarily to ALA. Up to 20 percent of individuals with ALA will develop amoebic pulmonary disease, with symptoms similar to those of bacterial pneumonia, including chest pain, dyspnoea, and haemoptysis. The mortality associated with extra-hepatic spread of *E. histolytica* ranges from 20–75 percent (Theel and Pritt 2012).

Although diagnosis of *E. histolytica* infection is traditionally by light microscopic examination of faecal samples (or concentrates) for cysts or trophozoites, it must be differentiated from other intestinal protozoa, which is possible for some protozoa, but not always easy. However, such differentiation is not possible for morphologically identical protozoa, such as non-pathogenic *Entamoeba dispar*, and here isoenzymatic, immunologic or molecular methods are useful. A large number of serologic and molecular tests are available (Theel and Pritt 2012) and antibody detection is particularly useful for detection of invasive or disseminated amoebiasis, in which parasites have interacted sufficiently with the host's immune system and is the test of choice of diagnosis of ALA. Serologic testing is more useful in non-endemic countries, where *E. histolytica* exposure is minimal and seropositivity is rare. Of the tests available, IgG detection by ELISA is most commonly used in laboratory settings due to ease of use and high levels of specificity and sensitivity (Hira et al. 2001).

It is generally accepted that if *E. histolytica* infection is identified, patients should be treated, regardless of clinical presentation, in order to limit onward transmission, but prophylaxis in asymptomatic cases is not recommended (Theel and Pritt 2012). A combination of treatment regimens is usually recommended in order to ensure complete eradication. Although surgical intervention may be warranted in cases of intestinal perforation or obstruction, invasive procedures are not usually recommended due to elevated risk of accidental cyst/trophozoite contamination or bacterial infection.

Foodborne transmission

Any food matrix, including water, which can be contaminated with faecal matter from a person or animal infected with *E. histolytica,* has the potential to act as a vehicle of infection. Although trophozoites of *E. histolytica* die rapidly once excreted, the cysts are robust, surviving for up to 48 h on fresh produce, and for up to one month in aqueous matrices such as water or sewage (Theel and Pritt 2012). Prolonged survival of the transmission stage is obviously advantageous for foodborne transmission.

Between 1950–2006, there have been nine reported outbreaks of amoebiasis globally, mostly due to contamination of water supplies with sewage. An earlier outbreak of amoebiasis occurred in 1933 in Chicago, USA, in association with the World Fair, in which a fresh water tank was contaminated by sewage. In this outbreak, over 1,400 individuals were infected, of whom 58 died (Markell 1986). More recently, an outbreak of waterborne amoebiasis has been reported from Jiangshan City in China (reported by Baldursson and Karanis 2011).

Although no outbreaks of foodborne *E. histolytica* are documented, amoebiasis, often called amoebic dysentery, is frequently described as being a foodborne infection (e.g., Mackey-Lawrence and Petri 2011). Food-handlers, in particular, are often associated with contamination of food with *E. histolytica* (Theel and Pritt 2012) and surveys of food-handlers from endemic areas suggest that there is good reason to speculate that direct contamination of food with *E. histolytica* cysts is not unrealistic (see Table 5). Nevertheless, most of these surveys are reliant on microscopy; the survey from Tunisia (Ben Ayed et al. 2008) found 12 samples from 4,266 food-handlers to contain cysts (0.3 percent), but subsequent molecular characterisation revealed that most were non-pathogenic *E. dispar*, and that none were *E. histolytica*.

Table 5. Surveys of food-handlers for *Entamoeba histolytica*.

Survey location	**Prevalence of *E. histolytica* in survey cohort**	**Reference**
Niterói city, Brazil	48.5% of 33 samples from food-handlers in hospitals	Lourenço et al. 2004
Florianópolis city, Brazil	10.9% of 238 fast-food workers	Nolla and Cantos 2005
Uberlândia, Brazil	2% of 264 food-handlers in public primary schools	de Rezende et al. 1997
Solapur city, India	13.7% of 276 food-handlers in hotels and restaurants	Takalkar et al. 2010
Kisii Municipality, Kenya	14.9% of 168 food-handlers	Nyarango et al. 2008
Abeokuta, Nigeria	72% of 100 food vendors on streets and in schools	Idowu and Rowland 2006
Al-Medinah, Saudi Arabia	23% of 13,216 food-handlers in a hospital setting	Ali et al. 1992
Khartoum, Sudan	4.3% of 1,500 food-handlers	Babiker et al. 2009
Tunisia	0 of 4,266 samples from food-handlers	Ben Ayed et al. 2008
Manisa, Turkey	69.9% of 8,895 food-handlers	Gündüz et al. 2008

There have also been some surveys of fresh produce for *E. histolytica* cysts, but, again, the data available must be treated cautiously due to lack of information on whether the cysts detected are actually *E. histolytica*, or another species that is morphologically identical to *E. histolytica*.

While there is no doubting the potential severity of infection with *E. histolytica*, the confusion resulting from the identical morphologies of this pathogen and non-pathogenic *E. dispar*, along with the variation in pathogenicity of different *E. histolytica* genotypes, has led to reduced focus on this parasite. Nevertheless, whilst cases of foodborne transmission have yet to be documented, the potential for this route of transmission should not be underestimated.

Giardia duodenalis

Introduction

Although the parasite that we now know as *Giardia duodenalis* was first discovered in the 1680s, its taxonomy and its public and veterinary health impacts have been, and continue to be, matters of controversy. *Giardia* are diplozoic flagellated protozoa with paired organelles, and are principally distinguished from other members of the Hexamitidae family by the ventral disc. Morphological differences were first used to divide the *Giardia* genus into 3 groups, of which the group or species of public health significance is that of *G. duodenalis*. Although the names *G. intestinalis* and *G. lamblia* are often used as synonyms, it has been powerfully argued that the use of these two names is on the basis of preference, rather than taxonomy, and creates unnecessary confusion and controversy (Thompson and Monis 2011). Additionally, further confusion arises as *G. duodenalis* tends not to be considered as a uniform species, but is further delineated into different genetic groupings or assemblages, which have some degree of host specificity, as well as differences in metabolism, biochemistry, predilection site *in vivo*, and other characteristics (Thompson and Monis 2011). Currently these different genetic groupings are labelled alphabetically (A–H), of which only A and B are infectious for humans, but it has been suggested that the taxonomy should be revised, and different species names given for each of the main Assemblages (Thompson and Monis 2011). This is not widely accepted and increasing recognition of sub-groupings (genotypes) within each group, suggest that sub-structuring will be developed before another system of nomenclature becomes established. For the purposes of this chapter, reference to *Giardia duodenalis* or *Giardia* generally refers to those genotypes that are considered to be infectious for humans, unless otherwise stated.

The lifecycle of *G. duodenalis* is generally considered as simple and direct. Infection occurs when cysts are ingested, and excyst in the small intestine, releasing two trophozoites per cyst. These are non-invasive, remaining in the lumen of the small intestine, either free or attached to the intestinal wall and multiply asexually by longitudinal binary fission. As the trophozoites transit towards the colon, they encounter conditions that trigger encystation. The resultant cysts are excreted in the host faeces, and are the infective and disseminating stage. Cysts are infectious upon excretion. Although the cyst is the stage found most commonly in faeces, trophozoites may also be excreted in some infections. However, their survival in the environment is limited. Although replication in *Giardia* is traditionally considered as exclusively asexual, this has been challenged by some recent studies (Cacciò and Sprong 2011) and it has been discussed that genetic exchange may occur, although infrequently. Although the Assemblages and genotypes of *G. duodenalis* that are infectious to humans have been identified in different animals, including various domestic animals (dogs, cats, cattle, sheep) and wildlife, the importance of giardiasis as a zoonosis remains unresolved. It seems that the majority of *Giardia* infections in animals pose little or no risk to public health.

Giardia duodenalis is globally distributed, and usually considered as the most common intestinal parasite of humans (Cook and Lim 2012). Prevalence depends on location, but has been estimated to be somewhere around 5 percent in industrialized countries and usually around five times that in developing countries. Giardiasis is particularly associated with pre-school and school-age children in developing countries (Boeke et al. 2010).

The patient

Although human infection with *G. duodenalis* is usually associated with diarrhoea, usually with malabsorption, it can be asymptomatic, or associated with a broad range of clinical symptoms, ranging from acute to chronic (Robertson et al. 2010). The infectious dose of *Giardia* is known to be low, theoretically a single cyst; in early infection studies a dose of 10 cysts was reported to result in infection in two out of two volunteers (Rendtorff 1954). However, not all human-source isolates are equally infectious to all people (Nash et al. 1987) and both human and parasite factors may be responsible for the broad spectrum of symptoms that may occur with infection. Different studies have provided contrasting results when trying to associate particular genotypes with particular symptoms (see summary in Robertson et al. 2010).

Nausea, vomiting, and weight loss are more usual symptoms, and consequences such as retarded growth and development, poor cognitive function, and detrimental effects on nutritional status particularly in children, have been reported from some studies. A few infections have been associated with other unusual manifestations, including pruritis, urticaria, uveitis, sensitisation towards food antigens, and synovitis. The wide spectrum of symptom patterns and the occurrence of treatment failures have long been recognized; while some patients experience only a mild, self-resolving illness or one that responds rapidly to treatment, other patients suffer a severe, prolonged illness, which may respond poorly to treatment or for which sequelae persist even after the parasite has been finally eliminated. Chronic infection is usually associated with diarrhoea and intestinal malabsorption, resulting in steatorrhoea, lactase deficiency and vitamin deficiencies. Potential mechanisms for this include epithelial transport and barrier dysfunction. Following an outbreak of waterborne giardiasis in Bergen, Norway in 2004 (Robertson et al. 2006b), with over 1,200 laboratory-confirmed cases, a prospective cohort study demonstrated that over 30 percent of patients with persistent symptoms had chronic *Giardia* infection, with mean disease duration of seven months and with inflammation demonstrated in duodenal biopsies (Hanevik et al. 2007). In addition, two yr after the Bergen outbreak, out of 1,017 people who had been infected and successfully treated, up to 41 percent reported fatigue and abdominal symptoms similar to those associated with irritable bowel syndrome (Mørch et al. 2009).

Diagnosis of giardiasis is traditionally by demonstration of cysts, or less frequently trophozoites, in stool samples. However, intermittent excretion and low cyst concentrations mean that sensitivity is greatly improved if a concentration technique is used (Escobedo et al. 2010), and use of stains such as trichrome or iodine will also improve detection rate. Even greater sensitivity is achieved using IFAT, but requires the use of a fluorescence microscope, which may be impractical in many smaller diagnostic labs. Faecal antigen tests are also commercially available, and are popular in terms of rapidity of obtaining a result, but they are also expensive and some studies have indicated that when number of cysts are low, the sensitivity of such tests may be limited (Strand et al. 2008). Although molecular techniques have also been developed, they are currently not widely used for routine diagnostic application, and in countries where giardiasis is of most public health significance, are usually impractical because of infrastructure and expense considerations (Escobedo et al. 2010).

Although at least six different classes of drugs, with different mechanisms, are available for treatment of giardiasis, 5-nitroimidazole compounds are usually the agents of choice (Robertson et al. 2010, Escobedo et al. 2010). The alternative treatments are advantageous if 5-nitroimidazole therapy fails or for patients for whom this therapy is unsuitable (e.g., pregnant women during the first trimester).

Foodborne transmission

As with *Cryptosporidium,* various factors in the biology of *G. duodenalis* favour a foodborne or waterborne route of transmission (Robertson and Lim 2011). These include the high cyst excretion rate from infected humans and animals, with outputs in excess of 100,000 cysts per gram over periods of days or longer, the low infectious dose (as described in the previous section) and the highly robust transmission stage (the cyst), which can survive for several months in the environment provided that they are not desiccated or exposed to freeze-thaw fracture cycles. The robustness of *Giardia* cysts is considered to be due to the filamentous cyst wall with strong interchain interactions (Gerwig et al. 2002). Two additional factors that may contribute to successful foodborne and waterborne transmission are: 1) the potential for zoonotic infections such that contamination by either animals or humans may be of public health significance, and 2) the possibility of onward contamination by cysts being moved by transport hosts, such as birds or insects (Robertson and Lim 2011).

Many outbreaks of waterborne giardiasis have been documented, with over 132 reported from the period between World War I and 2003, of which 103 were associated with contaminated drinking water systems (Karanis et al. 2007). A review which considered more recent (2004–2010) waterborne outbreaks, reported a further 70 that were considered to be caused by *G. duodenalis* (Baldursson and Karanis 2011). Most of these outbreaks are reported from developed countries where detection and monitoring systems are more likely to be in place. However, it is in less developed countries, where giardiasis is more likely to be endemic and where infrastructures related to water supply, sewage disposal, catchment control and public health may be sub-optimal, that the population is probably at greatest risk of waterborne disease transmission (Robertson and Lim 2011). If the country-by-country distribution of outbreaks of giardiasis associated with drinking water are compared, in association with time of report, some interesting patterns can be seen (Table 6), with New Zealand, for example being responsible for less than 1 percent of the outbreaks up until 2003, but over 85 percent after 2003. This probably reflects changes in reporting and monitoring rather than anything else, and it should be noted that the outbreaks reported from New Zealand are all relatively limited in size (Baldursson and Karanis 2011); the 61 outbreaks since 2004 being responsible in total for just 229 cases, considerably fewer than the 2,500 estimated cases from the single outbreak from Bergen, Norway (Baldursson and Karanis 2011). This latter outbreak, one of the largest in recent years, was considered to be due to contamination of the water source during heavy rainfall and insufficient water treatment to inactivate the parasite (Robertson et al. 2006b). Indeed, for most outbreaks of giardiasis related to drinking water, the reasons are reported as deficiencies in water treatment, including insufficient barriers or poorly operated treatment and disinfection systems, or distribution system deficiencies. The change in numbers of outbreaks of giardiasis reported from Canada is perhaps more interesting, being responsible for well over 10 percent of the outbreaks in the first time

Table 6. Country distribution of documented outbreaks of giardiasis associated with drinking water from 1950 until 2003, and from 2004 until 2010 (data derived from Karanis et al. 2007 and Baldursson and Karanis 2011).

From ca. 1950 until 2003 (data from Karanis et al. 2007)			**From 2004–2010** (data from Baldursson and Karanis 2011)		
Country	**No. outbreaks**	**%**	**Country**	**No. outbreaks**	**%**
USA	81	< 79	New Zealand	61	< 88
Canada	14	< 14	USA	4	< 6
UK	3	< 3	Norway	2	< 3
Sweden	2	< 2	Finland	1	< 2
Germany	2	< 2	Malaysia	1	< 2
New Zealand	1	< 1	Turkey	1	< 2
Finland	0	0	Sweden	0	0
Norway	0	0	UK	0	0
Malaysia	0	0	Germany	0	0
Turkey	0	0	Canada	0	0
TOTAL	**103**	**100**	**TOTAL**	**70**	**100**

period, but none in the second. Of course, the second period is considerably shorter, but it might be suggested that the measures implemented in Canada following the earlier outbreaks have been effective at preventing further outbreaks.

As with *Cryptosporidium* and other protozoa, food products can be contaminated with *Giardia* cysts, and thus serve as vehicles of transmission, at any step of the field-to-fork continuum. Nevertheless, very few outbreaks of foodborne giardiasis have been documented (less than ten from 1960 until today, with the most recent from over ten years ago and, in all, accounting for only around 200 cases of infection; see Robertson and Lim 2011, Cook and Lim 2012). Additionally, for some of these documented foodborne outbreaks, there is now some doubt about whether the infection route described is actually correct. For example, an outbreak recorded in 1991 suggests that a small number of people became infected with *Giardia* as a result of having soup made from sheep tripe (Karabiber and Aktas 1991). The authors suggest that the infection could have originated in the animal, and been protected during cooking in the crevices of the intestine. Although this theory cannot be disproven, our current knowledge suggests that *Giardia* in sheep is most likely to be of Assemblage E, and thus non-infective to humans (e.g., see Robertson 2009) and if the soup was boiled, it seems unlikely that the cysts would be infective anyway; thus, it is possible that a food-handler might be a more likely common source. However, an earlier outbreak, in which three people were apparently infected with *Giardia* as a result of eating Christmas pudding (Conroy 1960), quite remarkably extensive analyses were undertaken by the manufacturer of the pudding. In these analyses, sub-portions of centrifuged sediment were stained with Lugol's iodine and examined by microscopy, and large quantities of protozoan cysts, resembling *Giardia* cysts, were found. Contamination of the pudding with animal faeces was assumed, and, in this instance, it seems that this could have been the likely route of contamination and subsequent infection.

Despite the lack of evidence that foodborne outbreaks of giardiasis occur commonly, a case-control study in England indicated that consumption of lettuce was associated with an increased risk of giardiasis (Stuart et al. 2003). A similar study performed in Germany provided similar findings, with a significant association demonstrated between daily consumption of green salad and the risk of symptomatic giardiasis (Espelage et al. 2010).

The development of validated methods for analysing water samples for *Giardia* cysts (e.g., US EPA Method 1623, ISO Method 15553), has resulted in the development of methods for analysing food for this parasites. As IMS and IFAT are available for both *Cryptosporidium* and *Giardia*, analysis for one parasite is often conducted simultaneously with analysis for the other, as it does not require any additional work. Thus, as with *Cryptosporidium*, the method developed for analysing food samples for *Giardia* are based upon the four steps used in water analysis; elution, concentration, purification (usually by IMS), and detection (usually by IFAT) and again modifications may be used, particularly regarding the elution steps, as dictated by particular food matrix characteristics. As for *Cryptosporidium*, the development of an ISO method for analysing fresh fruit and vegetables is currently in progress, but less data are available for *Giardia* than *Cryptosporidium*. Indeed, when the method developed by Cook et al. (2007) for detecting *Giardia* cysts on lettuce was tested by eight experienced laboratories in a collaborative trial in UK, detection reproducibility was less than 50 percent, perhaps reflecting a reduced expertise in detecting *Giardia* cysts in environmental samples and these results were not published in a scientific journal (Cook and Lim 2012).

Table 7 provides an overview of development and use of different, but similar, methods for analysing different types of fresh produce for *Giardia* cysts, including recovery efficiencies.

As for *Cryptosporidium*, shellfish have also been analysed for *Giardia* cysts and again digestion of shellfish tissue in acidified pepsin appears to provide higher, more consistent recoveries than elution techniques, as well as enabling analysis of a larger representative sample (Robertson and Gjerde 2008).

As with *Cryptosporidium*, method development has resulted in surveys being conducted for *Giardia* in different food matrices. For fresh produce, contamination with *Giardia* cysts has been detected in water spinach, lettuce, sprouted seeds, potatoes, carrots, cabbage and cilantro, and surveys have been conducted in all regions of the world, including countries in Africa,

Table 7. Published methods used for the analysis of fresh produce for *Giardia* cysts and reported recovery efficiencies.

Reference	Outline of method	Food matrices tested	Recovery efficiencies
Robertson and Gjerde 2000	Washing in elution buffer in a rotating drum, concentration by centrifugation and IMS, detection by IFAT	Assorted lettuces, strawberries, cabbage, carrots, bean sprouts	Approximately 70% for all matrices, apart from bean sprouts for which recoveries were lower and more variable.
Di Benedetto et al. 2006	Method as in Robertson and Gjerde 2000	Leafy vegetables (lettuce, chicory) & berry vegetables (tomatoes, peppers)	Approximately 70% for leafy vegetables and 43% for berry vegetables
Cook et al. 2007	Stomaching in 1M glycine, concentration by centrifugation and IMS, detection by IFAT	Lettuce	Approximately 45% using seeded samples, and approximately 35% using internal controls.
Amorós et al. 2010	Method as in Cook et al. 2007	Lettuce, cabbage.	Approximately 17% using internal controls.

Asia, and South America where giardiasis is perhaps considered endemic (e.g., Morocco, Eritrea, Costa Rica, Brazil, Cambodia), but also in European countries such as UK, Norway, and Spain. Interestingly, such surveys have not yet been reported from USA. However, viability, infectivity, and genotype have seldom been investigated. Again, irrigation water could be a likely route for contamination of fresh produce, and surveys which have analysed irrigation water have generally supported this theory. Results from these studies emphasize the need for improvement in sanitary conditions, particularly in traditional growing areas and also care in the use of fertilizers (Cook and Lim 2012).

There have been fewer surveys for *Giardia* in shellfish than for *Cryptosporidium* and, indeed, it appears that *Giardia* seems to be less frequently detected in shellfish. Of five surveys for *Giardia* in shellfish published between 1997 and 2007, only three reported detection of *Giardia* cysts (Robertson 2007). However, four further surveys that have been published since then, have all reported the occurrence of *Giardia* cysts in shellfish, and, additionally, some studies have investigated the genotype of the *Giardia* cysts found (Robertson 2013). *Giardia* from Assemblage A and Assemblage B have been reported in different studies, indicating public health significance. However, although further studies are required to determine whether *Giardia* cysts remain viable and infectious in naturally contaminated bivalve molluscs, it has been concluded in different reviews of the literature that shellfish should be considered as potential sources for *Giardia* infection (Robertson 2007, Gómez-Couso and Ares-Mazás 2012).

As contamination with *Giardia* cysts arises ultimately from a human or animal source, the most effective means to control contamination from occurring on fresh produce is application of good agricultural practice during primary production, good manufacturing practice during processing, and good hygienic practice before consumption (Dawson 2005). Good agricultural practice includes use of clean water for irrigation, fertilizer application and washing, and exclusion of wild animals from growing areas. During processing (e.g., washing, chopping, packaging), it is also essential that water of potable standard is used, and that water that is reused should be subjected to effective disinfection procedures (Cook and Lim 2012). During many food manufacturing processes, elimination of microbial pathogens from foods is achieved by a variety of methods, including (most commonly) heat and chemical disinfection, and also irradiation or high pressure. The most commonly used sanitizer for fresh produce is chlorine, and commonly-used industry practice for treatment of fresh vegetables is to use 100 ppm hypochlorite, which yields 30–40 ppm free chlorine, depending upon the organic load, at pH 6.8–7.1 for a contact time of 2 min at 4°C (Seymour 1999). Contact with 1.5 ppm chlorine for less than 10 minutes at 25°C has been reported to result in a 99 percent reduction in *Giardia* viability, but 8 ppm is necessary to achieve the same effect at 5°C (Jarroll et al. 1981). However, one problem for assessing inactivation effects of different treatment protocols is that such a study requires an appropriate method for assessing viability or infectivity of the cysts. Unlike bacteria, *in vitro* cultivation of *Giardia* is frequently difficult, and animal model infectivity is also difficult for some isolates, as well as incorporating ethical issues. Some studies have employed the use of vital dyes for assessing viability, but these often over-estimate viability.

The potential severity of giardiasis and the relatively high potential for *Giardia* cysts to contaminate a range of foodstuffs means that *Giardia* is generally regarded as a significant foodborne pathogen. Cases of foodborne giardiasis may occur sporadically, thereby contributing to the endemicity of this pathogen, but it seems that foodborne outbreaks are not frequent. Robust detection methods may assist food safety management, but appropriate guidelines for control of contamination of foods and drinking water sources may be more effective in limiting the threat to public health from foodborne *Giardia*.

Toxoplasma gondii

Introduction

Toxoplasma is a coccidian parasite belonging to the phylum Apicomplexa and is infectious to practically all warm-blooded animals, including humans, livestock, birds, and marine mammals. Notably, there is only one species in the *Toxoplasma* genus, *Toxoplasma gondii*. However, based on molecular analyses, in conjunction with mouse virulence information, *T. gondii* has been classified into 3 genetic types (I, II, III), of which type I isolates are 100 percent lethal to mice, irrespective of dose, while types II and III are generally avirulent for mice. Strains that did not fall into these three clonal types were previously considered atypical, but a 4th clonal type has been recently recognized, mostly in wildlife (Khan et al. 2011). It has been suggested that certain genetic types of *T. gondii* may be associated with clinical toxoplasmosis in humans, with type I isolates, or recombinants of types I and III, or atypical strains more likely to result in clinical toxoplasmosis, but genetic characterization has been essentially limited to isolates from patients with toxoplasmosis. However, type II strains are largely predominant in both benign and severe congenital toxoplasmosis in France, suggesting no apparent association between severity of disease and genotype (Fekkar et al. 2011). Additionally, in immunocompromised patients, the genotype of *T. gondii* strain (type II vs. non-type II) is not associated with the clinical presentation or the severity of infection, but is strongly linked to the geographical area of infection

(Guy et al. 2012). Indeed, the information regarding the genetic diversity of *T. gondii* isolates circulating in the general human population is scarce, and therefore, any claims of an association between parasite genotypes and disease presentations should be treated cautiously in the absence of more complete knowledge of the *T. gondii* genotypes in human populations.

The overall lifecycle of *Toxoplasma* contains two distinct cycles; the sexual enteroepithelial cycle and the asexual cycle. The definitive hosts of *T. gondii* are members of the cat family (Felidae), thus the sexual cycle of the parasite occurs only within the intestinal epithelial cells of felids.

In the enteroepithelial cycle of *T. gondii*, the parasites pass through various morphologically distinct stages before gametogeny occurs. Gametes can be found between three and 15 d after infection, occurring throughout the small intestine, although with apparent predilection for the ileum. After fertilisation of the female macrogamete by the male microgamete, oocysts develop. The oocysts are the zygotic stage of the lifecycle, and are excreted unsporulated in the faeces of the cat. Speed of oocyst sporulation in the environment depends on factors such as temperature and humidity, but usually takes around three d. A sporulated oocyst (approximately 11μm x 13μm in size, and sub-spherical) contains two sporocysts each holding four sporozoites. The oocysts are environmentally robust, and can retain infectivity in a cool damp environment for months (Guy et al. 2012).

The asexual cycle occurs when consumption of tissue cysts (see below) or oocysts results in infection of the intestine, and the tachyzoite form of the parasite multiplies asexually in the cells of lamina propria by repeated divisions until the cells rupture. Tachyzoites from ruptured cells are released into surrounding tissues resulting in systemic infection. Circulating tachyzoites infect new cells throughout the body, and all cells, with the exception of mature erythrocytes seem to be susceptible, but cardiac and skeletal muscle and the central nervous system are more often infected. After several more rounds of asexual division, tissue cysts are formed and these remain intracellular. Tissue cysts of *T. gondii* range from 5 μm to over 100 μm in size and contain bradyzoites.

These bradyzoites are infectious when ingested with the tissue surrounding them. If ingested by a felid, the sexual enteroepithelial cycle will occur, or if ingested by any other host, the asexual cycle, as described in the previous paragraph will occur. In addition, if a female host is pregnant when infected, then circulating tachyzoites may move through the placenta to the foetus (intrauterine or congenital transmission), and lactogenic transmission of tachyzoites may also occur from an infected mother to a suckling infant.

Toxoplasma gondii is perhaps the most widespread protozoan parasite affecting humans, and it has been estimated that between 1 and 2 billion of the world's population is infected at any one time (Montoya and Liesenfeld

2004). Although infection in humans occurs worldwide, prevalence varies significantly between populations living in different geographical areas. For example, between 11 and 40 percent of adults have been found to be seropositive in USA and UK, but in other countries in western Europe, typical seroprevalence rates vary from 11–28 percent in Scandinavia, to 42 percent in Italy, an up to 67 percent in Belgium (Guy et al. 2012). In some regions of Brazil, infection rates over 70 percent have been reported, while rates of around 40 percent have been reported from different African countries. In Asia, infection rates vary from less than 10 percent to over 70 percent, depending on location (Guy et al. 2012). Although comparison of seroprevalence rates is hampered by population characteristics and serological tests used, foci of high infection rates can, nevertheless, be identified (Guy et al. 2012).

The patient

The clinical picture of infection with *Toxoplasma* is greatly influenced by the immune status of the patient. In the immunocompetent, *T. gondii* infection is usually asymptomatic, but may cause a mild to moderate illness, in which typical symptoms include low grade fever, lymphadenopathy, fatigue, muscle pain, sore throat, and headache. In some cases, ocular toxoplasmosis may occur, which may be accompanied by partial or total loss of vision. However, the rate of ocular toxoplasmosis seems to differ according to unknown factors, probably associated with both the host and the parasite. Although latent *Toxoplasma* infection is generally accepted as being benign in the immunocompetent, some studies have suggested that the parasite may affect behaviour (Flegr 2007), perhaps being a contributory, or even causative, factor in various psychiatric disorders including depression, anxiety, and schizophrenia (Henriquez et al. 2009). It has been proposed that *Toxoplasma* may affect dopamine levels within the brain, resulting in alterations in CNS function. Should the association between *Toxoplasma* infection and psychiatric dysfunction be proven, the overall burden of disease and risk to health and well-being due to this parasite would need to be reevaluated (Guy et al. 2012).

In the immunocompromised and immunodeficient, for example HIV-patients and/or those receiving profound immunosuppressive therapy, severe or life-threatening disease can result either from acute *Toxoplasma* infection or reactivation of a previously latent infection. Here, encephalitis is the most clinically-significant manifestation, but retinochoroiditis, pneumonitis, and other systemic disease may also occur. In patients with acquired immunodeficiency syndrome (AIDS), toxoplasmic encephalitis is the most common cause of intracerebral mass lesions and ranks high on the list of diseases resulting in the death of AIDS patients.

Congenital toxoplasmosis is another serious potential manifestation of *T. gondii* infection. In an immunocompetent mother, it is generally accepted that *Toxoplasma* is passed on to the foetus from an infection acquired immediately before or during pregnancy, i.e., prior to onset of the latent phase of infection. However rare cases of transplacental infection have been reported in which the mother has had a previous latent infection. The risk of transplacental infection increases throughout pregnancy, but the risk of severe disease or foetal death decreases. Symptoms commonly associated with transplacental infection include spontaneous termination, foetal death, ventricular dilatation, and intracranial calcification (Guy et al. 2012). Neonates may present with hydrocephalus, seizures, retinochoroiditis, spasticity, deafness, hepatosplenomegaly, jaundice and/or rash and children that are asymptomatic at birth, may suffer from mental retardation or retinochoroidal lesions later in life. Children who have been infected late on in the pregnancy are usually asymptomatic or have only mild complications.

Diagnosis of *T. gondii* infection is based principally upon serological testing for specific IgG, with acute phase infection confirmed by detection of IgM and IgA. In pregnant women that are positive for *Toxoplasma*-specific IgM, it is critical for management and assessment of risk to the foetus to determine whether the infection was acquired before or after conception. In such cases, measurement of IgG avidity may provide a more precise estimate of duration of infection. Other methods used to investigate possible active or reactivated *Toxoplasma* infection include nucleic acid amplification tests (NAAT), examination of tissues for any changes consistent with toxoplasmosis, enhanced immunohistological staining for detection of *Toxoplasma*, and culture.

In the immunocompetent patient, toxoplasmosis is essentially a self-limiting disease and so treatment is usually unnecessary. However, for the immunodeficient and immunosuppressed, active infection may be life-threatening and treatment is usually a synergistic combination of pyrimethamine and sulphadiazine, whose action is based partially upon disruption of the biochemical pathway for production of folic acid. As mammalian cells are able to utilize folinic acid as a precursor for folic acid, this may be provided as a supplement. As this treatment is not effective against the bradyzoite form of *Toxoplasma*, the parasite within tissue cysts cannot be inactivated. Various agents have been proposed for treatment of bradyzoites, but efficacy appears to be limited (Guy et al. 2012).

Foodborne transmission

There are three potentially infectious stages of *Toxoplasma*: tachyzoites, bradyzoites, and oocysts. With regard to foodborne transmission, two of these stages are of particular relevance—bradyzoites and oocysts. Bradyzoites may be ingested with the tissue of an infected intermediate host, while oocysts may be ingested with any produce that has the potential to be contaminated with the faeces of an infected felid. Thus, unlike the other protozoa of focus in this chapter, *T. gondii* has completely dissimilar potential routes of foodborne infection that must be considered separately. In addition, though probably of less significance, tachyzoites excreted in the milk of infected animals might result in milkborne infection, and for breastfed infants is also a human-to-human, foodborne infection route (Camossi et al. 2011). Although it has been suggested that tachyzoites should be destroyed by contact with gastric acid, it is possible that the neutral pH of the milk has a protective function, and that passage of milk through the stomach is relatively rapid also minimizes their degradation (Camossi et al. 2011). Indeed, various outbreaks of toxoplasmosis associated with consumption of unpasteurized goats' milk have been reported in the literature (Guy et al. 2012) and consumption of such milk has been considered a risk factor for acquisition of *T. gondii* infection in USA (Jones et al. 2009).

Human infection via bradyzoites in meat is dependent on various factors, including the prevalence of *Toxoplasma* infection in meat animals, cultural factors regarding eating of meat and meat preparation, and individual factors (such as age and immunological status) of the person exposed. Parasite factors concerned with infectivity and genetic type, are probably of relevance also. Virtually all edible portions of an animal can harbour viable *T. gondii* tissue cysts, which can survive in food animals for years. Most species of livestock, including sheep, goats, and pigs, are susceptible to infection with *T. gondii* (Dubey 2009a, Guy et al. 2012). In some countries sheep and goats are the most important hosts of *T. gondii*, in terms of human exposure and high prevalence (Dubey 2009b) and in different case-control studies consumption of lamb/mutton has been identified as a risk factor for pregnant women contracting toxoplasmosis (Kijlstra and Jongert 2008). In addition, in some countries, such as France, raw or undercooked lamb is considered a delicacy. However, in other countries, for example in USA, lamb and mutton are considered relatively minor food commodities (Guy et al. 2012). Of the major meat animal species investigated in USA to date, pig is the only species that has been found to frequently harbour the parasite (Dubey and Jones 2008), although prevalence has declined dramatically in areas where they are predominantly raised indoors (Guy et al. 2012). The risk of acquiring toxoplasmosis from beef also demonstrates regional variability with some European studies suggesting that it can be a significant contributor to human infection (Cook et al. 2000, Opsteegh et al. 2011) and consuming raw beef has also been noted as a risk factor for infection in a study from USA (Jones et al. 2009). However, in a survey of

698 retail outlets in 28 metropolitan statistical areas, covering 80 percent of the population of USA, no beef samples were positive by bioassay or by ELISA (Dubey et al. 2005). Although poultry are also susceptible to infection with *T. gondii*, and may theoretically pose a source of infection to humans, the relatively limited lifespan of poultry and the fact that they tend to be well-cooked before consumption, limits their importance as sources of infection for humans (Kijlstra and Jongert 2008). Chickens have not been indicated as a source of human infection in USA, despite high-rates of infection occurring in some flocks (Guy et al. 2012).

Well-documented outbreaks of toxoplasmosis associated with the meatborne transmission route are few, presumably due to the low proportion of infections that result in significant disease, the varying nature of symptoms between individuals, and the fact that meatborne infections are likely to affect relatively few individuals (compared with, for example, waterborne infections). Nevertheless, various meatborne outbreaks have been documented in the literature and these have been associated with beef, mutton, pork, and venison (Guy et al. 2012). Indeed, game animals are considered to be potentially important sources of meatborne toxoplasmosis, particularly as such meat is often consumed undercooked (Opsteegh et al. 2011), with wild-boar and venison particularly implicated in Europe (Kijlstra and Jongert 2008). However, in other parts of the world, other game meats may be of equal or greater importance; for example, kangaroos are considered to be highly susceptible to *T. gondii* infection and consumption of kangaroo has been linked to an outbreak in Australia (Kijlstra and Jongert 2008), while in Arctic regions consumption of undercooked game meat, particularly from marine mammals, seems to be an important risk factor for human infection (Davidson et al. 2011).

Currently, there are no programmes for the slaughter inspection of pigs or other meat animals for toxoplasmosis. Detection of microscopic *T. gondii* tissue cysts by visual inspection is not possible, and due to the high prevalence of infection in many areas it may not even be particularly useful. It is probably better that all meat animals are considered as potentially infective. However, for survey purposes in assessing prevalence, the most definitive methods for detecting tissue cysts in pigs and other livestock are bioassay of tissues in mice or cats, of which cat bioassay is considered the "gold standard". Indirect methods are available in which detection of anti-*Toxoplasma* antibody or parasite DNA is presumed to indicate infection, therefore the presence of tissue cysts (Guy et al. 2012). Of these, serum ELISA is considered to be the most sensitive, and direct PCR the least (Guy et al. 2012).

If it is assumed that all meat animals are potential sources of infection with *T. gondii*, it is important to be aware of methods of inactivation. Although *T. gondii* tissue cysts have been shown to retain viability in pork at temperatures slightly below freezing, they are inactivated rapidly at lower temperatures and there is no evidence that different Types of *T. gondii* have different freezing susceptibilities (Guy et al. 2012). Thorough cooking is also effective at inactivating *Toxoplasma* cysts, although it should be noted that when times and temperatures are quoted, they refer to the whole of the meat product—uneven heating throughout the meat may mean that some cysts avoid inactivation and remain infective, as has been found to occur with microwaves (Guy et al. 2012). Other treatments such as curing, irradiation, and high pressure processing have all been shown to be effective at inactivating *Toxoplasma* cysts under particular described conditions (Guy et al. 2012).

Human infection via oocysts occurs when a person ingests something that has been contaminated with faeces from an infected cat. Oocysts from an infected felid may be shed over a period of several weeks and at peak production, most commonly at about one week post infection, over a million oocysts may be excreted per day (Guy et al. 2012), giving a high potential for environmental contamination. Unlike *Cryptosporidium* oocysts, *Toxoplasma* oocysts are not infective at excretion, requiring a period of some days in the environment to sporulate, and thus direct infection from handling an infected cat is not particularly likely. However, the oocysts are very hardy (and, unlike bradyzoites, can survive freezing), so contamination of produce or water may provide a route for transmission. It is possible that the importance of the oocyst infection route has been generally under-estimated previously, perhaps due to the alternative, well-known transmission route, and also because most resultant infections are probably asymptomatic. However, there have been a few well-documented waterborne outbreaks that have resulted in considerable morbidity, as well as subsequent congenital infection due to pregnant women being infected in the outbreak (Karanis et al. 2007, Baldursson and Karanis 2011, Bowie et al. 1997, Balasundaram et al. 2010). Due to the high numbers of people infected at one time, these waterborne outbreaks are more likely to be identified. A recent study using a technique that identified sporozoites, and thus indicated infection with oocysts rather than via bradyzoites in meat, suggested that undetected contamination of food and water by *Toxoplasma* oocysts frequently causes human infections in North America (Boyer et al. 2011). Additionally, studies using berries contaminated with *T. gondii* oocysts under laboratory conditions have demonstrated that oocysts can adhere to berries and can be recovered by bioassays in mice. It was also shown that raspberries retain more inoculated oocysts than blueberries, presumably due to the different surfaces of these two berry species (Kniel et al. 2002).

Although methods have been developed for detection of *Toxoplasma* oocysts in water, they have not been widely adopted. For fresh produce, consumption of which has not been associated with any outbreak of toxoplasmosis, there has been even less focus on development of methods and surveying. Indeed, the first survey of fresh produce for *Toxoplasma* contamination was published in 2011 and describes analysis of samples of carrots, radishes, lettuces and strawberries in Poland using a calcium carbonate flocculation method followed by quantitative real-time PCR (Lass et al. 2012). Although the minimum levels of detection in this study were rather high, being estimated to be 100 oocysts on radishes and 1,000 oocysts on strawberries, the authors found *Toxoplasma* DNA on almost 10 percent of samples (21/216) and concluded that contaminated fruits and vegetables may contribute to the prevalence of *T. gondii* infection in Poland (Lass et al. 2012). There has been relatively little research on methods of inactivating *Toxoplasma* oocysts on fresh produce, and general kitchen hygiene probably remains the main protective measure. However, high pressure processing (at a minimum of 340 MPa) of fresh fruits and vegetables renders contaminating oocysts non-infectious (Lindsay et al. 2008). Other methods of inactivation of oocysts on fresh produce includes irradiation at 0.5 KGy, prolonged (> 24 h) freezing at –20°C, or boiling (Guy et al. 2012).

As for other protozoa that may be transmitted by contamination of food or produce, there has been increasing interest in recent years in the potential for shellfish to act as transmission vehicles, although cases of human toxoplasmosis directly associated with consumption of shellfish have not been reported. However, *T. gondii* oocysts are apparently widespread in the marine environment (Conrad et al. 2005, Forman et al. 2007). Also, as significantly higher levels of *T. gondii* seroprevalence were also reported in a species of porpoise with a high predilection for shellfish consumption than in a species feeding principally on fish (Forman et al. 2007), it was concluded that consumption of shellfish was a likely risk factor for infection with *T. gondii*. Both oysters and mussels have been shown, experimentally, to concentrate *Toxoplasma* oocysts, and some experiments have also demonstrated that *T. gondii* oocysts can sporulate in seawater and remain infectious (Arkush et al. 2003, Lindsay et al. 2001, 2004, Robertson 2007). More recently, natural contamination of shellfish has been investigated via surveys (Table 8). Although, not all these surveys found evidence of *Toxoplasma* contamination in shellfish, the possibility is supported by a case-control study of recently-infected (< 6 mon) adults in which consumption of raw oysters, clams, or mussels were identified as risk factors for *T. gondii* infection (Jones et al. 2009).

Although toxoplasmosis is estimated to have a disease burden and economic impact comparable to campylobacteriosis and salmonellosis, there are presently no explicit monitoring programmes to screen animals entering the food-chain, and reporting of human toxoplasmosis in different countries is not standardized (Guy et al. 2012). The Biological Safety Panel of the European Food Safety Authority has stated that there is an urgent

Table 8. Surveys of shellfish for contamination with *Toxoplasma gondii*.

Survey location	Shellfish	Analysis method	Results	Reference
Central California, USA	California mussels (*Mytilus californianus*)	TaqMan PCR method on pelletized haemocytes and digestive gland.	1 mussel positive out of 1109 sampled. Genotyped as Type X.	Miller et al. 2008
Nunavik, Quebec, Canada	Blue mussels (*Mytilus edulis*)	PCR analysis of 50 g aliquots of pooled whole mussel tissue.	All results negative.	Lévesque et al. 2010
Cananéia region, São Paulo State, Brazil	Mangrove oysters (*Crassostrea rhizophorae*) and mangrove mussels (*Mytella guyanensis*)	Aliquots from groups (5 oysters or 15 mussels)—bioassay in mice or PCR-RFLP. Assay sensitivity approximately 100 oocysts.	All mussel samples negative. Two of 60 groups of oysters positive by PCR—but not be mouse bioassay. Genotyping unsuccessful.	Esmerini et al. 2010
Varano Lagoon, Apulia, Italy	Pacific cupped oysters (*Crassostrea gigas*), Mediterranean mussels (*Mytilus galloprovincialis*), true clams, (*Tapes decussatus*) and Manila clams (*Tapes philippinarum*)	Hamolymph, digestive gland and gill aliquots from grouped samples analysed by nested-PCR and a fluorescent amplicon generation real-time PCR.	One pooled sample of gills from *C. gigas* and one pooled sample of haemolymph from *T. decussatus* positive by both techniques.	Putignani et al. 2011

requirement for reference materials and reagents to be made widely available so the suitability of different assays can be determined using field trials (EFSA 2007, Guy et al. 2012).

Concluding Remarks

The intention of this chapter was to provide an overview over those protozoan parasites that are of public health significance and that may be transmitted by the foodborne route. Due to space limitations, this overview is relatively cursory in nature, but hopefully demonstrates to the reader that despite protozoa usually being considered as pathogens of secondary importance, in comparison with bacteria or viruses they have a considerable impact on global public health. Also, it is emphasized, that the impact of these protozoa is not limited to tropical countries, or countries with poor infrastructure; even in the most developed and advanced nations, foodborne protozoa may compromise the health and well-being of the population.

Nevertheless, the burden of diseases caused by foodborne protozoa remains largely unknown and the data that indicate any trends in foodborne infectious intestinal disease are limited to a few industrialized countries, and even fewer pathogens. Epidemiological evidence suggests that *Cryptosporidium*, *Cyclospora*, *Giardia*, and *Toxoplasma* are the protozoa with the largest epidemiological risks for foodborne transmission. However, as described in the different sections in this chapter, the biology and ecology of these protozoa mean that their prevention and control in food manufacturing and food service operations are far from simple. Other protozoan infections may also be transmitted to humans via the foodborne route, and the type of foods associated with a particular protozoan species may vary according to a range of factors, including the infectious stages, the lifecycle, zoonotic factors, and human behavioural and cultural issues (Núñez and Robertson 2012). Although parasitic protozoa do not, in general, multiply in foods, they often have a transmission stage that is environmentally robust, perhaps being resistant to temperature extremes, disinfectants, desiccation and other stress and may survive in or on moist foods for months in appropriate environments.

The patterns of foodborne protozoan disease described in this chapter should not be considered as static. A wide range of factors contribute to changing trends in foodborne diseases, including globalisation, travel, climate change, changes in eating habits and farming practices, and demographic shifts (Broglia and Kapel 2011, Núñez and Robertson 2012).

Although identifying trends in foodborne protozoan disease is challenging, various messages are apparent. More effective food safety education programmes for food-handlers and consumers are needed, and control strategies should consider food safety-related trends, including large-scale production and wide distribution of food, globalization of the food supply, eating outside of the home, emergence of new pathogens, and escalating populations of at-risk consumers. Maintaining dialogue and collaboration across sectors and disciplines, such as between public health officials, veterinarians, and food safety experts, is essential for successful monitoring changing trends in foodborne protozoan diseases, for detecting emerging pathogens, for understanding transmission routes better and for developing effective control strategies (Núñez and Robertson 2012).

References

Alakpa, G.E., S.C. Clarke and A.F. Fagbenro-Beyioku. 2003. *Cyclospora cayetanensis* infection: vegetables and water as possible vehicles for its transmission in Lagos, Nigeria. Br. J. Biomed. Sci. 60: 113–114.

Alfano-Sobsey, E.M., M.L. Eberhard, J.R. Seed, D.J. Weber, K.Y. Won, E.K. Nace and C.L. Moe. 2004. Human challenge pilot study with *Cyclospora cayetanensis*. Emerg. Infect. Dis. 10: 726–728.

Ali, I.K.M., C.G. Clark and W.A. Petri. 2008. Molecular epidemiology of amebiasis. Infect. Genet. Evol. 8: 698–707.

Ali, I.K.M., U. Mondal, S. Roy, R. Haque, W.A. Petri and C.G. Clark. 2007. Evidence for a link between parasite genotype and outcome of infection with *Entamoeba histolytica*. J. Clin. Microbiol. 45: 285–289.

Ali, S.I., K. Jamal and S.M. Qadri. 1992. Prevalence of intestinal parasites among food handlers in Al-Medinah. Ann. Saudi Med. 12: 63.

Amorós, I., J.L. Alonso and G. Cuesta. 2010. *Cryptosporidium* oocysts and *Giardia* cysts on salad products irrigated with contaminated water. J. Food Prot. 73: 1138–1140.

Anonymous. 2010a. Import Policy for Fresh Guatemalan Raspberries and Blackberries [Online]. Available at http://www.inspection.gc.ca/english/fssa/frefra/safsal/guat/berguatemalane.shtml.

Anonymous. 2010b. *Cyclospora* outbreak investigation closed locally. Community Health Services Department, County of Lambston, Connecticut.

Arkush, K.D., M.A. Miller, C.M. Leutenegger, I.A. Gardner, A.E. Packham, A.R. Heckeroth, A.M. Tenter, B.C. Barr and P.A. Conrad. 2003. Molecular and bioassay-based detection of *Toxoplasma gondii* oocyst uptake by mussels (*Mytilus galloprovincialis*). Int. J. Parasitol. 33: 1087–1097.

Ayala-Gaytan, J.J., C. Diaz-Olachea, P. Riojas-Montalvo and C. Palacios-Martinez. 2004. Cyclosporidiosis: clinical and diagnostic characteristics of an epidemic outbreak. Rev. Gastroenterol. Mex. 69: 226–229.

Babiker, M.A., M.S. Ali and E.S. Ahmed. 2009. Frequency of intestinal parasites among food-handlers in Khartoum, Sudan. East. Mediterr. Health J. 15: 1098–1104.

Balasundaram, M.B., R. Andavar, M. Palaniswamy and N. Venkatapathy. 2010. Outbreak of acquired ocular toxoplasmosis involving 248 patients. Arch. Ophthalmol. 128: 28–32.

Baldursson, S. and P. Karanis. 2011. Waterborne transmission of protozoan parasites: review of worldwide outbreaks—an update 2004–2010. Water Res. 45: 6603–6614.

Ben Ayed, S., A.R. Ben, M. Mousli, K. Aoun, M. Thellier and A. Bouratbine. 2008. Molecular differentiation of *Entamoeba histolytica* and *Entamoeba dispar* from Tunisian food handlers with amoeba infection initially diagnosed by microscopy. Parasite 15: 65–68.

Blackburn, B.G., J.M. Mazurek, M. Hlavsa, J. Park, M. Tillapaw, M. Parrish, E. Salehi, W. Franks, E. Koch, F. Smith, L. Xiao, M. Arrowood, V. Hill, A. daSilva, S. Johnston and J.L. Jones. 2006. Cryptosporidiosis associated with ozonated apple cider. Emerg. Infect. Dis. 12: 684–686.

Blans, M.C., B.U. Ridwan, J.J. Verweij, M. Rozenber-Arska and J. Verhoef. 2005. Cyclosporiasis outbreak, Indonesia. Emerg. Infect. Dis. 11: 1453–1455.

Boeke, C.E., M. Mora-Plazas, Y. Forero and E. Villamor. 2010. Intestinal protozoan infections in relation to nutritional status and gastrointestinal morbidity in Colombian school children. J. Trop. Pediatr. 56: 299–306.

Botero-Garces, J., M.N. Montoya, J.I. Barguil and A. Castano-Gonsalez. 2006. An outbreak of *Cyclospora cayetanensis* in Medellín, Colombia. Revista Salud Publica 8: 258–268.

Bowie, W.R., A.S. King, D.H. Werker, J.L. Isaac-Renton, A. Bell, S.B. Eng and S.A. Marion. 1997. Outbreak of toxoplasmosis associated with municipal drinking water. Lancet 350: 173–177.

Boyer, K., D. Hill, E. Mui, K. Wroblewski, T. Karrison, J.P. Dubey, M. Sautter, A.G. Noble, S. Withers, C. Swisher, P. Heydemann, T. Hosten, J. Babiarz, D. Lee, P. Meier, R. McLeod and other members of the Toxoplasmosis Study Group. 2011. Unrecognized ingestion of *Toxoplasma gondii* oocysts leads to congenital toxoplasmosis and causes epidemics in North America. Clin. Infect. Dis. 53: 1081–1089.

Broglia, A. and C. Kapel. 2011. Changing dietary habits in a changing world: emerging drivers for the transmission of foodborne parasitic zoonoses. Vet. Parasitol. 182: 2–13.

Caceres, V.M., R.T. Ball, S.A. Somerfedlt, L. Mackey, S.E. Nichols, W.R. MacKenzie and B.L. Herwaldt. 1998. A foodborne outbreak of cyclosporiasis caused by imported raspberries. J. Fam. Practice. 47: 231–234.

Cacciò, S.M. and H. Sprong. 2011. Chapter 2: Epidemiology of giardiasis in humans. *In:* H. Lujan and S. Svärd (eds.). Giardia—A Model Organism. 2011. Springer Wien New York. pp. 17–28.

Camossi, L.G., H. Greca-Júnior, A.P. Corrêa, V.B. Richini-Pereira, R.C. Silva, A.V. Da Silva and H. Langoni. 2011. Detection of *Toxoplasma gondii* DNA in the milk of naturally infected ewes. Vet. Parasitol. 177: 256–261.

CDC. 1997. Update: Outbreaks of cyclosporiasis—United States and Canada, (1997). MMWR Surveill. Summ. 46: 521–523.

CDC. 1998. Outbreak of cyclosporiasis—Ontario, Canada, May (1998). MMWR Surveill. Summ. 47: 806–809.

CDC. 2004. Outbreak of cyclosporiasis associated with snow peas—Pennsylvania (2004). MMWR Surveill. Summ. 53: 876–878.

CDC. 2010a. CDC Vessel Sanitation Program [Online]. Available at http://www.cdc.gov/nceh/vsp/surv/gilist.htm.

CDC. 2010b. Health of resettled Iraqi refugees—San Diego County, California, October 2007 –September 2009. MMWR Surveill. Summ. 59: 1614–1618.

Chalmers, R.M., G. Robinson, K. Elwin, S.J. Hadfield, L. Xiao, U. Ryan, D. Modha and C. Mallaghan. 2009. *Cryptosporidium* sp. rabbit genotype, a newly identified human pathogen. Emerg. Infect. Dis. 15: 829–830.

Chappell, C.L., P.C. Okhuysen, R. Langer-Curry, G. Widmer, D.E. Akiyoshi, S. Tanriverdi and S. Tzipori. 2006. *Cryptosporidium hominis:* experimental challenge of healthy adults. Am. J. Trop. Med. Hyg. 75: 851–857.

Collins, M.V., G.J. Flick, S.A. Smith, R. Fayer, R. Croonenberghs, S. O'Keefe and D.S. Lindsay. 2005a. The effect of high-pressure processing on infectivity of *Cryptosporidium parvum* oocysts recovered from experimentally exposed Eastern oysters (*Crassostrea virginica*). J. Eukaryot. Microbiol. 52: 500–504.

Collins, M.V., G. Flick, S.A. Smith, R. Fayer, E. Rubendall and D.S. Lindsay. 2005b. The effects of E-beam irradiation and microwave energy on Eastern Oysters (*Crassostrea virginica*) experimentally infected with *Cryptosporidium parvum*. J. Eukaryot. Microbiol. 52: 484–488.

Conrad, P.A., M.A. Miller, C. Kreuder, E.R. James, J. Mazet, H. Dabritz, D.A. Jessup, F. Gulland and M.E. Grigg. 2005. Transmission of *Toxoplasma*: clues from the study of sea otters as sentinels of *Toxoplasma gondii* flow into the marine environment. Int. J. Parasitol. 35: 1155–1168.

Conroy, D.A. 1960. A note on the occurrence of *Giardia* sp. in a Christmas pudding. Rev. Iber. Parasitol. 20: 567–571.

Cook, A.J., R.E. Gilbert, W. Buffolano, J. Zufferey, E. Petersen, P.A. Jenum, W. Foulon, A.E. Semprini and D.T. Dunn. 2000. Sources of toxoplasma infection in pregnant women: European multicentre case-control study. European Research Network on Congenital Toxoplasmosis. BMJ. 321(7254): 142–147.

Cook, N. and Y.A.L. Lim. 2012. Chapter 4: *Giardia duodenalis*: contamination of fresh produce. *In:* L.J. Robertson and H.V. Smith (eds.). Foodborne Protozoan Parasites. Nova Science Publishers, New York, pp. 107–132.

Cook, N., R.A. Nichols, N. Wilkinson, C.A. Paton, K. Barker and H.V. Smith. 2007. Development of a method for detection of *Giardia duodenalis* cysts on lettuce and for simultaneous analysis of salad products for the presence of *Giardia* cysts and *Cryptosporidium* oocysts. Appl. Environ. Microbiol. 73: 7388–7391.

Cook, N., C.A. Paton, N. Wilkinson, R.A.B. Nichols, K. Barker and H.V. Smith. 2006. Towards standard methods for the detection of *Cryptosporidium parvum* on lettuce and raspberries. Part 1: development and optimization of methods. Int. J. Food Microbiol. 109: 215–221.

Davidson, R., M. Simard, S.J. Kutz, C.M. Kapel, I.S. Hamnes and L.J. Robertson. 2011. Arctic parasitology: why should we care? Trends Parasitol. 27: 239–245.

Dawson, D. 2005. Foodborne protozoan parasites. Int. J. Food Microbiol. 103: 207–227.

Deng, M.Q. and D.O. Cliver. 2001. Inactivation of *Cryptosporidium parvum* oocysts in cider by flash pasteurization. J. Food Prot. 64: 523–527.

de Rezende, C.H., J.M. Costa-Cruz and M.L. Gennari-Cardoso. 1997. Enteroparasitoses in food handlers of the public schools in Uberlândia (Minas Gerais), Brazil. Rev. Panam. Salud Publica 2: 392–397.

Di Benedetto, M.A., F. Di Piazza, R. Oliveri, G. Cerame, R. Valenti and A. Firenze. 2006. Development of a technique for recovering *Giardia* cysts and *Cryptosporidium* oocysts from fresh vegetables. Annali di igiene: medicina preventiva e di comunità 18: 101–107.

Dillingham, R.A., A.A. Lima and R.L. Guerrant. 2002. Cryptosporidiosis: epidemiology and impact. Microbes Infect. 4: 1059–1066.

Döller, P.C., K. Dietrich, N. Filipp, S. Brockmann, C. Dreweck, S. Vonthein, C. Wagner-Wiening and A. Wiedenmann. 2002. Cyclosporiasis outbreak in Germany associated with the consumption of salad. Emerg. Infect. Dis. 8: 992–994.

Dubey, J.P. 2009a. Toxoplasmosis of Animals and Humans, 2nd ed. CRC Press, Boca Raton, FL, USA.

Dubey, J.P. 2009b. Toxoplasmosis in sheep—the last 20 years. Vet. Parasitol. 163: 1–14.

Dubey, J.P., D.E. Hill, J.L. Jones, A.W. Hightower, E. Kirkland, J.M. Roberts, T. Lehmann, C. Sreekumar, M.C.B. Vianna, O.C.H. Kwok, S.K. Shen and H.R. Gamble. 2005. Prevalence of viable *Toxoplasma gondii* in beef, chicken and pork from retail meat stores in the United States. J. Parasitol. 91: 1082–1093.

Dubey, J.P. and J.L. Jones. 2008. *Toxoplasma gondii* infections in humans and animals in the United States. Int. J. Parasitol. 38: 1257–1278.

Escobedo, A.A., P. Almirall, L.J. Robertson, R.M.B. Franco, K. Hanevik, K. Mørch and S. Cimerman. 2010. Giardiasis: the ever-present threat of a neglected disease. Infect. Disord. Drug Targets 10: 329–348.

Esmerini, P.O., S.M. Gennari and H.F. Pena. 2010. Analysis of marine bivalve shellfish from the fish market in Santos city, São Paulo state, Brazil, for *Toxoplasma gondii*. Vet. Parasitol. 170: 8–13.

Espelage, W., M. An der Heiden, K. Stark and K. Alpers. 2010. Characteristics and risk factors for symptomatic *Giardia lamblia* infections in Germany. BMC Public Health 10: 41–49.

European Food Safety Authority. 2007. Surveillance and monitoring of *Toxoplasma* in humans, food and animals: Scientific Opinion of the Panel on Biological Hazards. EFSA Journal 583: 1–64.

Fekkar, A., D. Ajzenberg, B. Bodaghi, F. Touafek, P. Le Hoang, J. Delmas, P.Y. Robert, M.L. Dardé, D. Mazier and L. Paris. 2011. Direct genotyping of *Toxoplasma gondii* in ocular fluid samples from 20 patients with ocular toxoplasmosis: predominance of type II in France. J. Clin. Microbiol. 49: 1513–1517.

Flegr, J. 2007. Effects of *Toxoplasma* on human behaviour. Schizophr. Bull. (Bp) 33: 757–760.

Fleming, C.A., D. Caron, J.E. Gunn and A. Barry. 1998. A Foodborne outbreak of *Cyclospora cayetanensis* at a wedding. Arch. Intern. Med. 158: 1121–1125.

Forman, D., N. West, M. Powell, J. Francis and E. Guy. 2007. *Toxoplasma* in cetaceans around the British Isles. Vet. Rec. 161: 279.

Gargala, G. 2008. Drug treatment and novel drug target against *Cryptosporidium*. Parasite 15: 275–281.

Gerwig, G.J., J.A. van Kuik, B.R. Leeflang, J.P. Kamerling, J.F. Vliegenthart, C.D. Karr and E.L. Jarroll. 2002. The *Giardia intestinalis* filamentous cyst wall contains a novel beta(1-3)-N-acetyl-D-galactosamine polymer: a structural and conformational study. Glycobiology 12: 499–505.

Gómez-Couso, H. and M.E. Ares-Mazás. Chapter 4: *Giardia duodenalis*: contamination of bivalve molluscs. *In:* L.J. Robertson and H.V. Smith (eds.). Foodborne Protozoan Parasites. Nova Science Publishers, New York, pp. 133–150.

Gómez-Couso, H., F. Méndez-Hermida, J.A. Castro-Hermida and E. Ares-Mazás. 2006. Cooking mussels (*Mytilus galloprovincialis*) by steam does not destroy the infectivity of *Cryptosporidium parvum*. J. Food Prot. 69: 948–950.

Goodgame, R.W., R.M. Genta, A.C. White and C.L. Chappell. 1993. Intensity of infection in AIDS-associated cryptosporidiosis. J. Infect. Dis. 167: 704–709.

Gündüz, T., M.E. Limoncu, S. Cümen, A. Ari, E. Serdağ and Z. Tay. 2008. The prevalence of intestinal parasites and nasal *S. aureus* carriage among food handlers. J. Environ. Health 70: 64–65.

Guy, E., J.P. Dubey and D.E. Hill. 2012. Chapter 6: *Toxoplasma gondii*. *In:* L.J. Robertson and H.V. Smith (eds.). Foodborne Protozoan Parasites. Nova Science Publishers, New York, pp. 167–188.

Hanevik, K., T. Hausken, M.H. Morken, E.A. Strand, K. Mørch, P. Coll, L. Helgeland and N. Langeland. 2007. Persisting symptoms and duodenal inflammation related to *Giardia duodenalis* infection. J. Infect. 55: 524–530.

Harp, J.A., R. Fayer, B.A. Pesch and G.J. Jackson. 1996. Effect of pasteurization on infectivity of *Cryptosporidium parvum* oocysts in water and milk. Appl. Environ. Microbiol. 62: 2866–2868.

Henriquez, S.A., R. Brett, J. Alexander, J. Pratt and C.W. Roberts. 2009. Neuropsychiatric disease and *Toxoplasma gondii* infection. Neuroimmunomodulation 16: 122–133.

Herwaldt, B.L. 2000. *Cyclospora cayetanensis*: a review, focusing on the outbreaks of cyclosporiasis in the 1990s. Emerg. Infect. Dis. 31: 1040–1057.

Herwaldt, B.L. and M.-L. Ackers. 1997. An outbreak in 1996 of cyclosporiasis associated with imported raspberries. The Cyclospora Working Group. N. Engl. J. Med. 336: 1548–1556.

Hira, P.R., J. Iqbal, F. Al-Ali, R. Philip, S. Grover, E. D'Almeida and A.A. Al-Eneizi. 2001. Invasive amebiasis: challenges in diagnosis in a non-endemic country (Kuwait). Am. J. Trop. Med. Hyg. 65: 341–345.

Ho, A.Y., A.S. Lopez, M.G. Eberhart, R. Levenson, B.S. Finkel, A.J. daSilva, J.M. Roberts, P.A. Orlandi, C.C. Johnson and B.L. Herwaldt. 2002. Outbreak of cyclosporiasis associated with imported raspberries, Philadelphia, Pennsylvania. Emerg. Infect. Dis. 8: 783–788.

Hoang, L., M. Fyee, C. Ong, J. Harb, S. Champagne, B. Dixon and J. Isaac-Renton. 2005. Outbreak of cyclosporiasis in British Columbia associated with imported Thai basil. Epidemiol. Infect. 131: 23–27.

Hunter, P.R., S. Hughes, S. Woodhouse, N. Raj, Q. Syed, R.M. Chalmers, N.Q. Verlander and J. Goodacre. 2004. Health sequelae of human cryptosporidiosis in immunocompetent patients. Clin. Infect. Dis. 39: 504–510.

Idowu, O.A. and S.A. Rowland. 2006. Oral fecal parasites and personal hygiene of food handlers in Abeokuta, Nigeria. Afr. Health Sci. 6(3): 160–164.

Insulander, M., B. Svenungsson, M. Lebbad, L. Karlsson and B. de Jong. 2010. A foodborne outbreak of *Cyclospora* infection in Stockholm, Sweden. Foodborne Pathog. Dis. 7: 1585–1587.

Jarroll, E.A., A.K. Bingham and E.A. Meyer. 1981. Effect of chlorine on *Giardia lamblia* cyst viability. Appl. Environ. Microbiol. 41: 483–487.

Jones, J.L., V. Dargelas, J. Roberts, C. Press, J.S. Remington and J.G. Montoya. 2009. Risk factors for *Toxoplasma gondii* infection in the United States. Clin. Infect. Dis. 49: 878–884.

Karabiber, N. and F. Aktas. 1991. Foodborne giardiasis. Lancet 377: 376–377.

Karanis, P., C. Kourenti and H. Smith. 2007. Waterborne transmission of protozoan parasites: a worldwide review of outbreaks and lessons learnt. J. Water Health 5: 1–38.

Katz, D., S. Kumar, J. Malecki, M. Lowdermilk, E.H. Koumans and R. Hopkins. 1999. Cyclosporiasis associated with imported raspberries, Florida, (1996). Public Health Rep. 114: 427–438.

Khan, A., J.P. Dubey, C. Su, J.W. Ajioka, B.M. Rosenthal and L.D. Sibley. 2011. Genetic analyses of atypical *Toxoplasma gondii* strains reveals a fourth clonal lineage in North America. Int. J. Parasitol. 41: 645–655.

Kijlstra, A. and E. Jongert. 2008. Control of the risk of human toxoplasmosis transmitted by meat. Int. J. Parasitol. 38: 1359–1370.

Kniel, K.E., D.S. Lindsay, S.S. Sumner, C.R. Hackney, M.D. Pierson and J.P. Dubey. 2002. Examination of attachment and survival of *Toxoplasma gondii* oocysts on raspberries and blueberries. J. Parasitol. 88: 790–793.

Kniel, K.E., A.E.H. Shearer, J.L. Cascarino, G.C. Wilkins and M.C. Jenkins. 2007. High hydrostatic pressure and UV light treatment of produce contaminated with *Eimeria acervulina* as a *Cyclospora cayetanensis* surrogate. J. Food Prot. 70: 2837–2842.

Kniel, K.E., S.S. Sumner, D.S. Lindsay, C.R. Hackney, M.D. Pierson, A.M. Zajac, D.A. Golden and R. Fayer. 2003. Effect of organic acids and hydrogen peroxide on *Cryptosporidium parvum* viability in fruit juices. J. Food Prot. 66: 1650–1657.

Koumans, E.H., D.J. Katz, J.M. Malecki, S. Kumar, S.P. Wahquist, M.J. Arrowood, A.W. Hightower and B.L. Herwaldt. 1998. An outbreak of cyclosporiasis in Florida in (1995): a harbinger of multistate outbreaks in 1996 and 1997. Am. J. Trop. Med. Hyg. 59: 235–242.

Lass, A., H. Pietkiewicz, B. Szostakowska and P. Myjak. 2012. The first detection of *Toxoplasma gondii* DNA in environmental fruits and vegetables samples. Eur. J. Clin. Microbiol. Infect. Dis. 31: 1101–1108.

Lévesque, B., C. Barthe, B.R. Dixon, L.J. Parrington, D. Martin, B. Doidge, J.F. Proulx and D. Murphy. 2010. Microbiological quality of blue mussels (*Mytilus edulis*) in Nunavik, Quebec: a pilot study. Can. J. Microbiol. 56: 968–977.

Lindsay, D.S., M.V. Collins, S.M. Mitchell, C.N. Welch, A.C. Rosypal, G.J. Flick, A.M. Zajac, A. Lindquist and J.P. Dubey. 2004. Survival of *Toxoplasma gondii* oocysts in eastern oysters (*Crassostrea virginica*). J. Parasitol. 90: 1054–1057.

Lindsay, D.S., D. Holliman, G.J. Flick, D.G. Goodwin, S.M. Mitchell and J.P. Dubey. 2008. Effects of high pressure processing on *Toxoplasma gondii* oocysts on raspberries. J. Parasitol. 94: 757–758.

Lindsay, D.S., K.K. Phelps, S.A. Smith, G. Flick and J.P. Dubey. 2001. Removal of *Toxoplasma gondii* oocysts from seawater by eastern oysters (*Crassostrea virginica*). J. Eukaryot. Microbiol. 48: 197S–198S.

Lopez, A.S., D.R. Dodson, M.J. Arrowood, P.A. Orlandi Jr., A.J. da Silva, J.W. Bier, S.D. Hanauer, R.L. Kuster, S. Oltman, M.S. Baldwin, K.Y. Won, E.M. Nace, M.L.Eberhard and B.L. Herwaldt. 2001. Outbreak of cyclosporiasis associated with basil in Missouri in 1999. Clin. Infect. Dis. 32: 1010–1017.

Lourenço, A.E., C.M. Uchoa and O.M. Bastos. 2004. Hospital food handlers in Niterói, RJ, Brazil: intestinal parasitism. Arch. Latinoam. Nutr. 54(4): 395–401.

Mackey-Lawrence, N.M. and W.A. Petri Jr. 2011. Amoebic dysentery. Clin. Evid. (Online). pii: 0918.

Manuel, D.G., R. Shahin, W. Lee and M. Grmusa. 1999. The first reported cluster of food-borne cyclsporiasis in Canada. Can. J. Public Health 90: 399–402.

Markell, E.K. 1986. The 1933 Chicago outbreak of amebiasis. West. J. Med.144: 750.

Miller, M.A., W.A. Miller, P.A. Conrad, E.R. James, A.C. Melli, C.M. Leutenegger, H.A. Dabritz, A.E. Packham and D. Paradies, M. Harris, J. Ames, D.A. Jessup, K. Worcester and M.E. Grigg. 2008. Type X *Toxoplasma gondii* in a wild mussel and terrestrial carnivores from coastal California: new linkages between terrestrial mammals, runoff and toxoplasmosis of sea otters. Int. J. Parasitol. 38: 1319–1328.

Milord, F., E. Lampon-Goulet, M. St. Amour, E. Levac and D. Ramsey. 2011. *Cyclospora cayetanensis*: a description of clinical aspects of an outbreak in Quebec, Canada. Epidemiol. Infect. 27: 1–7.

Montoya, J. and O. Liesenfeld. 2004. Toxoplasmosis. Lancet 363: 965–976.

Moriarty, E.M., G. Duffy, J.M. McEvoy, S. Caccio, J.J. Sheridan, D. McDowell and I.S. Blair. 2005. The effect of thermal treatments on the viability and infectivity of *Cryptosporidium parvum* on beef surfaces. J. Appl. Microbiol. 98: 618–623.

Mørch, K., K. Hanevik, G. Rortveit, K.A. Wensaas, G.E. Eide, T. Hausken and N. Langeland. 2009. Severity of *Giardia* infection associated with post-infectious fatigue and abdominal symptoms two years after. BMC Infect. Dis. 9: 206.

Nash, T.E., D.A. Herrington, G.A. Losonsky and M.M. Levine. 1987. Experimental human infections with *Giardia lamblia*. J. Infect. Dis. 156: 974–984.

Nolla, A.C. and G.A. Cantos. 2005. Relationship between intestinal parasites in food handlers and epidemiological factors in the city of Florianópolis, Santa Catarina, Brazil. Cad. Saude Publica. 21(2): 641–645.

Núñez, F.A. and L.J. Robertson. 2012. Chapter 11: Trends in foodborne protozoan disease, including deaths. *In:* L.J. Robertson and H.V. Smith (eds.). Foodborne Protozoan Parasites. Nova Science Publishers, New York, pp. 293–307.

Nyarango, R.M., PA Aloo, E.W. Kabiru and B.O. Nyanchongi. 2008. The risk of pathogenic intestinal parasite infections in Kisii Municipality, Kenya. BMC Public Health 8: 237.

Nydam, D.V., S.E. Wade, S.L. Schaaf and H.O. Mohammed. 2001. Number of *Cryptosporidium parvum* oocysts or *Giardia* spp. cysts shed by dairy calves after natural infection. Am. J. Vet. Res. 62: 1612–1615.

Okhuysen, P.C., C.L. Chappell, J.H. Crabb, C.R. Sterling and H.L. DuPont. 1999. Virulence of three distinct *Cryptosporidium parvum* isolates for healthy adults. J. Infect. Dis. 180: 1275–1281.

Opsteegh, M., S. Prickaerts, K. Frankena and E.G. Evers. 2011. A quantitative microbial risk assessment for meatborne *Toxoplasma gondii* infection in The Netherlands. Int. J. Food Microbiol. 150: 103–114.

Ortega, Y.R., R.H. Gilman and C.R. Sterling. 1994. A new coccidian parasite (Apicomplexa: Eimeriidae) from humans. J. Parasitol. 80: 625–629.

Ortega, Y.R., C.R. Roxas, R.H. Gilman, N.J. Miller, L. Cabrera, C. Taquiri and C.R. Sterling. 1997. Isolation of *Cryptosporidium parvum* and *Cyclospora cayetanensis* from vegetables collected in markets of an endemic region in Peru. Am. J. Trop. Med. Hyg. 57: 683–686.

Puente, S., A. Morente, T. García-Benayas, M. Subirats, J. Gascón and J.M. González-Lahoz. 2006. Cyclosporiasis: a point source outbreak acquired in Guatemala. J. Travel Med. 13: 334–337.

Putignani, L., L. Mancinelli, F. Del Chierico, D. Menichella, D. Adlerstein, M.C. Angelici, M. Marangi, F. Berrilli, M. Caffara, D.A. Frangipane di Regalbono and A. Giangaspero. 2011. Investigation of *Toxoplasma gondii* presence in farmed shellfish by nested-PCR and real-time PCR fluorescent amplicon generation assay (FLAG). Experimental Parasitol. 2: 409–417.

Rendtorff, R.C. 1954. The experimental transmission of human intestinal protozoan parasites. II. *Giardia lamblia* cysts given in capsules. Am. J. Hyg. 59: 209–220.

Robertson, L.J. 2007. The potential for marine bivalve shellfish to act as transmission vehicles for outbreaks of protozoan infections in humans: a review. Int. J. Food Microbiol. 120: 201–216.

Robertson, L.J. 2009. *Giardia* and *Cryptosporidium* infections in sheep and goats: a review of the potential for transmission to humans via environmental contamination. Epidemiol. Infect. 137: 913–921.

Robertson, L.J. 2013. Chapter 5: Protozoan infections of public health importance: significance of molluscan shellfish as a transmission vehicle. *In:* J.M.M. Oliveira (ed.). Bivalve Molluscs: Pollution and Public Health. In production.

Robertson, L.J. and R.M. Chalmers. 2013. Foodborne cryptosporidiosis: is there really more in Nordic countries? Trends Parasitol. 29: 3–9.

Robertson, L.J. and R. Fayer. 2012. Chapter 2: *Cryptosporidium*. *In:* L.J. Robertson and H.V. Smith (eds.). Foodborne Protozoan Parasites. Nova Science Publishers, New York, pp. 32–64.

Robertson, L.J., T. Forberg, L. Hermansen, B.K. Gjerde, J.O. Alvsvåg and N. Langeland. 2006a. *Cryptosporidium parvum* infections in Bergen, Norway, during an extensive outbreak of waterborne giardiasis in autumn and winter 2004. Appl. Environ. Microbiol. 72: 2218–2220.

Robertson, L.J. and B. Gjerde. 2000. Isolation and enumeration of *Giardia* cysts, *Cryptosporidium* oocysts and *Ascaris* eggs from fruits and vegetables. J. Food Prot. 63: 775–778.

Robertson, L.J. and B. Gjerde. 2008. Development and use of a pepsin digestion method for analysis of shellfish for *Cryptosporidium* oocysts and *Giardia* cysts. J. Food Prot. 71: 959–966.

Robertson, L.J., B. Gjerde and A.T. Campbell. 2000. Isolation of *Cyclospora* oocysts from fruits and vegetables using lectin-coated paramagnetic beads. J. Food Prot. 63: 1410–1414.

Robertson, L.J., K. Hanevik, A.A. Escobedo, K. Mørch and N. Langeland. 2010. Giardiasis—why do the symptoms sometimes never stop? Trends Parasitol. 26: 75–82.

Robertson, L.J., L. Hermansen, B.K. Gjerde, E. Strand, J.O. Alvsvåg and N. Langeland. 2006b. Application of genotyping during an extensive outbreak of waterborne giardiasis in Bergen, Norway, during autumn and winter 2004. Appl. Environ. Microbiol. 72: 2212–2217.

Robertson, L.J. and Q. Huang. 2012. Analysis of cured meat products for *Cryptosporidium* oocysts following possible contamination during an extensive waterborne outbreak of cryptosporidiosis. J. Food Prot. 75: 982–988.

Robertson, L.J. and Y. Lim Ai Lian. 2011. Chapter 3: Waterborne and environmentally-borne giardiasis. *In:* H. Lujan and S. Svärd (eds.). Giardia—A Model Organism. 2011. Springer, Wien, NewYork. pp. 29–70.

Robertson, L.J. and H.V. Smith. 2012. Foodborne Protozoan Parasites. Nova Science Publishers, New York.

Sathyanarayanan, L. and Y. Ortega. 2006. Effects of temperature and different food matrices on *Cyclospora cayetanensis* oocyst sporulation. J. Parasitol. 92: 218–222.

Scallan, E., R.M. Hoekstra, F.J. Angulo, R.V. Tauxe, M.-A.Widdowson, S.L. Roy, J.L. Jones and P.M. Griffin. 2011. Foodborne illness acquired in the United M-A; States—major pathogens. Emerg. Infect. Dis. 17: 7–15.

Schuster, F.L. and L. Ramirez-Avila. 2008. Current world status of *Balantidium coli*. Clin. Microbiol. Rev. 21: 626–638.

Sewell, A.M. and J.M. Farber. 2001. Foodborne outbreaks in Canada linked to produce. J. Food Prot. 64: 1863–1877.

Seymour, I.J. 1999. Review of current industry practice on fruit and vegetable decontamination. CCFRA Review 14. Chipping Campden, Gloucestershire: Campden and Chorleywood Food Research Association (CCFRA).

Shields, J.M. and Y.R. Ortega. 2012. Chapter 3: *Cyclospora cayetanensis*. *In:* L.J. Robertson and H.V. Smith (eds.). Foodborne Protozoan Parasites. Nova Science Publishers, New York, pp. 65–103.

Stanley, S. 2003. Amoebiasis. Lancet. 361: 1025–1034.

Strand, E.A., L.J. Robertson, K. Hanevik, J.O. Alvsvåg, K. Mørch and N. Langeland. 2008. Sensitivity of a *Giardia* antigen test in persistent giardiasis following an extensive outbreak. Clin. Microbiol. Infect. 14: 1069–1071.

Stuart, J.M., H.J. Orr, F.G. Warburton, S. Jeyakanth, C. Pugh, I. Morris, J. Sarangi and G. Nichols. 2003. Risk factors for sporadic giardiasis: a case-control study in southwestern England. Emerg. Infect. Dis. 9: 229–233.

Takalkar, A.A., N.S. Madhekar, A.P. Kumavat and S.M. Bhayya. 2010. Prevalence of intestinal parasitic infections amongst food handlers in hotels and restaurants in Solapur city. Indian J. Public Health 54: 47–48.

Theel, E. and B.S. Pritt. 2012. Chapter 1: *Balantidium coli* and *Entamoeba histolytica*. *In:* L.J. Robertson and H.V. Smith (eds.). Foodborne Protozoan Parasites. Nova Science Publishers, New York, pp. 5–31.

Thompson, R.C.A. and P.T. Monis. 2011. Chapter 1: Taxonomy of *Giardia* species. *In:* H. Lujan and S. Svärd (eds.). Giardia—A Model Organism. 2011. Springer Wien, NewYork. pp. 1–16.

Torres-Stimming, P.A., C.C. Mundaca, M. Moran, J. Quispe, O. Colina, D.J. Bacon, A.G. Lescan, R.H. Gilman and D.L. Blazes. 2006. Outbreak of cyclosporiasis at a naval base in Lima, Peru. Am. J. Trop. Med. Hyg. 75: 546–548.

Tram, N.T., L.M.N. Hoang, P.D. Cam, P.T. Chung, M.W. Fyfe, J.L. Isaac-Renton and C.S.L. Ong. 2008. *Cyclospora* spp. in herbs and water samples collected from markets and farms in Hanoi, Vietnam. Trop. Med. Int. Health 13: 1415–1420.

Uga, S., J. Matsuo, E. Kono, K. Kimura, M. Inoue, S.K. Rai and K. Ono. 2000. Prevalence of *Cryptosporidium parvum* infection and pattern of oocyst shedding in calves in Japan. Vet. Parasitol. 94: 27–32.

Verdier, R.-I., D.W. Fitzgerald, W.D. Johnson and J.W. Pape. 2000. Trimethoprim–Sulfamethoxazole compared with Ciprofloxacin for treatment and prophylaxis of *Isospora belli* and *Cyclospora cayetanensis* infection in HIV-infected patients. A randomized, controlled trial. Ann. Intern. Med. 132: 885–888.

Walzer, P.D., F.N. Judson, K.B. Murphy, G.R. Healy, D.K. English and M.G. Schultz. 1973. Balantidiasis outbreak in Truk. Am. J. Trop. Med. Hyg. 22: 33–41.

Warren, C.A. and R.L. Guerrant. 2008. Clinical disease and pathology. *In:* R. Fayer and L. Xiao (eds.). Cryptosporidium and Cryptosporidiosis. 2nd edition. CRC Press, Boca Raton. pp. 235–253.

WHO. 2008. The Global Burden of Disease, 2004 Update. World Health Organization. WHO Press, Geneva, Switzerland.

Xiao, L. and R. Fayer. 2008. Molecular characterisation of species and genotypes of *Cryptosporidium* and *Giardia* and assessment of zoonotic transmission. Int. J. Parasitol. 38: 1239–1255.

Xiao, L. and Y. Feng. 2008. Zoonotic cryptosporidiosis. FEMS Immunol. Med. Microbiol. 52: 309–323.

CHAPTER 14

Salmonella: Disease Burden and Sources of Infection

Tine Hald[#]

Salmonella is an important cause of foodborne disease in humans throughout the world and is a significant cause of morbidity, mortality and economic loss (Roberts and Sockett 1994, Mead et al. 1999, Adak et al. 2002, Voetsch et al. 2004, Schroeder et al. 2005). Four forms of *Salmonella* infections exist: enteritis, septicaemia, enteric fever and healthy carrier. Enteritis is the most common form. The incubation period is commonly 6 to 72 hours and the symptoms include nausea, vomiting, and diarrhea. Fever, abdominal cramps, myalgias and headache are also common. *Salmonella* can induce chronic conditions, such as aseptic reactive arthritis, Reiter's syndrome, and ankylosing spondylitis (Murray et al. 2002, Montvill et al. 2012). Irritable bowel syndrome has also been identified as outcomes of salmonellosis (Doorduyn et al. 2008, Haagsma et al. 2010).

The majority of infections are transmitted from healthy carrier animals to humans via contaminated food. The main reservoir of zoonotic *Salmonella* is food animals and the main sources of infections in industrialised countries are animal derived products, notably eggs and fresh meat. In developing countries contaminated water, vegetables and human contacts are believed to contribute to a comparatively larger proportion of the human cases (Acha and Szyfres 2001).

[#] National Food Institute, Technical University of Denmark, Mørkhøj Bygade 28, Building H, 2860 Søborg, Denmark.
Email: tiha@food.dtu.dk

Despite many efforts to prevent and control foodborne salmonellosis during the last two decades, the incidence of human salmonellosis in many industrialised countries has remained high (EFSA 2010).

In this paper, the recent research on the burden and sources of human salmonellosis with a focus on the European situation, is discussed.

Disease Incidence and Burden of Human Salmonellosis

Human salmonellosis is the second ranking foodborne disease in EU and most European countries, only exceeded by campylobacteriosis. A total of 99,020 confirmed cases were reported from 27 EU Member States in 2010 through the European Surveillance System (TESSy). The EU notification rate was 21.5 cases per 100,000 population, ranging from 1.9 in Portugal to 91.1 confirmed cases per 100,000 population in Slovakia. As in previous years, *S.* Enteritidis and *S.* Typhimurium were the most frequently reported serovars (67.4% of all reported cases where the information on serovars was provided) (EFSA 2012a). This proportion is continuously decreasing (it was 75.6% in 2009). During the past five years, a decreasing trend of human salmonellosis has been observed, which particularly is explained by a decrease in the number of *S.* Enteritidis infections (EFSA 2012a).

Statistics for the incidence of human salmonellosis (and other foodborne infections) are notoriously difficult to compare between countries and sometimes even within a country, as they depend on the definition of a case, the diagnostic method used and how the information is collected and analysed. In addition, the subjective reactions of the patients and general practitioners will determine whether a case will be diagnosed and reported. There have been relatively few attempts to calibrate *Salmonella* surveillance data at national surveillance institutes, but some research groups have attempted to equate disease in the population to what appears in official statistics (Wheeler et al. 1999, Mead et al. 1999, Gallay et al. 2000, de Wit et al. 2001, Voetsch et al. 2004, van Pelt et al. 2003). The studies suggest that for every reported case of salmonellosis, between 3.8 and 38 persons in the population fell ill (Mølbak et al. 2006).

Underreporting factors for human salmonellosis in the EU MSs has recently been estimated employing information on the risk from Swedish travellers in the EU in 2009 (EFSA 2011a). The methodologies used for estimating the underreporting factors are described in (Havelaar et al. 2012). The risk of salmonellosis in returning Swedish travellers in the EU was 8.44 per 100,000 travels (95% CI: 8.18–8.71), ranging between 0.13 for travel to Finland to 94.3 for travel to Bulgaria. Based on these risk estimates, the true incidence of salmonellosis in 2009 was estimated to be 6.2 (95% CI: 1.0–19) million cases and the underreporting factor at the EU-level was 57.5 (95% CI: 9.0–172). The disease burden of salmonellosis and its sequelae was estimated in the same study to be 0.23 (0.05–0.6) million disability-adjusted life years (DALYs) per year and total annual costs were estimated at 2 (0.3–4) billion EURO (EFSA 2011a).

Another recent study had the ambitious aim of estimating the global burden of human salmonellosis (Majowicz et al. 2010). By synthesizing existing data from multiple studies and surveillance reports (including prospective population-based studies, "multiplier studies," disease notifications, and returning travellers data), incidence estimates were calculated for each of the 21 Global Burden of Disease (GBD) regions, which were then summed to provide a global number of cases. It was estimated that 61.8–131.6 million (ranging from 40/100,000 cases in Pacific and Central Asia to 3,980/100,000 in East and Southeast Asia) cases of gastroenteritis due to *Salmonella* occur globally each year, with 39,000–303,000 deaths. The study obviously involves a lot of uncertainty. For instance, for 11 of the 21 GBD regions the risk estimates applied were based on returning Swedish travellers data (Ekdahl et al. 2005) in a similar manner as the study at the EU level described above. Both studies, therefore, assume that the incidence rate in the local population is comparable with that of Swedish travellers acquiring their infection in that region. This assumption is debatable, as travellers (mostly tourists) may consume and behave very differently than the local population and may also be more susceptible to infection due to a general low exposure level in their home country. Still, the resulting estimates at least for the EU MSs are believed to provide a better reflection of the true human salmonellosis incidence than the actual reported number of cases.

Epidemiology and Disease Transmission in Humans

Salmonella can affect mammals, birds, reptiles and insects (Mølbak et al. 2006). There are numerous transmission pathways through which humans can be exposed to *Salmonella* including a wide range of domestic and wild animals and a variety of foodstuff including food both of animal and plant origin. Infected animals will carry *Salmonella* in the faeces and the usual route of infection is through faecal-oral transmission. The epidemiology of *Salmonella* is, therefore, primarily due to direct or indirect faecal contamination of live animals, food or humans. Most clinical infections are transmitted from healthy carrier animals to humans through food, but other types of transmission, e.g., direct contact with live animals and environmental transmission, can also be responsible for human disease (O'Reilly et al. 2007, Baker et al. 2007). Person-to-person transmission occurs occasionally.

Despite the many efforts to prevent and control food borne salmonellosis during the last twenty years, this pathogen continues to be one of the leading causes of human gastroenteritis. Many factors that contribute to this development exist. Among these are the adaptive ability of the pathogen itself, the changing characteristics of the population, the increasing globalisation of the food trade and changes in industrial structure and in consumer behaviours.

The occurrence of antimicrobial resistance among zoonotic *Salmonella* is an increasing problem. Antimicrobial-resistant *Salmonella* involved in disease among humans are spread mostly through foods, predominantly poultry meat, eggs, pork and beef (Hald et al. 2007). Meat is recognised as a source of human exposure to fluoroquinolone-resistant *Salmonella* spp. and high levels of extended spectrum beta lactamase (ESBL)-producing *Salmonella* have also been reported particularly in poultry meat in some EU MSs (EFSA 2012b). Such resistant strains may be associated with a significant level of human infection, depending on the pathogenicity of the strains involved and the opportunity for them to contaminate the food chain (Butaye et al. 2006, EFSA 2011b, de Jong et al. 2012, Rodriguez et al. 2012). The control of antimicrobial-resistant bacteria in food is complicated by the fact that resistance mechanisms can be located on mobile genetic elements such as plasmids and thereby be transferred between different bacterial species, for instance between generally apathogenic *E. coli* and *Salmonella* spp.

The use of antimicrobials for food animals is a major contributing factor for the selection and dissemination of resistance *Salmonella* (Emborg et al. 2007, van den Bogaard and Stobberingh 1999), but also the increasing use of antimicrobials, particularly fluoroquinolones, in humans has recently been shown to be associated with an increased incidence of infections caused by drug-resistant *Salmonella* (Koningstein et al. 2010). Compared with patients infected with susceptible *Salmonella* strains, patients with multi-drug resistant infections may be more likely to have a protracted course of disease that in addition is more severe, often requires hospitalization and lead to excess mortality (Helms et al. 2002, Varma et al. 2005).

Children and elderly people are considered to be more at risk of an infection with *Salmonella* than the average adult. It is generally accepted, that immunocompromised people suffering from underlying diseases, e.g., cancer, AIDS or chronic bowel disorders, are more prone to an infection than people in good health (Helms et al. 2006). People receiving antacids have also been reported as having an increased risk of infection due to the increased pH-level in the ventricle (Neal et al. 1994). Since the group of both elderly and chronically diseased people is growing, this may also contribute to the explanation of the continuing high level of human salmonellosis.

A rapidly growing international trade in live animals (incl. breeding animals), animal feed, raw materials and processed foods has facilitated the introduction of new *Salmonella* types in importing countries and resulted in an increasing length and complexity of the food chain. Concurrently, there has been an increase in the consolidation of food industries, including the primary production and mass distribution. This trend towards greater geographic distribution of products from large centralised food processors carries a risk for more widespread outbreaks affecting more people. The dissemination of *S.* Enteritidis in the table-egg industry (Thorns 2000, EFSA 2010) and the occurrence of multistate and international foodborne outbreaks (Isaacs et al. 2005, Werber et al. 2005, Pezzoli et al. 2008) are examples of this.

Traditionally, foods implicated in foodborne outbreaks have been poultry products including eggs, red meats and unpasteurised milk. In recent years, new types of food previously thought to be safe are emerging as sources of outbreaks. These include in particular fresh produce, which may be contaminated with animal faeces during growth, harvest and distribution. In particular, alfalfa sprouts have been implicated in large multistate or international outbreaks (Mahon et al. 1997, van Beneden et al. 1999), and sprouts are recognised as a special problem because of the potential for pathogen growth during the sprouting process (EFSA 2011c).

International travel has grown rapidly during the last decades. In countries with a low prevalence of *Salmonella* in their domestic livestock and food, this fact influences the national human statistics markedly. In Sweden and Norway, for instance, it is estimated that approximately 70–80% of all human *Salmonella* infections are acquired abroad (Kapperud and Hasseltvedt 1999, Ekdahl et al. 2005, EFSA 2010). Overall, around 64% of human *Salmonella* cases in EU were reported to be acquired domestically and 8% abroad in 2008. The proportion of cases with an unknown location of origin represented around 28% of confirmed cases, but in some MSs this proportion is 100%. Data on domestic versus travel-related cases are, therefore, often incomplete, but should to the extent possible be accounted for, as a high proportion of travel-associated cases is likely to reduce the expected effect of national intervention strategies.

Classification and Subtypes

Currently the genus *Salmonella* is divided into two species: *S.* enterica and *S.* bongori. The species *S.* enterica consists of six subspecies: *S.* enterica, *S.* salamae, *S.* arizonae, *S.* diarizonae, *S.* houtenae and *S.* indica, whereas no subspecies has been assigned to *S.* bongori (Su and Chiu 2007). Within the species and subspecies, more than 2,600 different serovars have been identified (Guibordenche et al. 2010). The vast majority of zoonotic serovars associated with human illness belong to *S.* enterica ssp. enterica.

All *Salmonella* serovars are considered potentially pathogenic for humans, but the degree of host adaptation varies, which affects the pathogenicity. Some serovars: *S.* Typhi, *S.* Paratyphi and *S.* Sendai, are highly adapted to man (Mølbak et al. 2006). They cause severe systemic illness in humans characterised by fever and abdominal symptoms (enteric/(para) typhoid fever (Miller et al. 1995). These serovars are usually not pathogenic to animals and not considered to have a zoonotic potential.

Non-typhoid, ubiquitous serovars, such as *S.* Typhimurium and *S.* Infantis, affect a wide range of animals and humans. Although these serovars in principle are non-host-adapted, strong associations between certain serovars or subtypes within a serovar and a given animal reservoir may occur, e.g., *S.* Enteridis in laying hens. In contrast, there exist a group of serovars that are highly adapted to an animal host, e.g., *S.* Cholerasuis in pigs, *S.* Dublin in cattle, *S.* Abortus-ovis in sheep, and *S.* Gallinarum in poultry. These serovars infect humans only occasionally, where they may produce no, mild or serious disease (Acha and Szyfres 2001, Mølbak et al. 2006). The non-host-adapted serovars are those with principal zoonotic significance and the ability of these to infect animals and eventually infect humans via food seems to vary (Hald et al. 2006, Pires and Hald 2010).

A subtyping system based on lysis of *Salmonella* from a panel of *Salmonella* bacteriophages (phage typing) is available. Phage typing is in some countries routinely used for the more common serovars, *S.* Enteritidis and *S.* Typhimurium (EFSA 2012a). Phage typing subdivides serovars into phage types (PT) in *S.* Enteritidis or definitive types (DT) in *S.* Typhimurium. Antimicrobial susceptibility testing may also be used to characterize *Salmonella* isolates and studies have indicated that there seems to be a strong association between some phage types and antimicrobial resistance patterns in particular *S.* Typhimurium strains (Emborg et al. 2007).

Molecular methods based on characterisation of the bacterial DNA (e.g., Pulsed Field Gel Electrophoresis (PFGE)) have a considerably higher discriminatory power than the above-mentioned phenotypic methods. The most recently developed methods typically target specific areas or genes of the genome and include Multiple-Locus Variable number tandem repeat Analysis (MLVA), where the numbers of repeat elements in specific loci are measured, and multi-locus sequence typing (MLST), where DNA sequences of specific genes are determined (Lindstedt et al. 2004, Dingle et al. 2001). Particularly for the sequence-based methods, a whole new research area has become available as the latest technology makes it possible to perform large scale sequencing at an affordable price.

Subtyping of *Salmonella* is used in epidemiological investigations. The high differentiation of strains obtained from genotyping is particularly useful in the investigation of outbreaks, as it helps to define groups of cases that have been infected with the same strain from the same source

(Mølbak et al. 2006, Torpdahl et al. 2007). Subtyping is also increasingly being applied for tracing the sources of sporadic *Salmonella* cases, i.e., for source attribution purposes (Pires et al. 2009) (see below).

Tracing the Sources of Human Salmonellosis—Source Attribution

Source attribution is defined as the partitioning of the human disease burden of one or more foodborne infections to specific sources, where the term *source* includes animal reservoirs and vehicles, e.g., foods. Source attribution methods attempt to attribute the burden of disease at the population level, and do not describe causation of disease at the individual level. Methods for source attribution of foodborne diseases include microbiological approaches, epidemiological approaches, intervention studies and expert elicitations. For a thorough review of source attribution methods see (Pires et al. 2009). In the following, recent source attribution studies for human salmonellosis are presented and discussed.

Source attribution using microbial subtyping

The microbial subtyping approach involves characterization of isolates of the pathogen by phenotypic and/or genotypic subtyping methods. The principle is to compare the distribution of subtypes in potential sources (e.g., animals and food) with the subtype distribution in humans and it is enabled by the identification of strong associations between some of the dominant subtypes and a specific reservoir or source, providing a heterogeneous distribution of subtypes among the sources. Subtypes exclusively or almost exclusively isolated from one source are regarded as indicators for the human health impact of that particular source, assuming that all human infections with these subtypes originate only from that source. Human infections caused by subtypes found in several reservoirs are then distributed proportionally to the prevalence of the indicator types (Hald et al. 2004). This approach requires a collection of temporally and spatially related isolates from various sources and humans and is consequently facilitated by an integrated foodborne disease surveillance programme focused on the collection of isolates from the major food animal reservoirs of foodborne diseases and from humans (Pires et al. 2009). The data quality and requirements are considered the biggest limitation of this approach.

The principle of comparing the distribution of subtypes found in animal and food sources with those found in humans to make inferences about the most important sources of human disease has been applied by several research groups (Van Pelt et al. 1999, Sarwari et al. 2001). A Bayesian model developed to attribute human salmonellosis in Denmark (Hald et al. 2004) has regularly being improved to include data on antimicrobial susceptibility (Hald et al. 2007) as well as data from multiple years (Pires and Hald 2010). The model attributes domestically acquired laboratory-confirmed human *Salmonella* infections caused by different *Salmonella* subtypes (e.g., serotypes, phage types, antimicrobial resistant profiles) as a function of the prevalence of these subtypes in animal and food sources and the amount of each food source consumed. This approach has proved to be a valuable tool in focusing food safety interventions to the appropriate animal reservoir in Denmark (Fig. 1) (Wegener et al. 2003, Korsgaard et al. 2009), and the model has recently been adapted to attribute human salmonellosis in other EU countries (Pires et al. 2008 (Fig. 2), Wahlström et al. 2011, Valkenburgh et al. 2007), as well as in the United States (Guo et al. 2011), New Zealand (Mullner et al. 2009), and Japan (Toyofuku et al. 2011).

The results of the study presented in Fig. 2 indicate that the most important sources of human salmonellosis varied between countries presumably reflecting different control strategies and consumption patterns. The study also showed that international travel is an important risk factor in all countries. It is emphasized that the data used in these models represents different years, so the estimates provided do not necessarily reflect the

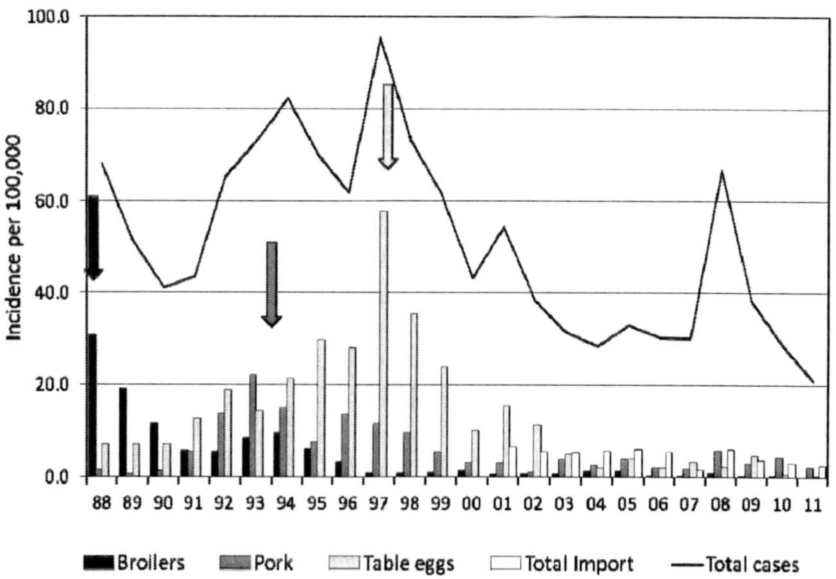

Figure 1. Effects of *Salmonella* control programmes in Denmark as estimated by application of the microbiological subtyping approach on an annual basis. The arrows indicate the initiation of a new control programme in broiler chickens, in pigs and pork, and laying hens, respectively. Remaining cases were attributable to beef, imported food products, infections acquired while travelling abroad, and unknown sources.

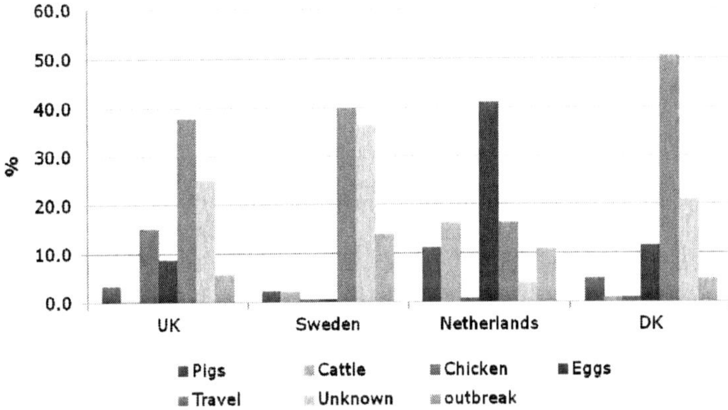

Figure 2. Estimated proportion of cases attributed to specific sources in the UK, Sweden, The Netherlands and Denmark (%) (Pires et al. 2008).

current situation. For instance in UK, a lot of effort had gone in to control *Salmonella* in broilers in recent years, so the number of cases associated with the consumption of chickens is expected to be lower than when this study was performed. This implies that if this approach is applied on a regular basis, it also allows for the analysis of the dynamic spread and trends of the most important sources of disease over time (Fig. 1).

The microbial subtyping approach has also been adapted to accommodate data from several European Union (EU) countries in a model that utilized data from the European Center for Disease Control (ECDC) and the European Food Safety Authority (EFSA) (Pires et al. 2011). The model was applied to data from 24 Member States (MSs) and attributed human sporadic salmonellosis to four animal reservoirs: pigs, broilers, layers and turkeys. Overall estimates for the EU showed that, the laying hen reservoir (eggs) was estimated to be the most important source of salmonellosis (43.8% of cases), followed by pigs (26.9%). Turkeys (4.0%) and broilers (3.4%) were estimated to be less important sources of *Salmonella*. Around 9.2% of all salmonellosis cases were reported as being travel-related, 3.6% of were reported as being part of outbreaks with unknown source, and 9% of the cases could not be attributed to any source included in the model (Table 1).

Results varied substantially according to EU region (Table 1), revealing differences in the epidemiology of *Salmonella* among regions, in the relative contribution of sources for disease and potentially in the efficiency of surveillance systems and data availability and representativeness. Layers were the most important source in Eastern, Northern and Southern Europe, contributing between 30% and 59.4% of human salmonellosis, whereas pigs were the major source of salmonellosis in Southern Europe (43.6%). Turkeys

Table 1. Proportion (%) of *Salmonella* cases attributed to food sources in the EU regions[(a)], 2007–2009, median and 95% Credibility Interval (%) (adapted from Pires et al. 2011).

	EU	Eastern EU	Northern EU	Western EU	Southern EU
Broilers	3.4 (3.1–3.7)	7.0 (6.4–7.6)	1.2 (1.0–1.4)	2.1 (1.8–2.5)	3.1 (2.6–3.6)
Pigs	26.9 (26.3–27.6)	22.7 (21.5–23.9)	10.6 (10.0–11.1)	34.1 (33.5–34.7)	43.6 (42.5–44.8)
Turkey	4.0 (3.8–4.3)	2.2 (2.0–2.5)	7.4 (6.9–8.0)	4.1 (3.8–4.3)	7.6 (6.8–8.4)
Layers	43.8 (43.2–44.4)	59.4 (58.1–60.6)	30.0 (29.4–30.6)	41.8 (41.3–42.3)	28.4 (27.5–29.3)
Outbreak[(b)]	3.6 (n.a) [(c)]	5.4 (n.a.)	4.0 (n.a.)	2.2 (n.a.)	4.2 (n.a.)
Travel	9.2 (n.a.)	0.8 (n.a.)	34.5 (n.a.)	4.8 (n.a.)	0.7 (n.a.)
Unknown	9.0 (8.7–9.3)	2.5 (1.9–3.1)	12.4 (11.8–13.0)	10.9 (10.5–11.4)	12.5 (11.4–13.5)

(a) EU regions as defined by the United Nations (Pires et al. 2010b). Eastern Europe: Czech Republic, Hungary, Poland and Slovakia. Northern Europe: Denmark, Estonia, Finland, Ireland, Latvia, Lithuania, Sweden and the United Kingdom. Southern Europe: Cyprus, Greece, Italy, Portugal, Slovenia, Spain. Western Europe: Austria, Belgium, France, Germany, Luxembourg and the Netherlands.

(b) Includes outbreaks with unknown source. Outbreak cases for which the source was identified were assigned to the correspondent animal sources.

(c) n.a.—not applicable: The proportions of outbreak- and travel-related cases were derived directly from the reported data (i.e., they were not estimated and consequently no Credibility Intervals were calculated).

and broilers contributed to varying but lower proportions of reported cases. In Northern EU, a large proportion of the reported *Salmonella* infections was reported as acquired abroad.

The methods currently applied to subtype *Salmonella* isolates, mostly phenotypic methods, have limitations in their power to identify the origin of a given isolate and consequently to assist source attribution studies. Molecular methods based on characterisation of the bacterial DNA (e.g., MLVA and MLST) have a considerably higher discriminatory power than the phenotypic methods and are increasingly being applied in outbreak investigation for pinpointing a particular source (Torpdahl et al. 2007). Still, the DNA-based methods value for source attribution of human salmonellosis has not been examined and will undoubtedly challenge the optimal strategy: "one typing method that fits all needs". Very discriminatory methods are not necessarily the best solution for source attribution, where we are not looking for a single source for a particular outbreak, but rather want to relate groups of *Salmonella* strains with particular reservoirs/sources and then attribute human sporadic cases to these sources. Such a process must allow for some genetic diversity between strains from human and food sources even if they are epidemiologically related. It is, therefore, expected that serotyping, phage typing and susceptibility testing will remain for some time a useful tool for source attribution and strengthening the global *Salmonella* surveillance in general.

Source attribution using outbreak data

Another way of trying to assess the proportion that is likely to be foodborne and the foods implicated in causing human disease, is to use data from outbreak investigations. One advantage is that these data are observed at the public health endpoint. A simple descriptive analysis or summary of outbreak data is useful for attributing illnesses to foods, but often the implicated food is a "complex" food containing several food items, where any of the items could be the actual source of the infection.

An alternative method for conducting an analysis of data from outbreak investigations was developed in the United States. In this method, food items are categorized into a hierarchical scheme, according to their ingredients (Painter et al. 2009). Foods that contain ingredients that are members of a single food category are considered 'simple foods', while foods that contain ingredients that are members of multiple food categories are considered 'complex foods'. As an example, steak is a simple food, whereas meat loaf is a complex food. Each implicated food is assigned to one or more mutually exclusive food categories, according to its ingredients. For outbreaks that have implicated a simple food item, all illnesses are attributed to that single category. For outbreaks that have implicated a complex food item, illnesses are partitioned to each category in the complex food according to the proportion of illnesses attributed to each of those categories in outbreaks caused by simple foods. As a result, illnesses in an outbreak due to a complex food item are attributed to a category in the implicated complex food, only if that category has been implicated in at least one outbreak due to a simple food. The number of illnesses attributed to each category are then summed and used to determine the percentage of disease attributed to each category (Painter et al. 2009).

This method has been adapted to attribute human salmonellosis in Europe (Pires et al. 2010). Based on foodborne outbreak data collected by EFSA for the reporting years 2005 and 2006, the authors estimated that the most important food sources were eggs (32%) and meat and poultry-meat (15%), but also that a large proportion of cases could not be linked to any source. Among illnesses that could be attributed to a source, 58% of salmonellosis cases were attributed to eggs.

Limitations of using of outbreak data for attribution include that the quality of evidence varies between data sources and classification schemes for the data are not used consistently. Also, large outbreaks, those associated with point sources, those that have short incubation periods and others that cause serious illness, are more likely to be investigated. Likewise, certain food vehicles are more likely to be associated with reported outbreaks than others, which can lead to an overestimation of the proportion of human illnesses attributed to a specific food. An important factor to consider is that illnesses included in data from outbreak investigations may not be representative of all food-borne illnesses. The fraction of the burden of foodborne disease that is associated with outbreaks, varies between pathogens but is typically smaller than that corresponding to sporadic disease. Consequently, the extrapolation of source attribution estimates obtained through an analysis of data from outbreaks to the overall burden of disease should be made with care. The authors of the above study, however, concluded that the approach seemed useful for attributing human salmonelllosis, but not campylobacteriosis. The latter because there are relatively few reported outbreaks of campylobacteriosis and the relative importance of the implicated sources seem to differ between outbreaks (e.g., water) and sporadic cases (e.g., poultry meat).

Discussion on the Important Sources

In general, the relative importance of different sources varies between countries and regions depending on differences in prevalences, consumption patterns and preferences and animal and food production systems. In EU, the overall incidence of human salmonellosis has been decreasing from 2006 to 2010, which is mainly explained by a decrease in the number of *S.* Enteritidis infections presumably as a result of an improved surveillance and control of *S.* Enteritidis in laying hens in many MSs (EFSA 2012a, Korsgaard et al. 2009). In contrast, the incidence of *S.* Typhimurium infections has changed little indicating the need for improved monitoring and control of *Salmonella* in the major sources of these infections, particular pigs and pig meat.

Despite the decreasing trend of *S.* Enteritidis cases, eggs from laying hens are still considered one of the most important source of *S.* Enteritidis infections in many EU MS. This is supported by the source attribution analysis based on subtyping data (Pires et al. 2011) and outbreak data (Pires et al. 2010). A certain proportion of human *S.* Enteritidis infections are also assessed to be attributable to broilers, particularly in countries with a high *S.* Enteritidis prevalence in broiler flocks.

Human *S.* Typhimurium infections represents between ca. 10–20% of all cases and the majority of human *S.* Typhimurium cases are likely to be associated with pig meat consumption (Pires et al. 2011). Certainly broilers and beef also contribute to these infections, but the contribution is assessed to be low due to low prevalences and/or lower impact through the food production chain. The latter is derived from the fact that some of the dominant *S.* Typhimurium phage types in broilers occur only in low frequencies in humans.

Travel appears to be an important "source" of sporadic salmonellosis particularly in Northern Europe. However, travel data is lacking from many countries, so the role may be underestimated in other parts of EU.

Only a few countries collect data on *Salmonella* in imported food systematically. Experience from Denmark suggests that the relative importance of this source increases when efforts to control domestic sources are successful. This will, however, depend on the amount and origin of the imported food.

Fresh produce is described as an increasing source of particular foodborne outbreaks in US, also for salmonellosis. In Europe, we have also seen multistate outbreaks caused by fresh produce (e.g., Pezzoli et al. 2008, EFSA 2011d), but based on the attribution study using outbreak data, fresh produce does not seem to be a major source in EU in general. The reasons for this need further investigations.

In conclusion, salmonellosis continues to represent a considerable burden in the majority of countries and is the second ranking reported foodborne and zoonotic illness in EU. There is considerable underreporting, and the true number of cases of illness is likely to be much higher than what is reported. To estimate the burden of disease and provide a baseline for evaluating the human health effects of interventions, a harmonized and active surveillance of salmonellosis in EU MSs could be established.

The successful control of *Salmonella* (and other zoonotic pathogens) requires knowledge about the most important sources or reservoirs as well as the principal routes of transmission, and these are very likely to vary between countries and regions. The identification of sources should, whenever possible, be based on several source attribution approaches, as the combined results will increase our confidence in the results. In addition to evaluating the trends and dynamics of sources to human infections, the results will support risk managers in their decision of allocating resources in order to achieve the highest possible benefit. The fact that we are living in a "global village" furthermore calls for food-safety surveillance and efforts to reduce transmission of *Salmonella* by food and other routes to be implemented on a global scale.

Overall in EU, the efforts to control *S.* Enteritidis in particularly laying hens appear to be effective. In contrast, the incidence of *S.* Tymphimurium infections remains high and the sources for the majority of these infections are likely to be associated with the pig reservoir. Further research is now needed to identify the most cost-effective control measures in pig production.

Finally, it is stressed that the successful surveillance and control of *Salmonella* (and other zoonoses) require collaboration between all experts in the food-production chain i.e. between microbiologists and epidemiologists, across veterinary and public health borders and between food-safety authorities, scientists and the industry.

References

Acha, P.N. and B. Szyfres. 2001. *Salmonella. In:* Zoonoses and communicable diseases common to man and animals, 3rd ed., P.N. Acha and B. Szyfres (eds.). Washington,D.C.: PAHO, ©2001.3 vol.—(Scientific and Technical Publication No. 580), pp. 233–246. http://www.scribd.com/doc/39011784/Zoonoses-and-Communicable-Diseases-Common-to-Man-and-Animals.

Adak, G.K., S.M. Long and S.J. O'Brien. 2002. Trends in indigenous foodborne disease and deaths, England and Wales: 1992 to 2000. Gut 51: 832–841.

Baker, M.G., C.N. Thornley, L.D. Lopez, N.K. Garrett and C.M. Nicol. 2007. A recurring salmonellosis epidemic in New Zealand linked to contact with sheep. Epidemiol. Infect. 135: 76–83.

Butaye, P., G.B. Michael, S. Schwarz, T.J. Barrett, A. Brisabois and D.G. White. 2006. The clonal spread of multidrug-resistant non-typhi Salmonella serotypes. Microbes Infect. 8: 1891–1897. Available from http://www.ncbi.nlm.nih.gov/pubmed/16714135.

de Jong, A., B. Stephan and P. Silley. 2012. Fluoroquinolone resistance of Escherichia coli and Salmonella from healthy livestock and poultry in the EU. J. Appl. Microbiol. 112: 239–245. Available from http://www.ncbi.nlm.nih.gov/pubmed/22066763.

de Wit, M.A., M.P. Koopmans, L.M. Kortbeek, W.J. Wannet, J. Vinjé, F. van Leusden, A.I. Bartelds and Y.T. van Duynhoven. 2001. Sensor, a population-based cohort study on gastroenteritis in the Netherlands: incidence and etiology. Amer. J. Epidemiol. 154: 666–674.

Dingle, K.E., F.M. Colles, D.R. Wareing, R. Ure, A.J. Fox, F.E. Bolton, H.J. Bootsma, R.J. Willems, R. Urwin and M.C. Maiden. 2001. Multilocus sequence typing system for *Campylobacter jejuni*. J. Clin. Microbiol. 39: 14–23.

Doorduyn, Y., W. van Pelt, C.L. Siezen, F. van der Horst, Y.T. van Duynhoven, B. Hoebee and R. Janssen. 2008. Novel insight in the association between salmonellosis or campylobacteriosis and chronic illness, and the role of host genetics in susceptibility to these diseases. Epidemiol. Infect. 136: 1225–1234.

EFSA. 2010. EFSA Panel on Biological Hazards (BIOHAZ); Scientific Opinion on a quantitative estimate of the public health impact of setting a new target for the reduction of *Salmonella* in laying hens. EFSA Journal 8(4): 1546. 86 pp.

EFSA. 2011a. EFSA Panel on Biological Hazards (BIOHAZ); Scientific Opinion on a quantitative estimation of the public health impact of setting a new target for the reduction of Salmonella in broilers. EFSA Journal 9(7): 2106, 94 pp.

EFSA. 2011b. EFSA Panel on Biological Hazards (BIOHAZ); Scientific Opinion on the public health risks of bacterial strains producing extended-spectrum β-lactamases and/or AmpC β-lactamases in food and food-producing animals. The EFSA Journal 9(8): 95 pp.

EFSA. 2011c. EFSA Panel on Biological Hazards (BIOHAZ); Scientific Opinion on the risk posed by Shiga toxin-producing Escherichia coli (STEC) and other pathogenic bacteria in seeds and sprouted seeds. EFSA Journal 9(11): 2424, 101 pp.

EFSA. 2011d. Shiga toxin-producing E. coli (STEC) O104:H4 2011 outbreaks in Europe: Taking Stock. EFSA Journal 9(10): 2390, 22 pp.

EFSA. 2012a. The European Union Summary Report on Trends and Sources of Zoonoses, Zoonotic Agents and Food-borne Outbreaks in 2010. EFSA Journal 10(3): 2597, 442 pp.

EFSA. 2012b. The European Union Summary Report on antimicrobial resistance in zoonotic and indicator bacteria from humans, animals and food in 2010. EFSA Journal 10(3): 233 pp.

Ekdahl K., B. de Jong, R. Wollin and Y. Andersson. 2005. Travel-associated non-typhoidal salmonellosis: geographical and seasonal differences and serotype distribution. Clin. Microbiol. Infect. 11: 138–144

Emborg, H.D., H. Vigre, V.F. Jensen, A.R. Vieira, D.L. Baggesen and F.M. Aarestrup. 2007. Tetracycline consumption and occurrence of tetracycline resistance in *Salmonella* typhimurium phage types from Danish pigs. Microb. Drug Resist. 13: 289–294.

Gallay, A., V. Vaillant, P. Bouvet, P. Grimont and J.C. Desenclos. 2000. How many foodborne outbreaks of *Salmonella* infection occurred in France in 1995? Application of the capture-recapture method to three surveillance systems. Am. J. Epidemiol. 152: 171–177.

Guibourdenche, M., P. Roggentin, M. Mikoleit, P.I. Fields, J. Brockemühl, P.A. Grimont and F.X. Weill. 2010. Supplement 2003–2007 (No. 47) to the White-Kauffmann-Le Minor scheme. Res. Microbiol. 161: 26–29.

Guo, C., R.B. Hoekstra, C.M. Schroeder, S.M. Pires, K.L. Ong, E. Hartnett, A. Naugle, J. Harman, P. Bennett, P. Cieslak, E. Scallan, B. Rose, K.G. Holt, B. Kissler, E. Mbandi, R. Roodsari, F.J. Angulo and D. Cole. 2011. Application of Bayesian Techniques to Model the Burden of Human Salmonellosis Attributable to U.S. Food Commodities at the Point of Processing: Adaptation of a Danish Model Foodborne Pathogens and Disease 8(4): 509–516.

Haagsma, J.A., P.D. Siersema, N.J. de Wit and A.H. Havelaar. 2010. Disease burden of post-infectious irritable bowel syndrome in The Netherlands. Epidemiol. Infect. 138: 1650–1656.

Hald, T., D. Vose, H.C. Wegener and T. Koupeev. 2004. A Bayesian approach to quantify the contribution of animal-food sources to human salmonellosis. Risk Anal. 24: 255–269.

Hald, T., A. Wingstrand, T. Brøndsted and D.M.A. Lo Fo Wong. 2006. Human health impact of *Salmonella* contamination in imported soybean products: A semiquantitative risk assessment. Foodborne Pathog. Dis. 3: 422–431.

Hald, T., D. Lo Fo Wong and F.M. Aarestrup. 2007. The attribution of human infections with antimicrobial resistant *Salmonella* bacteria in Denmark to sources of animal origin. Foodborne. Pathog. Dis. 4: 313–326.

Havelaar, A.H., S. Ivarsson, M. Löfdahl and M.J. Nauta. 2012. Estimating the true incidence of campylobacteriosis and salmonellosis in the European Union, 2009. Epidemiology and Infection, Published online 13 April, 2012.

Helms, M., J. Simonsen and K. Mølbak. 2006. Foodborne bacterial infection and hospitalization: a registry-based study. Clin. Infect. Dis. 42: 498–506.

Helms, M., P. Vastrup, P. Gerner-Smidt and K. Mølbak. 2002. Short and long term mortality associated with foodborne bacterial gastrointestinal infections: registry based study. BMJ. 326(7385): 357.

Isaacs, S., J. Aramini, B. Ciebin, J.A. Farrar, R. Ahmed, D. Middleton, A.U. Chandran, L.J. Harris, M. Howes, E. Chan, A.S. Pichette, K. Campbell, A. Gupta, L.Y. Lior, M. Pearce, C. Clark, F. Rodger, F. Jamieson, I. Brophy and A. Ellis. Salmonella Enteritidis PT30 outbreak investigation working group. 2005. An international outbreak of salmonellosis associated with raw almonds contaminated with a rare phage type of *Salmonella* Enteritidis. J. Food Protect. 68: 191–198.

Kapperud, G. and V. Hasseltvedt. 1999. Status og udviklingstendenser for fødevarebårne zoonoser i Norge. Zoonose-Nyt 5: 9–14.

Koningstein, M., J. Simonsen, M. Helms and K. Mølbak. 2010. The interaction between prior antimicrobial drug exposure and resistance in human *Salmonella* infections. J. Antimicrob. Chemother. 65: 1819–1825.

Korsgaard, H., M. Madsen, N.C. Feld, J. Mygind and T. Hald. 2009. The effects, costs and benefits of *Salmonella* control in the Danish table-egg sector. Epidemiol. Infect. 137: 828–836.

Lindstedt, B.-A., T. Vardund, L. Aas and G. Kapperud. 2004. Multiple-locus variable-number tandem-repeats analysis of *Salmonella* enterica subsp. enterica serovar Typhimurium using PCR multiplexing and multicolour capillary electrophoresis. J. Microbiol. Meth. 59: 163–172.

Mahon, B.E., A. Pönkä, W.N. Hall, K. Komatsu, S.E. Dietrich, A. Siitonen, G. Cage, P.S. Hayes, M.A. Lambert-Fair, N.H. Bean, P.M. Griffin and L. Slutsker. 1997. An international outbreak of *Salmonella* infections caused by alfalfa sprouts grown from contaminated seeds. J. Infect. Dis. 175: 876–882.

Majowicz, S.E., J. Musto, E. Scallan, F.J. Angulo, M. Kirk, S.J. O'Brien, T.F. Jones, A. Fazil, R.M. Hoekstra, and International Collaboration on Enteric Disease 'Burden of Illness' Studies. 2010. The global burden of nontyphoidal *Salmonella* gastroenteritis. Clin. Infect. Dis. 50: 882–889.

Mead, P.S., L. Slutsker, V. Dietz, L.F. McCaig, J.S. Bresee, C. Shapiro, P.M. Griffin and R.V. Tauxe. 1999. Food-related illness and death in the United States. Emerg. Infect. Dis. 5: 607–625.

Miller, S.I., E.L. Hohmann and D.A. Pegues. 1995. *Salmonella* (including *Salmonella* typhi). *In:* G.L. Mandell, J.E. Bennett and R. Dolin (eds.). Principles and practice of infectious diseases. Churchill Livingstone, New York. pp. 2013–2033.

Montville, T.J., K.R. Matthews and K.E. Kniel. 2012. Food Microbiology—an introduction, 3rd ed. Ch. 14. Salmonella species. ASM Press, Washington, DC. pp. 205–211.

Mullner, P., G. Jones, A. Noble, S.E. Spencer, S. Hathaway and N.P. French. 2009. Source attribution of food-borne zoonoses in New Zealand: a modified Hald model. Risk Anal. 29: 970–984.

Murray, P.R., K.S. Rosentahl, G.G. Kobayashi and M.A. Pfaller. 2002. Medical Microbiology, 4th ed. Ch. 24 Enterobacteriaceae. Mosby, St. Louis, London, Philadelphia, Sydney, Toronto. pp. 273–275.

Mølbak, K., J.E. Olsen and H.C. Wegener. 2006. *Salmonella* infections. *In:* H.P. Rieman and D.O. Cliver (eds.). Foodborne infections and intoxications. 3rd ed. Academic press, Elsevier, School of Veterinary Medicine, University of California, Davis.

Neal, K.R., S.O. Briji, R.C. Slack, C.J. Hawkey and R.F. Logan. 1994. Recent treatment with H2 antagonists and antibiotics and gastric surgery as risk factors for *Salmonella* infection. BMJ 308: 176.

O'Reilly, C.E., A.B. Bowen, N.E. Perez, J.P. Sarisky, C.A. Shepherd, M.D. Miller, B.C. Hubbard, M. Herring, S.D. Buchanan, C.C. Fitzgerald, V. Hill, M.J. Arrowood, L.X. Xiao, R.M. Hoekstra, E.D. Mintz, M.F. Lynch and Outbreak Working Group. 2007. A waterborne outbreak of gastroenteritis with multiple etiologies among resort island visitors and residents: Ohio, 2004. Clin. Infect. Dis. 44: 506–512.

Painter, J.A., T. Ayers, R. Woodruff, E. Blanton, N. Perez, R.M. Hoekstra, P.M. Griffin and C. Braden. Recipes for foodborne outbreaks: a scheme for categorizing and grouping implicated foods. Foodborne Pathog. Dis. 6: 1259–1264. doi:10.1089/fpd.2009.0350.

Pezzoli, L., R. Elson, C.L. Little, H. Yip, I. Fisher, R. Yishai, E. Anis, L. Valinsky, M. Biggerstaff, N. Patel, H. Mather, D.J. Brown, J.E. Coia, W. van Pelt, E.M. Nielsen, S. Ethelberg, E. de Pinna, M.D. Hampton, T. Peters and J. Threlfall. 2008. Packed with *Salmonella*—Investigation of an International Outbreak of *Salmonella* Senftenberg Infection Linked to Contamination of Prepacked Basil in 2007. Foodborne Pathog. Dis. 5: 661–668.

Pires, S.M., G. Nichols, H. Whalström, A. Kaesbohrer, J. David, H. Spitznagel, W. Van Pelt, A. Baumann and T. Hald. 2008. *Salmonella* source attribution in different European countries. Proceeding in FoodMicro 2008, Aberdeen, Scotland.

Pires, S.M., E.G. Evers, W. van Pelt, T. Ayers, E. Scallan, F.J. Angulo, A. Havelaar and T. Hald. 2009. Attributing the human disease burden of foodborne infections to specific sources. Foodborne Pathog. Dis. 6: 417–424.

Pires, S.M. and T. Hald. 2010. Assessing the differences in public health impact of *Salmonella* subtypes using a Bayesian microbial subtyping approach for source attribution. Foodborne Pathog. Dis. 7: 143–151.

Pires, S.M., H. Vigre, P. Makela and T. Hald. 2010. Using outbreak data for source attribution of human salmonellosis and campylobacteriosis in Europe. Foodborne Pathog. Dis. 7: 1351–1361.

Pires, S.M., L. de Knegt and T. Hald. 2011. Estimation of the relative contribution of different food and animal sources to human Salmonella infections in the European Union. Question No EFSA-Q-2010-00685. Published as an external scientific report on 28 July 2011: http://www.efsa.europa.eu/en/supporting/pub/184e.htm.

Roberts, J.A. and P.N. Sockett. 1994. The socio-economic impact of human *Salmonella* enteritidis infection. Int. J. Food Microbiol. 21: 117–129.

Rodriguez, I., S. Jahn, A. Schroeter, B. Malorny, R. Helmuth and B. Guerra. 2012. Extended-spectrum beta-lactamases in German isolates belonging to the emerging monophasic Salmonella enterica subsp. enterica serovar Typhimurium 4,[5],12:i:- European clone. J. Antimicrob. Chemother. 67: 505–508. Available from http://www.ncbi.nlm.nih.gov/pubmed/22058374.

Sarwari, A.R., L.S. Magder, P. Levine, A.M. McNamara, S. Knower, G.L. Armstrong, R. Etzel, J. Hollingsworth and J.G. Morris. 2001. Serotype distribution of *Salmonella* isolates from food animals after slaughter differs from that of isolates found in humans. J. Infect. Dis. 183: 1295–1299.

Schroeder, C.M., A.L. Naugle, W.D. Schlosser, A.T. Hogue, F.J. Angulo, J.S. Rose, E.D. Ebel, W.T. Disney, K.G. Holt and D.P. Goldman. 2005. Estimate of illness from *Salmonella* Enteritidis in eggs, United States, 2000. Emerg. Infect. Dis. 11: 113–115.

Su, L.H. and C.H. Chiu. 2007. *Salmonella*: Clinical importance and evolution of nomenclature. Chang Gung Med. J. 30(3): 210–219.

Thorns, C.J. 2000. Bacterial food-borne zoonoses. Review Science and Technology Off. Int. Epiz. 19: 226–239.

Torpdahl, M., G. Sørensen, B.-A. Lindstedt and E.M. Nielsen. 2007. Tandem repeat analysis for surveillance of human *Salmonella* Typhimurium infections. Emerg. Infect. Dis. 13: 388–395.

Toyofuku, H., S.M. Pires and T. Hald. 2011. Salmonella source attribution in Japan by a microbiological subtyping approach EcoHealth (ISSN: 1612-9202) (DOI:), 7(Suppl. 1): S22–S23.

Valkenburgh, S., R. Oosterom, O. Van Stenvers, M. Aalten, M. Braks, B. Schimmer, A. Giessen, W. Van Pelt and M. Langelaar. 2007. Zoonoses and zoonotic agents in humans, food, animals and feed in the Netherlands 2003–2006. RIVM rapportnummer:330152001 ISBN-13, 978-90-6960-184-7.

van Beneden, C.A., W.E. Keene, R.A. Strang, D.A. Werker, A.S. King, B. Mahon, K. Hedberg, A. Bell, M.T. Kelly, V.K. Balan, W.R. Mac Kenzie and D. Fleming. 1999. Multinational outbreak of *Salmonella* enterica serotype Newport infections due to contaminated alfalfa sprouts. JAMA 281: 158–162.

van den Bogaard, A.E. and E.E. Stobberingh. 1999. Antibiotic usage in animals: impact on bacterial resistance and public health. Drugs 58: 589–607.

Van Pelt, W., A.W. Van De Giessen, W.J. Leeuwen, W. Wannet, A.M. Henken and E.G. Evers. 1999. Oorsprong, omvang en kosten van humane salmonellose. Deel 1. Oorsprong van humane salmonellose met betrekking tot varken, rund, kip, ei en overige bronnen. Infectieziekten Bulletin 240–243.

Van Pelt, W., M.A.S. de Wit, W.J.B. Wannet, E.J.J. Ligtvoet, M.A. Widdowson and Y.T.H van Duynhoven. 2003. Laboratory Surveillance of Bacterial Gastroenteric Pathogens in The Netherlands, 1999–2001. Epidemiol. Infect. 130: 431–41.

Varma, J.K., K. Molbak, T.J. Barrett, J.L Beebe, T.F. Jones, T. Rabatsky-Ehr, K.E. Smith, D.J. Vugia, H.G. Chang and F.J. Angulo. 2005. Antimicrobial-resistant non-typhoidal *Salmonella* is associated with excess bloodstream infections and hospitalizations. J. Infect. Dis. 191: 554–61.

Voetsch, A.C., T.J. Van Gilder, F.J. Angulo, M.M. Farley, S. Shallow, R. Marcus, P.R. Cieslak, V.C. Deneen, R.V. Tauxe and Emerging Infections Program FoodNet Working Group. 2004. FoodNet estimate of the burden of illness caused by nontyphoidal *Salmonella* infections in the United States. Clin. Infect. Dis. 38 (Suppl. 3): S127–134.

Wegener, H.C., T. Hald, D. Lo Fo Wong, M. Madsen, H. Korsgaard, F. Bager, P. Gerner-Smidt and K. Mølbak. 2003. *Salmonella* control programs in Denmark. Emerg. Infect. Dis. 9: 774–780.

Werber, D., J. Dreesman, F. Feil, U. van Treeck, G. Fell, S. Ethelberg, A.M. Hauri, P. Roggentin, R. Prager, I.S.T. Fisher, S.C. Behnke, E. Bartelt, E. Weise, A. Ellis, A. Siitonen, Y. Andersson, H. Tschäpe, M.H. Kramer and A. Ammon. 2005. International outbreak of *Salmonella* Oranienburg due to German chocolate. BMC Infectious Diseases. 5:7 doi:10.1186/1471-2334-5-7

Wahlström, H., Y. Andersson, L. Plym-Forshell and S.M. Pires. 2011. Source attribution of human Salmonella cases in Sweden. Epidemiology and Infection 139: 1246–53.

Wheeler, J.G., D. Sethi, J.M. Cowden, P.G. Wall, L.C. Rodrigues, D.S. Tompkins, M.J. Hudson and P.J. Roderick on behalf of the Infectious Intestinal Disease Study Executive. 1999. Study of Infectious Intestinal Disease in England: rates in the community, presenting to General Practice and reported to national surveillance. Brit. Med. J. 318: 1046–1050.

CHAPTER 15

Shigella spp.

Swapan Kumar Niyogi[#]

Abstract

Shigella are one of the leading causes of bacterial foodborne illnesses and responsible for causing outbreaks. *Shigella* species are members of the family *Enterobacteriaceae* and are Gram-negative, non motile rods. Four subgroups exist based on O- antigen structure and biochemical properties; *S. dysenteriae* (Group A), *S. flexneri* (Group B), *S. boydii* (Group C) and *S. sonnei* (Group D). It has a low infectious dose which ranges from 10 to 100 organisms and therefore, can be transmitted easily through the faecal-oral route, commonly through food and water. The ability of *Shigella* to invade and colonize the intestinal epithelium is a key determinant of the disease. In all *Shigella*, a 180–220 kilobase pair plasmid carries most of the genes necessary for invasion, intracellular replication, and intracellular and intercellular spread. Clinical manifestations include mild to severe diarrhea with blood and mucus in stool, fever and abdominal pain. A combination of oral rehydration and antibiotics leads to the rapid resolution of infection. However, the emergence and dissemination of multidrug-resistant strains of *Shigella* is complicating the therapeutic management of shigellosis. Both traditional microbiological methods and molecular approaches are being used for detection, identification and analysis of *Shigella* from food. For subtyping *Shigella*, a combination of phenotypic and genotypic methods are used. Hand washing before handling food and cooking all food thoroughly before eating decreases the risk of getting *Shigella*. A comprehensive national surveillance system for foodborne diseases is essential to improve control of the diseases.

[#] National Institute of Cholera & Enteric Diseases, P33 CIT Road, Schm XM, P.O. Beliaghata, Kolkata, Pin 700010. India.
Email: niyogisk@yahoo.co.in

Introduction

Infections caused by foodborne bacterial pathogens continue to be a major public health issue around the world. The ingestion of pathogenic and virulent microorganisms generally results in foodborne illnesses affecting population in both developed and developing countries (WHO 2003). Though the global incidence of foodborne disease is difficult to estimate, it has been reported that in 2005 alone 1.8 million people died from diarrhoeal diseases and a great proportion of these cases can be attributed to contamination of food and drinking water (WHO 2007). The burden of diseases caused by food-borne pathogens remains largely unknown and are substantially underestimated, particularly in developing countries whereas there are several reported outbreaks of foodborne illnesses from industrialized countries. It was estimated that each year, major known pathogens acquired in the United States caused 9.4 million episodes of food-borne illness, resulting in 55,961 cases of hospitalization and 1,351 deaths. (Scallan et al. 2011). Bacteria are the causative agents of foodborne illness in 60% of cases requiring hospitalization (Mead et al. 1999). A variety of bacterial pathogens transmitted by food and water, including *Campylobacter* spp., *Clostridium perfringes*, *Cryptosporidium* spp., *Legionella* spp., and *Shigella* spp. and various serotypes of *Salmonella enterica* can infect humans (Tauxe et al. 2010). *Shigella* is the third most common pathogen transmitted through food. In FoodNet surveillance areas in 2008, the rate of *Shigella* food poisoning was 6.6 per 100,000 population, exceeded only by Salmonella (15.2/100,000) and Campylobacter (12.7/100,000) (CDC 2009).

The Pathogen

Shigella, the causative agent of shigellosis or "bacillary dysentery," was first discovered by a Japanese scientist Kiyoshi Shiga, during a large epidemic of dysentery in Japan in 1896 (Shiga 1898). In the years immediately following Shiga's discovery of the dysentery bacillus, similar organisms were reported by other investigators (Flexner 1900, Kruse 1900) and over the next 40 years three additional groups of related organisms were defined ultimately and taxonomically placed in the genus *Shigella* and named *S. dysenteriae*, *S. flexneri*, *S. boydii*, and *S. sonnei* to honour the lead workers, Shiga, Flexner, Boyd, and Sonne (Hale 1991).

The genus *Shigella* belongs to the tribe *Escherichia* in the family *Enterobacteriacae* and are nearly identical genetically to *Escherichia coli*. It is Gram negative, non-motile rods, 0.3 to 1 µm in diameter and 1 to 6 µm in length are non-sporeforming, facultatively anaerobic, oxidase negative, ferment glucose and other carbohydrates without producing gas. In the 1940s, *Shigella* strains were put in a different genus from *Escherichia coli* because of their medical significance. The new genus *Shigella* with four species was formally recognized, and the traditional classification follows the recommendations of Ewing (Ewing 1949). The genus *Shigella* is divided into four species: *S. dysenteriae* (Group A), *S. flexneri* (Group B), *S. boydii* (Group C) and *S. sonnei* (Group D). Each of these species, with the exception of *S. sonnei*, are subdivided into serotypes based on the O antigen component of lipopolysaccharide present on the outer membrane of the cell wall: *S. dysenteriae* (15 serotypes), *S. flexneri* (6 serotypes and two variants), *S. boydii* (20 serotypes). Serogoups A, B, and C are very similar physiologically while *S. sonnei* can be differentiated from the other serogroups by positive β-D-galactosidase and ornithine decarboxylase biochemical reactions. The identification of *Shigella* by species in the clinical laboratory is usually accomplished by slide agglutination using commercially available, absorbed rabbit antisera.

Virulence Factors and Pathogenesis of *Shigella*

Shigella infection is generally limited to the intestinal mucosa. The ability of *Shigella* to invade and colonize the intestinal epithelium is a key determinant of the disease. Cellular invasion and spread of infection is complex and a good example of multiple gene actions. The process can be arbitrarily divided into at least four stages: (1) cell invasion; (2) intracellular multiplication; (3) intra and intercellular spread; and (4) host cell killing (Sansonetti 1992, Parsot 2005). The genetic machinery responsible for invasion resides on a large (180–220 KDa) virulence plasmid. The virulence plasmid contains invasion plasmid antigen (*ipa*) genes which encode four highly immunogenic polypeptides: IpaA, IpaB, IpaC and IpaD. The organism can enter both enterocytes and M cells, which are specialized epithelial cells overlying mucosal lymphoid follicles. The infection process involves multiple steps including macropinocytosis, escape into the cytosol followed by multiplication and passage to the adjacent cells. The Ipa proteins, which mediate macropinocytosis, are encoded on the "virulence plasmid". The *Shigella* virulence gene is a complicated regulatory cascade and is not completely understood (Dorman and Porter 1998). *Shigella* invades epithelial cells by reorganizing the cytoskeleton, starting with the type III secretion system (Sansonetti and Egile 1998) apparently controlled by GTPases (Mounier et al. 1999). The ultimate result is infectious diarrhoea with loss of water and solutes, associated with abdominal cramps, fever and the loss of blood and mucus in the stools.

Toxins

The production of enterotoxins by the bacteria while they are in the small bowel probably causes the diarrhoea that preeceeds dysentery. *Shigella* strains produce 3 distinct enterotoxins: (a) chromosome encoded *Shigella* enterotoxin 1 (SHET1) which is present in all *S. flexneri* 2a (Venkatessan et al. 1991, Yavzori et al. 2002, Niyogi et al. 2004 a) but rarely found in other *Shigella* serotypes (Noriega et al. 1995), (b) *Shigella* enterotoxin 2 (SHET2) which is located on a large plasmid associated with virulence of *Shigella* (Nataro et al. 1995). SHET2 was found in many, but not all, *Shigella* of different serotypes and also in enteroinvasive *Escherichia coli* (Nataro et al. 1995, Vargas et al. 1999). The soluble toxins, SHET1 and SHET2, show significant enterotoxic activity *in vitro* when tested on rabbit ileal loops and Ussing chambers and (c) phage-borne Shiga toxin by *S. dysenteriae*.

Shiga toxin (Stx, also referred to as verotoxin or verocytotoxin) is produced by the Shiga bacillus *Shigella dysenteriae* type 1 (Sd1), which was first described by Kiyoshi Shiga in Japan in 1898 (Shiga 1898). Shiga toxin is neurotoxic, cytotoxic, and enterotoxic. The Shiga toxin inhibits protein synthesis in eukaryotic cells via inactivation of ribosomal RNA, leading to cell death. It targets glomerular epithelial cells and central nervous system microvascular endothelial cells, causing an haemolytic-uremic syndrome and seizures (Cherla et al. 2003). The toxin has a molecular weight of ca 70Kda and consists of two polypeptide subunits A and B. The B subunit mediates binding of the toxin to surface receptors on target cells. The A subunit enters cells by endocytic transport, binding to 60s ribosome and inhibitory protein and DNA synthesis, leading to cell death (O'Brien and Holmes 1987). Shiga toxins are also produced by certain strains of *Escherichia coli*. Shiga toxins are encoded in bacteriophages and thus may be transmitted horizontally.

Food/Sources/Reservoirs of Infection

Unlike other bacterial human pathogens, which may have multiple animal or environmental niches, humans and higher primates are the principal reservoirs for *Shigella*. Most cases of shigellosis result from the ingestion of faecally contaminated food and water. *Shigella* has a low infectious dose, the ingestion of 10 to 100 organisms generally produces the disease which facilitates person to person spread (Morris 1986). Transmission by house flies has also been documented (Levine and Levine 1991). Although being a foodborne pathogen, *Shigella* are not considered to have a food animal reservoir. In underdeveloped countries, harvesting of seafood from faecally contaminated waters and use of unsanitary water in food preparation may be sources of the disease.

Shigella, although classically thought of as a waterborne pathogen, has been involved in an increasing number of food-borne outbreaks in both developed and developing countries (Smith 1987). An estimated 20% of the total number of cases of shigellosis involve food as the vehicle of transmission (Mead et al. 1999). Food products associated with *Shigella* outbreaks are most commonly subjected to hand processing or preparation, limited heat treatment, or served/delivered raw to the consumer (Wu et al. 2000). Any food may, in principle, be contaminated and a wide range has been implicated as vehicles of shigellosis. *Shigella* spp. are not associated with specific foods. Some foods that have been identified in *Shigella* outbreaks include salads (potato, shrimp, tuna, chicken, turkey, macaroni, fruit, and lettuce), chopped turkey, rice balls, beans, pudding, strawberries, spinach, raw oysters, fish, luncheon meat, ground beef, milk and raw vegetables (Smith 1987). Six separately reported outbreaks in the United States and Canada in 1998 involving *S. sonnei* were traced back to parsley grown in Mexico (Naimi et al. 2003). *Shigella* is also responsible for a substantial portion of foodborne outbreaks on cruise ships. In two incidents, one in October 1989 and the other in August 1994, passengers and crew members reported having gastrointestinal symptoms. A multi drug resistant strain of *Shigella flexneri type 2a* were isolated from several ill passengers and crew members in 1989, and the source of this outbreak was identified as German potato salad. Contamination was introduced by infected food handlers, initially in the country where the food was originally prepared and second, only by a member of the galley crew on the cruise ship. In August 1994, *Shigella flexneri type 2a* was isolated from patient and the suspected source of contamination was spring onions (Rooney et al. 2004). However, sanitation violations related to food handling and communicable disease have decreased substantially over the past years (Cramer et al. 2008). Improper refrigeration of the contaminated product often contributes to illness. After the initial infection from a contaminated food, the disease readily spreads from person to person by the faecal-oral route of transmission.

Foods can be contaminated in several ways. One of the striking features about foodborne outbreaks caused by *Shigella* is that in many situations, contamination of foods may not have originated at the processing plants but rather the source can be traced to either a food handler or water containing infected faecal matter used for irrigation. Epidemiological data clearly indicate that the major cause of shigellosis in developed countries is infected food handlers.

From infected carriers, *Shigella* are spread by several routes, including food, finger, faeces and flies. Improper storage of contaminated foods is the second most common factor contributing to food-borne outbreak of shigellosis. Other contributing factors are inadequate cooking, contaminated utensils and food obtained from unsafe sources. Foods are not routinely examined for the presence of *Shigella* unless epidemiological data suggest that it may be the source of an outbreak.

Survival

These laboratory data suggest that *Shigella* may survive for extended periods in foods and may grow under selected conditions. However, in practice, *Shigella* is rarely isolated from processed products. Manufacturers do not routinely test their products, raw materials, or processing environments for *Shigella*, and there is no evidence to suggest that routine testing is warranted. Most outbreaks result from contamination of raw or previously cooked foods during preparation in the home or in foodservice establishments. Generally, the source of contamination can be traced to a carrier whose personal hygiene is poor (Bryan 1978). *Shigella* are usually considered to be relatively fragile; i.e., they do not survive well outside the host. Like most other members of the family *Enterobacteriaceae*, they are readily killed by most heat treatments employed in the processing and preparation of foods and do not survive well at pH below 4.5. However, studies on the survival of *Shigella* suggest that under certain selected conditions, they can survive for extended periods in foods (Bryan 1978). For example, *S. sonnei* and *S. flexneri* have been reported to survive at 25°C (77°F) in flour and in pasteurized whole milk for more than 170 days; in eggs, clams, and shrimp for more than 50 days; in oysters for more than 30 days; and in egg whites for more than 20 days. At lower temperatures, such as 20°C (–4°F) and 0.5°C (32.9°F), survival for longer periods has been reported (Taylor and Nakamura 1964).

Clinical Manifestations

Individual susceptibility to *Shigella* infection varies according to age, nutritional status of the person and it is also dose related. Immunocompromised persons are of particular susceptibility and there is a great likelihood of bacteraemia, with consequent high mortality. Infection is rare, however, in neonates and very young children owing to lack of receptor sites on colonic epithelial cells. Food-borne shigellosis is characterized by a high attack rate, common-source epidemiology, and short incubation periods of 12–50 hr. Symptoms may appear a day or two after consuming contaminated food. The symptoms of *Shigella* food-borne illness range from short period of mild watery diarrhoea to severe bacillary dysentery with fever, abdominal pain, blood and mucus in stools. The presence of high number of leukocytes and *Shigella* in stool specimens occurs in the early stage of the disease. Later in infection bacteria spread from cell to cell, incurring more destruction and sloughing of the colonic mucosal cells as evidenced by the presence of blood, pus and mucus in stools. Shigellosis usually resolves in 5–7 days with the notably exception of *Shigella dysenteriae,* where it can be fatal. Frequently, an asymptomatic carrier state may develop during convalescence, lasting from a few days to several months. There have been only a few studies on the prevalence of such carriers. These carriers may be important in maintaining the transmission of *Shigella* organisms in the community. *Shigella dysenteriae* type 1 produces severe disease and may be associated with life-threatening complications. The highest incidence of shigellosis occur during the warmer months of the year. Members of the genus *Shigella* are responsible for mortality and/or morbidity in high risk populations such as children under five years of age, toddlers in daycare centers, patients in custodial institutions, homosexual men and, war and femine engulfed people.

Treatment

A combination of oral rehydration and antibiotics leads to the rapid resolution of infection. Antimicrobial therapy shortens the duration of the illness and reduces the excretion of infectious organisms in faeces (Bennish and Salam 1992). However, the emergence and dissemination of multidrug-resistant strains of *Shigella* is now an increasing global health problem that is complicating the therapeutic management of cases of shigellosis (Niyogi 2007). Thus, sulphonamides, tetracycline, ampicillin, co-trimoxazole and nalidixic acid are no longer effective for the treatment of shigellosis (Niyogi 2005). Frequent shifts in the prevalences of the different serogroups and changes in the resistance patterns is also a major concern (Niyogi et al. 2001). On the basis of its efficacy, safety and relatively low cost, ciprofloxacin has been recommended by the WHO as the first-line antibiotic for treatment of shigellosis, and use of nalidixic acid is not encouraged, even in areas where it is still effective against *Shigella* (WHO 2004). However, the emergence of fluoroquinolone resistance among *Shigella* spp. has been documented (Niyogi et al. 2004b). At present, fluoroquinolone-resistant *Shigella* strains remain susceptible to some other antimicrobial agents, e.g., the macrolides (azithromycin) or third-generation cephalosporins (cefixime/ceftriaxone). These agents have been reported to be effective for the treatment of moderate-to-severe shigellosis caused by multidrug-resistant *Shigella* strains (Ashkenazi et al. 2003). Currently, there is no licensed vaccine available for *Shigella.*

Detection and Identification from Food

In foodborne outbreak caused by *Shigella* spp., the initial recognition of shigellosis usually is based on the results from examining patients sample in the clinical laboratory. Unlike clinical samples of stool, with a high number of *Shigella*, foods pose significant obstacles to successful isolation and subsequent identification of this bacterium. An accurate diagnosis of the presence of *Shigella* in suspected food samples is usually inconclusive because of the low numbers and/or stressed/injured state of *Shigella* spp. Also it is further complicated by the composition and consistency of the food sample. Traditional microbiological methods for testing foods for the presence of pathogens rely on growth in culture media, followed by isolation, and biochemical and serological identification. The methods are laborious, time consuming and lack the specificity and sensitivity needed to detect the pathogens in food. The recovery may require special enrichment procedures for successful isolation of *Shigella* (Andrews et al. 1995). Suspected food samples are grown initially in enrichment broth supplemented with 0.5 to 3 µg/ml of novobiocin. This antibiotic appears to inhibit many Gram-negative bacteria but does not affect the growth of *Shigella*. The broth culture is grown at 42°C to 44°C for 20 hrs and then streaked onto MacConkey agar plate. After growth at 35°C for 20 hrs, suspected lactose-negative colonies are tested for specific biochemical reactions and subsequently tested by group-specific antiserum.

The use of chromogenic selective agar media for target organisms, in particular from specimens with high or mixed commensal microflora has been reported to offer higher culture rates. A novel chromogenic medium (Rainbow agar) which was compared with tryptic soy agar, xylose lysine desoxycholate agar (XLD), and *Salmonella Shigella* agar for the isolation and detection of *Shigella* spp. in foods, was a much more effective medium than was XLD for the isolation of *Shigella* spp. from foods (Zhang and Lampel 2010).

Even though the culture methods are time consuming and laborious, the isolation of microorganisms allows for further subtyping analysis and for storage in culture collections.

However, rapid detection of pathogens in food is essential for ensuring the safety of food for consumers and also facilitate the prevention of spread of this pathogen (Hoorfar 2011). The advent of biotechnology has introduced new technologies that led to the emergence of rapid diagnostic methods that reduce the analysis time, showed a higher diagnostic sensitivity and a faster result as compared to the conventional culture.

Nucleic acid-based detection methods such as the polymerase chain reaction (PCR) have been developed recently for the detection of *Shigella* spp. with greater specificity and sensitivity than conventional culture methods.

PCR has become a powerful diagnostic tool for detection of microorganisms in food and clinical samples (Naraveneni and Jamil 2005). PCR is an *in vitro* amplification system where bacterial detection is possible without the need for bacteria isolation. It is considered a fast, highly sensitive and specific assay that quickly amplifies specific sequences of the target DNA from bacterial pathogens such as *Shigella* spp. Also, enrichment PCR, where a primary culture broth is tested in PCR, is the most common approach in rapid testing. Recent reports show that it is possible both to enrich a sample and enumerate by pathogen-specific real-time PCR, if the enrichment time is short. This can be especially useful in situations where food producers ask for the level of pathogen in a contaminated product. The commercially available rapid detection methods, have substantially shortened the total time of the detection assay when compared to conventional methods.

However, PCR based assays do have their disadvantages. As in any PCR system, template DNA from live and dead cells can yield positive results. But in the case of *Shigella,* these bacteria are not indigenous to any food and it can be argued that a positive result from PCR indicates contamination with human faecal matter. A limitation of PCR assay that targets a plasmid-encoded virulence factor for detecting *Shigella* in foods is that the loss of the virulence plasmid renders methods ineffective. A major disadvantage of rapid methods over culture methods is that these methods involve damaging the cells. Therefore, viable cells for confirmation and further characterization can be obtained only by carrying out repeat analyses using standard culture procedures (Feng 2007).

In a recent report *Shigella* spp. isolated from food and stool samples by a combination of PCR and culture method and characterized by PFGE demonstrated different kinds of foods as vehicle of transmission of *Shigella,* which usually escape from detection in traditional culture methods (Mokhtari et al. 2012).

Other novel molecular approaches are being used for detection, identification and analysis of foodborne pathogens. Examples include microarrays, capillary sequencing, optical mapping and real-time sequencing (Pyrosequencing). These technologies are capable of rapidly delivering massive amounts of genetic information and are becoming routine mainstays of many laboratories.

Subtyping

Subtyping methods allow for differentiation of bacterial isolates beyond the species and subspecies level. The more conventional methods for subtyping of *Shigella* include the study of the phenotypic characteristics of the microorganisms. These phenotypic methods include biotyping (Vlajinac and Krajinovic 1988), serotyping (Seltmann 1972), bacteriophage typing (Bentley et al. 1996) and colicine typing (Aoki et al. 1967).

However, these phenotypic typing methods are limited since microorganisms are capable of suddenly altering their phenotypic characteristics due to environmental changes or genetic mutations. Serotyping is not always able to provide a correct species identification, due to cross-reactions or the absence of agglutination. New serotypes are regularly discovered and are sometimes found to cross-react with *Escherichia coli* strains. In addition, the discriminatory power of phenotypic tools and serotyping is limited and requires the manipulation of the live agent. The introduction of DNA-based molecular typing methods, such as ribotyping (Mendoza et al. 1996), plasmid profile analysis (Mendoza et al. 1988), restriction fragment length polymorphism (Litwin et al. 1991, Liu et al. 1995) and pulsed-field gel electrophoresis (PFGE) (Ribot et al. 2006) has greatly improved the ability of researchers to discriminate between epidemiologically related and unrelated isolates in outbreaks. In the United States, PulseNet, a network of laboratories implicated in food-borne disease surveillance (Swaminathan et al. 2001), uses PFGE typing coupled with strict quality control procedures in order to ensure interlaboratory reproducibility, but this approach remains labor-intensive for routine clinical strain typing, so cheaper alternatives are being pursued actively. Multilocus sequence typing (MLST) is a very powerful approach and it provides a clear view of the population structure (Yang et al. 2007, Choi et al. 2007). However, it is not yet appropriate for the routine, first-line genotyping of a large number of isolates. Recently, PulseNet members acknowledged that multilocus variable-number tandem-repeat (VNTR) analysis (MLVA) is a highly promising typing tool, likely to replace PFGE in the coming years (Hyytia-Trees et al. 2006). Amplified Fragment Length Polymorphism (AFLP) typing (Herrera et al. 2002) is also another reliable typing method. Multilocus variable-number tandem-repeat analysis is a molecular tool used extensively for phylogenetic analysis (Chiou et al. 2009).

There are plenty of molecular genotyping and subtyping methods. Currently, the most widely used technique for microbial source tracking is pulsed-field gel electrophoresis (PFGE), which is based in macrorestriction analysis of bacterial DNA and is usually carried out in reference or public health laboratories (Foley et al. 2009). Ribotyping is the only molecular typing technique that has been marketed in a completely automated format and allows for the large-scale characterization and fingerprinting of strains for epidemiological investigation. Ribotyping is a variant of the restriction fragment length polymorphism (RFLP) technique, which employs probes based on rDNA. Results can be obtained within 16 hours and riboprint patterns can be stored to create a unique database (Pavlic and Griffiths 2009).

Molecular typing of food borne pathogens is used to generate approximations of population variation, definition of specific clonal lineages, comparison of isolates of similar species from different geographical locations, and changes of types within the population over time. Thus, it can be used to confirm the identity of organisms responsible for sporadic cases or foodborne outbreaks, as well as facilitating trace back investigations and food product recalls.

Molecular subtyping and characterization methods may also facilitate the development of a novel framework for tracking, preventing, and regulating foodborne bacterial diseases, which is based on evolutionary relationships and genetic characteristics.

Control and Prevention

The control of shigellosis is best achieved by preventing food and water from becoming contaminated with the microbes. Thus, prevention and control of shigellosis requires either that infected humans not be permitted to handle foods or that they practice good personal hygiene such that, even if infected, the food does not become contaminated. Certainly, exclusion of infected individuals from food handling is desirable. Food handlers should be thoroughly trained in food hygiene, and if infected, they should not be permitted to work with food until they are demonstrably free of the disease. The spread of *Shigella* from an infected person to other persons can be stopped by frequent and careful hand washing. In child care settings, child care workers changing diapers need to take particular care in disposing of diapers and cleansing the changing areas with antibacterial wipes. However, routine testing of food workers for *Shigella* is neither practical nor necessary.

The traditional approach to address the problem of microbially contaminated foods in the processing plants is to inspect the final product. There are several drawbacks to this approach. Conventional bacteriological methods are often time consuming and laborious. An alternative to end product testing is the Hazard Analysis and Critical Control Points (HACCP) approach (Stannard 1997, Jasson et al. 2010), a process control system based on identification of points at which biological, chemical and physical hazards may be introduced into a food. The HACCP system identifies certain points of the processing system that may be most vulnerable to microbial contamination and other hazards. One of the most important sanitation steps to reduce the spread of shigellosis is the treatment of municipal water supplies. Water treatment for public consumption is a safe and highly effective preventative measure that has been in place for many years. Additionally, the treatment of sewage alleviates the spread of many disease-causing organisms, including *Shigella*. For this reason, use of municipal water supplies is recommended for all food handling facilities. Improved worker and facility hygiene at picking and packing facilities is a major step in preventing shigellosis caused by contaminated produce.

Basic food safety precautions and regular drinking water treatment prevent shigellosis. Products suspected of foodborne illness should be analyzed for *Shigella*. Prompt treatment with effective antimicrobial agents may shorten the duration of clinical symptoms and carriage and reduce the spread of infection. US government regulatory departments and agencies have implemented programmes to test imported and domestic products for the presence of selected bacterial agents as a measure to control the introduction of contaminated foods to the consumer.

Future Trends

Many factors have contributed to the increase in food borne diseases, such as industrialization, increase in international trade, globalization of food trade, increased travel, etc. Aggressive efforts are needed to identify, assess, and control microbiological hazards associated with each segment of the food production system. Safe food handling must be a national priority. Consumers have an important role to play in reducing their risk of foodborne illness. It may be noted that *Shigella*, like other bacteria responsible for food and waterborne diseases, has been classified as a potential biological threat agent due to its infection route and environmental stability (Rotz et al. 2002). Reduction of foodborne shigellosis may depend on the education of the employees and the consumers as to proper personal hygiene. A comprehensive national surveillance system for foodborne diseases is essential to improve control of the diseases.

References

Andrews, W.H., G.A June and P.S. Sherrod. 1995. *Shigella* in *Bacteriological Analytical Manual* 8th ed. AOAC International, Gathersburg. MD, 6.01–6.06.

Aoki, Y.T., T. Naito, S. Matsuo, N. Fujise, A. Aikeda, K. Miura and Y. Yakushili. 1967. Colicine typing of *Shigella sonnei*. Jpn J. Microbiol. 11: 73–85.

Ashkenazi, S., I. Levy, V. Kazaronovski and Z. Samra. 2003. Growing antimicrobial resistance of *Shigella* isolates. J. Antimicrob. Chemother. 51: 427–429.

Bennish, M.A. and M.A. Salam. 1992. Rethinking options for the treatment of shigellosis. J. Antimicrob. Chemother. 30: 243–247.

Bentley, C.A., J.A. Frost and B. Rowe. 1996. Phage typing and drug resistance of *Shigella sonnei* isolated in England and Wales. Epidemiol. Infect. 116: 295–302.

Bryan, F.L. 1978. Factors that contribute to outbreaks of foodborne disease. J. Food Prot. 41: 816.

Centers for Disease Control and Prevention. 2009. Preliminary FoodNet Data on the Incidence of Infection with Pathogens Transmitted Commonly Through Food—10 States, 2008. MMWR. 58(13): 333–337.

Cherla, R.P., SY. Lee and V.L. Tesh. 2003. Shiga toxins and apoptosis. FEMS Microbiol. Lett. 228: 159–166.

Chiou, C.S., H. Watanabe, Y.W. Wang, W.L. Wang, J. Terajima, K.L. Thong, D.C. Phung and S.K. Tung. 2009. Utility of multilocus variable-number tandem-repeat analysis as a molecular tool for phylogenetic analysis of *Shigella sonnei*. J. Clin. Microbiol. 47: 1149–1154.

Choi, S.Y., J. Yoon-Seong, J.H. Lee, B. Choi, S.H. Moon, L. Seidlein, J.D. Clemens, G. Dougan, J. Wain, J. Yu, J.C. Lee, S.Y. Seol, B.K. Lee, J.H. Song, M. Song, C. Czerkinsky, J. Chun and D.W. Kim. 2007. Multilocus sequence typing analysis of *Shigella flexneri* isolates collected in Asian countries J. Med. Microbiol. 56: 1460–1466.

Cramer, E.H., C.J. Blanton and C. Otto. 2008. Shipshape: sanitation inspections on cruise ships, 1990–2005, Vessel Sanitation Program, Centers for Disease Control and Prevention. J. Environ. Health. 70: 15–21.

Dorman, C.J. and M.E. Porter. 1998. The *Shigella* virulence regulatory cascade: a paradigm of bacterial gene control mechanisms. Mol. Microbiol. 29: 677–684.

Ewing, W.H. 1949. *Shigella* nomenclature. J. Bacteriol. 57: 633–638.

Feng, P. 2007. Rapid methods for the detection of foodborne pathogens: current and next-generation technologies. *In:* M.P. Doyle and L.R. Beuchat (eds.). Food microbiology, fundamentals and frontiers, 3rd edn. ASM Press, Washington, D.C. pp. 911–934.

Flexner, S. 1900. On the etiology of tropical dysentery. Bull. Johns Hopkins Hosp. 11: 231–242.

Foley, S.L., A.M. Lynne and R. Nayak. 2009. Molecular typing methodologies for microbial source tracking and epidemiological investigations of Gram-negative bacterial foodborne pathogens. Infect. Genet. Evol. 9: 430–440.

Hale, T.L. 1991. Genetic basis of virulence in *Shigella* species. Microbiol. Rev. 55: 206–224.

Herrera, S., R. Cabrera, M.M. Ramirez, M.A. Usera and M.A. Echeita. 2002. Use of AFLP, plasmid typing and phenotyping in a comparative study to assess genetic diversity of *Shigella flexneri* strains. Epidemiol. Infect. 129: 445–450.

Hoorfar, J. 2011. Rapid detection, characterization, and enumeration of foodborne pathogens. Acta Pathol. Microbiol. Immunol. Scand. 119 (Suppl. 133): 1–24.

Hyytia-Trees, E., S.C. Smole, P.A. Fields, B. Swaminathan and E.M. Ribot. 2006. Second generation subtyping: a proposed PulseNet protocol for multiple-locus variable-number tandem repeat analysis of Shiga toxin-producing *Escherichia coli* O157 (STEC O157). Foodborne Pathog. Dis. 3: 118–131.

Jasson, V., I. Jacxsens, P. Luning, A. Rajkovic and M. Uyttendaele. 2010. Review. Alternative microbial methods: An overview and selection criteria. Food Microbiol. 27: 710–730.

Kruse, W. 1900. Ueberdie Ruhr als Volkskrankheit und ihrer erreger. Dtsch. Med. Wochenschr. 26: 637–639.

Levine, O.S. and M.M. Levine. 1991. Houseflies (*Musca domestica*) as mechanical vectors of shigellosis. Rev. Infect. Dis. 13: 688–696.

Litwin, C.M., A.L. Storm, S. Chipowsky and K.J. Ryan. 1991. Molecular epidemiology of *Shigella* infections: plasmid profiles, serotype correlation, and restriction endonuclease analysis. J. Clin. Microbiol. 29: 104–108.

Liu, P.Y., Y.J. Lau, B.S. Hu, J.M. Shyr, Z.Y. Shi, W.S. Tsai, Y.H. Lin and C.Y. Tseng. 1995. Analysis of clonal relationships among isolates of *Shigella sonnei* by different molecular typing methods. J. Clin. Microbiol. 33: 1779–1783.

Mead, P.S., L. Slutsker, V. Dietz, L.F. McCaig, J.S. Bresee, C. Shapiro, P.M. Griffin and R.V. Tauxe. 1999. Food-related illness and death in the United States. Emerg. Infect. Dis. 5: 607–625.

Mendoza, M.C., A.J. Gonzalez, F.J. Mendez and. C. Hardisson. 1988. Plasmid typing of *Shigella sonnei* epidemic strains and molecular relationship of their R-plasmids. Eur. J. Epidemiol. 4: 158–163.

Mendoza, M.C., M.C Martin and M.A. Gonzalez-Hevia. 1996. Usefullness of ribotyping in molecular epidemiological study of shigellosis. Epidemiol. Infect. 116: 127–135.

Mokhtari, W., S. Nasaibia, D. Majouri, A.B. Hassen, A. Gharbi and M. Aouni. 2012. Detection and characterization of *Shigella* species isolated from food and human stool samples in Nabeul, Tunisia, by molecular methods and culture techniques. J. Appl. Microbiol. 113: 209–222.

Morris, G.K. 1986. *Shigella*. *In:* D.O. Cliver and B.A. Cochrane (eds.). "Progress in Food Safety," Food Res. Inst., Univ. of Wisconsin-Madison.

Mounier, J., V. Laurent, A. Hall, P. Fort, M.F. Carlier, P.J. Sansonetti and C. Egile. 1999. Rho family GTPases control entry of *Shigella flexneri* into epithelial cells but not intracellular motility J. Cell Sci. 112: 2069–2080.

Naimi, T.S., J.H. Wicklund, S.J. Olsen, G. Krause, J.G. Wells, J.M. Bartkus, D.J. Boxrud, M. Sullivan, H. Kassenborg, J.M. Besser, E.D. Mintz, M.T. Osterholm and C.W. Hedberg. 2003. Concurrent outbreaks of *Shigella sonnei* and enterotoxigenic *Escherichia coli* infections associated with parsley: Implications for surveillance and control of foodborne illness. J. Food Prot. 66: 535–541.

Naraveneni, R. and K. Jamil. 2005. Rapid detection of food borne pathogens by using molecular techniques. J. Med. Microbiol. 54: 51–54.

Nataro, J.P., J. Seriwatana, A. Fasano, D.R. Maneval, L.D. Guers, F. Noriega, F. Dubovsky, M.M. Levine and J.G. Morris, Jr. 1995. Identification and cloning of a novel plasmid-encoded enterotoxin of enteroinvasive *Escherichia coli* and *Shigella* strains. Infect. Immun. 63: 4721–4728.

Niyogi, S.K., U. Mitra and P. Dutta. 2001. Changing pattern of serotypes and antimicrobial susceptibilities of *Shigella* species isolated from children in Calcutta. Jpn. J. Infect. Dis. 54: 121–122.

Niyogi, S.K., M. Vargas and J. Vila. 2004a. Prevalence of the sat, set and sen genes among diverse serotypes of *Shigella flexneri* strains isolated from patients with acute diarrhoea. Clin. Microbiol. Infect. 10: 574–576.

Niyogi, S.K., K. Sarkar, P. Lalmalsawma, N. Pallai and S.K. Bhattacharya. 2004b. An outbreak of bacillary dysentery caused by quinolone resistant *Shigella dysenteriae* type 1 in a northeastern state of India. J. Health Popul. Nutr. 22: 97.

Niyogi, S.K. 2005. Shigellosis. J. Microbiol. 43: 133–143.

Niyogi, S.K. 2007. Increasing antimicobial resistance—an emerging problem in the treatment of shigellosis. Clin. Microbiol. Infect. 13: 1141–1143.

Noriega, F.R., F.M. Liao, S.B. Formal, A. Fasano and M.M. Levine. 1995. Prevalence of *Shigella* enterotoxin 1 among *Shigella* clinical isolates of diverse serotypes J. Infect. Dis. 172: 1408–1410.

O'Brien A.D. and R.K Holmes. 1987. Shiga and Shiga-like toxins. Microbiol. Rev. 51: 206–220.

Parsot, C. 2005. *Shigella* spp. and enteroinvasive *Escherichia coli* pathogenicity factors. FEMS Microbiol. Lett. 252: 11–18.

Pavlic, M. and M.W Griffiths. 2009. Principles, applications, and limitations of automated ribotyping as a rapid method in food safety. Foodborne Pathog. Dis. 6: 1047–1055.

Ribot, E.M., M.A. Fair, R. Gautom, D.N. Cameron, S.B. Hunter, B. Swaminathan and T.J. Barrett. 2006. Standardization of pulsed-field gel electrophoresis protocols for the subtyping of *Escherichia coli* O157:H7, *Salmonella*, and *Shigella* for PulseNet. Foodborne Pathog. Dis. 3: 59–67.

Rooney, R.M., E.H. Cramer, S. Mantha, G. Nichols, J.K. Bartram, J.M. Farber and P.K. Ben Embarek. 2004. A review of outbreaks of foodborne disease associated with passenger ships: evidence for risk management. Publ. Health Rep. 119: 427–434.

Rotz, L.D., A.S. Khan, S.R. Lillibridge, S.M. Ostroff and J.M. Hughes. 2002. Public health assessment of potential biological terrorism agents. Emerg. Infect. Dis. 8: 225–30.

Sansonetti, P.J. 1992. Molecular and cellular biology of *Shigella flexneri* invasiveness from cell assay system to shigellosis. Curr. Top. Microbiol. Immunol. 180: 1–19.

Sansonetti, P.J. and C. Egile. 1998. Molecular basis of epithelial cell invasion by *Shigella flexneri*. Antonie Van Leeuwenhoek. 74: 191–197.

Scallan, E., R.M. Hoekstra, F.J. Angulo, R.V. Tauxe, M.A. Widdowson, S.L. Roy, J.L. Jones and P.M. Griffin. 2011. Foodborne illnesses acquired in the United States—major pathogens. Emerg. Infect. Dis. 17: 7–15.

Seltmann, G. 1972. Antigen structure of *Shigella*. Chemical basis of *Shigella flexneri* antigens. Z. Allg. Mikrobiol. 12: 497–520.

Shiga, K. 1898. Ueber den Erreger der Dysenterie in Japan. Vorlaufige Mitteilung Zentralbal Bakteriol. Microbiol. Hyg. 23: 599–600.

Shiga, K. 1898. Ueber den Dysenterie-bacillus (Bacillus dysenteriae). Zentralbl. Bakteriol. 24: 817–828.

Smith, J.L. 1987. *Shigella* as a foodborne pathogen. J. Food Prot. 50: 788–801.

Stannard, C. 1997. Development and use of microbiological criteria for foods. Food Sci. Technol. Today 11: 137–177.

Swaminathan, B., T.J. Barrett, S.B. Hunter, R.V. Tauxe and CDC PulseNet Task Force. 2001. PulseNet: the molecular subtyping network for foodborne bacterial disease surveillance, United States. Emerg. Infect. Dis. 7: 382–389.

Taylor, B.C. and M. Nakamura. 1964. Survival of *Shigella* in food. J. Hyg. 62: 303–311.

Tauxe, R.V., M.P. Doyle, T. Kuchenmuller, J. Schlundt and C.E. Stein. 2010. Evolving public health approaches to the global challenge of foodborne infections. Int. J. Food Microbiol. 139 (Suppl. 1): S 16–28

Vargas, M., J. Gascon, M.T. Jimenez De Anta and J. Vila. 1999. Prevalence of *Shigella* enterotoxins 1 and 2 among *Shigella* strains isolated from patients with travelers diarrhea. J. Clin. Microbiol. 37: 3608–3611.

Venkatessan, M., C. Fernandez-Prada, J.M. Buysse, S.B. Formal and T.L. Hale. 1991. Virulence phenotype and genetic characteristics of the T32-ISTRATI *Shigella flexneri* 2a vaccine strain. Vaccine. 9: 358–363.

Vlajinac, H. and S. Krajinovic. 1988. Biotyping of *Shigella sonnei* (a note) Acta Microbiol. Hung. 35: 77–79.

World Health Organization. 2003. Food safety: issues—public health significance of food borne illnesses. Food Safety Programme. Regional Office for the Western Pacific.

World Health Organization. 2004. International notes on antibiotics in the management of shigellosis. Wkly Epidemiol. Rec. 79: 202.

World Health Organization. 2007. Food safety & food-borne illness. Fact sheet no. 237 (reviewed March 2007). Geneva: World Health Organization.

Wu, F.M., M.P. Doyle, L.R. Beuchat, J.G. Wells, E.D. Mintz and B. Swaminathan. 2000. Fate of *Shigella sonnei* on parsley and methods of disinfection. J. Food Prot. 63: 568–572.

Yang, J., H. Nie, L. Chen, X. Zhang, F. Yang, X. Xu, Y. Zhu, J. Yu and Q. Jin. 2007. Revisiting the molecular evolutionary history of *Shigella* spp. J. Mol. Evol. 64: 71–79.

Yavzori, M.D., D. Cohen and N. Orr. 2002. Prevalence of the genes for *Shigella* enterotoxins 1 & 2 among clinical isolates of *Shigella* in Israel. Epidemiol. Infect. 128: 533–535.

Zhang, G. and K.A. Lampel. 2010. Comparison of chromogenic Biolog Rainbow agar *Shigella*/ Aeromonas with xylose lysine desoxycholate agar for isolation and detection of *Shigella* spp. from foods. J. Food. Prot. 73: 1458–1465.

CHAPTER 16

Staphylococcal Food Poisoning

Marie-Louise Danielsson-Tham[#]

Introduction

Staphylococcal food poisoning, one of the most common food borne diseases in the world, is caused by ingesting staphylococcal enterotoxins, which are produced mainly by enterotoxigenic strains of *Staphylococcus aureus* (*S. aureus*) and sometimes by other *Staphylococcus* species. Currently, 22 different staphylococcal enterotoxins have been identified (Hennekinne et al. 2012).

The *Staphylococcus* genus includes at least 47 species and 24 subspecies (Euzéby 2012), many of which are harmless and reside normally on the skin and on mucous membranes of humans, animals, and birds (Hennekinne et al. 2012). The major criterion for classification is the organism's ability to clot plasma and thus the division into coagulase positive staphylococci and coagulase negative staphylococci (Tolan 2012): the most well known species is the coagulase positive *S. aureus*.

S. aureus is an important pathogen due to a combination of toxin-mediated virulence, invasiveness, and antibiotic resistance (Le Loir et al. 2003) and expresses a wide array of cell-associated and secreted virulence factors. The bacterium is the main cause of nosocomial infections and

[#] School of Hospitality, Culinary Arts and Meal Science, Örebro University, Box 1, S-712 02 Grythyttan, Sweden.
Email: marie-louise.danielsson-tham@oru.se

community-acquired diseases and staphylococcal infections ranging from pimples and furuncles to toxic shock syndrome and sepsis (Le Loir et al. 2003). *S. aureus* is a major food borne pathogen and a frequent contaminant of foodstuffs (Cretenet et al. 2011), therefore, staphylococcal food poisoning is one of the most common food borne diseases worldwide (Hennekinne et al. 2012).

The first description of food poisoning caused by staphylococci is considered to be by Vaughan, published in 1884 (Vaughan 1884), who investigated an outbreak of cheese poisoning in Michigan, US that involved nearly three hundred cases. This investigation was completely different from modern food poisoning investigations and Vaughan states the symptoms produced by the "sick" cheese were "Dryness of the throat, nausea, vomiting, diarrhea, nervous prostration, headache, and sometimes double vision. In short, the symptoms are those of a gastro-intestinal irritant, with marked secondary effects upon the nervous system. Notwithstanding the alarming symptoms, recovery follows. I have failed to find any record of a case terminating fatally…Five samples of the cheese which has proved harmful were examined… I covered some bits of the cheese with 90 per cent. alcohol, agitated thoroughly, and filtered…I ate a piece of this alcoholic extract, about the size of a hazel-nut….in ten or fifteen minutes there were marked constriction and dryness of the fauces, much the same as that experienced while under the effect of atropia. Later there was considerable nausea. This would indicate a chemical poison and not a bacteric one. However, this chemical poison might be generated by the agency of bacteria… Microscopial examination of the opalescent drops… revealed bacterium termo and spherical bodies from ¾ to 1 micromillimeter in diameter…They may be micrococci, or the resting spores of some bacillus… The production of this poisonous material is due to the rapid growth of some bacterium." (Vaughan 1884).

However, the association of staphylococci with food borne illness was not established until 1914, when Barber (Adams and Moss 2007) identified the signs and symptoms of the disease through consuming milk contaminated with a culture of *S. aureus*. In 1930, Dack (Adams and Moss 2007) established staphylococcal food poisoning is caused by a filterable enterotoxin.

Classification

Bergey's Manual of Systematic Bacteriology (the "dictionary" for the classification of bacteria, Kloos and Schleifer 1986) places staphylococci in the family *Micrococcaceae*, which includes the genera *Micrococcus*, *Stomatococcus*, *Planococcus* and *Staphylococcus* (Kloos and Schleifer 1986). The genus *Staphylococcus* is further subdivided into 46 species and 24 subspecies (Euzéby 2012). The term *Staphylococcus* is derived from the Greek terms *staphyle* meaning a bunch of grapes and *coccus* meaning a grain or berry (Kloos and Schleifer 1986): the type species is *Staphylococcus aureus*. Staphylococci are normally divided and identified according to their potential to produce the enzyme coagulase, which has the ability to clot blood plasma. As the coagulation of plasma around an organism both allows staphylococci to multiply and protects them from immediate phagocytosis, coagulase production is an important criterion for determining whether a staphylococcus is pathogenic or not. Although the classification distinguishes between coagulase positive staphylococci (CPS) and coagulase negative staphylococci (CNS) (Tolan 2012), several species of *Staphylococcus*, including both CPS and CNS, can be pathogenic and cause staphylococcal food poisoning. Typically, characterization has been through classic phenotypic tests; however, modern characterization uses molecular biological methods. Four biotypes of *S. aureus* are recognized, but the use of biotyping is limited as nearly all strains isolated from human sources belong to biotype A and although phage typing schemes have been used with *S. aureus*, they are now of minor importance (Adams and Moss 2007).

Bacteria of the genus *Staphylococcus* are facultative anaerobic Gram-positive cocci that are non-motile, non-spore-forming, and catalase-positive. The cells are spherical, 0.5–1.0 μm in diameter, and are arranged as single, paired, or grape-like clusters. Colonies are smooth, raised, glistening, circular with entire edge, and translucent and single colonies may reach a size of 6-8 mm on nonselective media (Kloos and Schleifer 1986). Although colonial pigment varies, most strains have some colony or cell pigmentation, ranging from gray or gray-white thorough yellow to orange (Kloos and Schleifer 1986). The cell wall contains peptidoglycan and teichoic acid and is resistant to lysozyme but sensitive to lysostaphin (Le Loir et al. 2003). Members of the *S. sciuri* group are oxidase-positive as they produce the enzyme cytochrome c oxidase: this group is the only clade within the genus *Staphylococcus* to possess this gene (Kloos and Schleifer 1986).

Virulence Factors and Toxins

Staphylocooci express a wide array of cell-associated and secreted virulence factors that in combination with antibiotic resistance make it a serious pathogen capable of a wide range of infections. The chief function of the secreted factors, including various enzymes, cytotoxins, exotoxins, and exfoliate toxins, is to turn host components into nutrients for the bacteria to use for growth (Pinchuk et al. 2010). The pathogenesis of staphylococcal infection is a chain reaction dependent upon the toxin produced, virulence factors involved, and the reaction of the tissues of the host. As staphylococci invade host tissues, a large number of extracellular proteins are produced, for example hemotoxins (coagulase α, β and γ toxins), phospholipase C, metalloprotease, exfoliative toxins, enterotoxins, protein A, lipase, protease, leucocidins, staphylokinase and fibrinolysin. In terms of food poisoning, enterotoxins are the toxins of interest.

In staphylococci, the *agr* system (accessory gene regulator) is the main regulator system controlling the expression of virulence factors, and gene regulation by *agr* can be transcriptional or translational. The *agr* system works in combination with the *sar* system (staphylococcal accessory regulator) (Hennekinne et al. 2012), and another important gene is *sae* (*S. aureus* exoprotein expression). Although not all staphylococcal toxins are regulated by *agr*, at least 15 genes are under *agr* control; therefore, the *agr* system is considered a "global regulator" (Montville et al. 2012).

Enterotoxins

The staphylococcal enterotoxins (SE) belong to the family of superantigens, also called the family of pyrogenic toxins, which originate from *Staphylococcus* and *Streptococcus* species. Pyrogenic toxins include staphylococcal enterotoxons, toxin shock staphylococcal toxin (TSST), exfoliatins A and B, and streptococcus pyrogenic toxins. These toxins share the some structure, functions, and sequence similarities and display phylogenetic relationships. Although pyrogenic toxins are involved in distinct pathologies, they have common biological activities, referred to as superantigenic activity: they are pyrogenic, cause immunosuppression, and non-specific T-cell proliferation. Besides these common features, some toxins cause other symptoms, and among superantigens, only staphylococcal enterotoxins have emetic activity: superantigen and emetic activity of staphylococcal enterotoxins are two separate functions localized on separate domains (Le Loir et al. 2003).

Nomenclature, and chemical and physical properties

Since the first characterization of staphylococcal enterotoxin A (SEA) and B (SEB) in 1959 to 1960 by Bergdoll (Bergdoll et al. 1959) and Casman (Casman 1960), 22 serologically distinct staphylococcal enterotoxins (SE) and enterotoxin-like (SE*l*) toxins have been identified (Hennekinne et al. 2012). These enterotoxins are designated A-V in alphabetical order, excluding F, which was later renamed TSST1 (Hennekinne et al. 2012): several variants of C, G, I, U, and V toxins have been described (Lawrynowicz-Paciorek et al. 2007). The toxins causing emesis after oral administration in laboratory monkeys are named staphylococcal enterotoxins (SE) and referred to as SEA, SEB, SEC etcetera. The other related toxins are named enterotoxin-like toxins (SE*l*), as they do not have emetic properties (SE*l*L, SE*l*K, SE*l*Q), or they have not yet been tested (SE*l*J, SE*l*M-SE*l*P, SE*l*R, SE*l*U, SE*l*V) (Lawrynowicz-Paciorek et al. 2007, Argudin et al. 2010).

All SE and SE*l* share similar properties, in that they are short, single-chain proteins with similar three-dimensional structures, secrete in media, are soluble in water and saline solutions (Lawrynowicz-Paciorek et al. 2007, Hennekinne et al. 2012), and have molecular weights ranging from 22 to 29 kDa (Hennekinne et al. 2012). Upon hydrolysis, SE and SE*l* yield 18 amino acids, with aspartic acid, glutamic acid, lysine, and tyrosine being the most abundant (Jay et al. 2005). They are highly stable, resistant to most proteolytic enzymes, such as pepsin and trypsin, and maintain their activity in the digestive tract after ingestion. They are also resistant to chymotrypsine, rennin and papain (Le Loir et al. 2003). SE and SE*l* are also resistant to conditions that easily destroy the bacteria producing them, e.g., heat treatment and low pH (Argudin et al. 2010); thus, food could be judged as "staphylococci not detected" but still cause food poisoning due to the presence of SE.

Molecular regulation of staphylococcal enterotoxin production

Genes encoding SE and SE*l* have different genetic supports, most of which are mobile genetic elements (Le Loir et al. 2003); therefore, horizontal transfer of genes between strains is not rare (Pinchuk et al. 2010). SE and SE*l* genes can be carried by plasmids (*seb, sed, sej, ser, ses* and *set*), phages (temperate for *sea*, defective for *see, selk, selp* and *selq*), and on pathogenicity islands (SaPI) (*seb*, sec, *selk, sell* and *selq*) or genomic islands (*seg, seh, seli, selk, sell, selm, seln, selo, selp, selq, selu* and *selv*). Some of the genes encoding SE and SE*l* have been found in several genetic elements such as *seb*, reported as located on the chromosome, a plasmid or a transposon, and *sec*, reported as located on a plasmid or a pathogenicity island (Cretenet et al. 2011).

Some, but not all, SE genes are controlled by the *agr* system. The *seb, sec,* and *sed* genes are *agr* dependent, whereas, *sea* and *sej* are *agr* independent. As *agr* expression is tightly linked to *quorum sensing*, the production of *agr*-dependent SE in food is dependent on the ability of *S. aureus* to produce a high cell density (estimated 10^6 cfu/g) in the food: therefore, environmental factors play an important role in SE gene expression (Le Loir et al. 2003).

Enterotoxin production

Staphylococcal enterotoxins such as SEB, SEC and SED are mainly expressed during the transition from log-phase to stationary phase of growth, whereas, staphylococcal enterotoxins such as SEA and SEJ are expressed predominately during the log-phase. SEA production is only slightly affected by culture conditions and is directly linked to the population level (Cretenet et al. 2011). The lowest number of *S. aureus* cells required to produce a minimum level of enterotoxin (1 ng/g) necessary to cause gastrointestinal syndromes in humans appears to differ among substrates and for the particular enterotoxin. Although detectable SEA with as low as 10^4 cfu per g have been found (Jay et al. 2005), the FDA (Food and Drug Administration) suggests that *S. aureus* levels of $> 10^5$ cells per g of food may produce sufficient enterotoxins to cause illness. Although lower cell populations are sometimes implicated, the typical range is considered as 10^5 to 10^8 cells per g (Montville et al. 2012). The production of many SE, e.g., SEC, is inhibited by growth in media containing glucose, as this affects *agr*. The optimal expression of the *agr* gene is at neutral pH, however, when staphylococci grow in media containing glucose, the pH is lowered, which directly reduces *agr* expression. The expression of *sec* and other *agr* target genes is also affected and glucose reduces *sec* expression in strains lacking *agr* (Montville et al. 2012).

Staphylococcal species known to Produce Enterotoxins

Although staphylococcal food poisoning is mainly associated with *S. aureus*, many other species of staphylococci, both coagulase positive (CPS) and coagulase negative (CNS), are capable of producing enterotoxins and causing food poisoning.

CPS strains that can produce enterotoxins include *S. hyicus* and *S. intermedius*: *S. intermedius* is a known CPS pathogen (pyodermitis) in dogs (Jay et al. 2005). At least 10 CNS species produce enterotoxins, *S. caprae*, *S. chromogens*, *S. cohnii*, *S. epidermidis*, *S. haemolyticus*, *S. lentus*, *S. saprophyticus*, *S. sciuri*, *S. warneri* and *S. xylosus*, many of which have been isolated from sheep and goats (Jay et al. 2005).

Nutritional Requirements for Growth

The nutrient medium "food" is heterogeneous, as they contain infinite arrays of ingredients, and these parameters, known as intrinsic factors, can influence the growth, both positively and negatively, of a certain bacterium. The ranges presented below are under optimal conditions.

Staphylococci are typical of other Gram-positive bacteria in requiring certain organic compounds in their nutrition (Jay et al. 2005). Nutritional requirements vary among strains but most staphylococci are auxotrophic for cystein, aspartate, glutamate, and some strains are even auxotroph for valine, leucine, glycine and proline (Cretenet et al. 2011), and vitamins, especially thiamine and nicotinic acid, are required (Jay et al. 2005).

Temperature

Staphylococci are mesophilic bacteria, with a growth range between 6–48°C: the optimal temperature is around 37°C (Hennekinne et al. 2012, Adams and Moss 2007). Enterotoxins are produced between 10 and 46°C, with an optimum between 35 and 45°C (Adams and Moss 2007, Jay et al. 2005).

pH

Most staphylococcal strains grow within a pH range of 4.0 to 9.8–10, with the optimum being pH 6–7 (Hennekinne et al. 2012, Adams and Moss 2007). Although the pH range over which enterotoxin production occurs is narrower, enterotoxin production is reported at pH 5.15 in homemade mayonnaise (Jay et al. 2005).

Water activity, the effect of NaCl

Staphylococci are unique in being able to grow at lower water activity values (a_w 0.83) than any other non-halophilic bacteria. Although most staphylococci grow well between 7–10% NaCl concentrations (Jay et al. 2005) and some strains can grow in 20% NaCl (a_w 0.83), SE production decreases with increasing salt concentration (Cretenet et al. 2011). SEE production is reported in media containing 10% NaCl, a concentration corresponding to about a_w 0.92 (Hennekinne et al. 2012).

Redox potential

Staphylococci are facultative anaerobic bacteria that grow best in the presence of oxygen: under anaerobic conditions, growth is much slower. Aerated cultures produce approximately 10-times more enterotoxin than cultures incubated in an atmosphere of 95% N_2 + 5% CO_2 (Hennekinne et al. 2012).

Bacterial antagonism

In general, staphylococci do not compete well with the normal biota of most foods, especially where conditions permit the growth of large numbers of lactic acid bacteria (LAB). Bacteria known to be antagonistic to staphylococci are *Acinetobacter*, *Aeromonas*, *Bacillus*, *Pseudomonas*, *S. epidermidis*, *Enterobacteriaceae*, *Lactobacillaceae*, and enterococci (Jay et al. 2005). Mathematical modeling of the interaction between *S. aureus* and LAB indicates that the critical parameter is not pH or lactic acid production but rather the LAB population itself (Cretenet et al. 2011).

Antibiotic Resistance

The origins of antibiotic-resistant *S. aureus* trace back to 1941, when penicillin was first introduced. Since the 1970s, *S. aureus* strains resistant to penicillinase-stable penicillin (cloxacillin, dicloxacillin, methicillin, nafcillin and oxacillin) have emerged. Resistance is due to a supplemental penicillin binding protein (PBP 2a) encoded by the chromosomal *mecA* gene. These strains are called methicillin resistant *S. aureus* (MRSA) and are resistant to all beta-lactamase agents (Turnidge et al. 2008). In beta-lactamase mediated-resistance the extracellular excretion of an enzyme by *S. aureus* inactivates the antibiotic through opening the beta-lactam ring (Turnidge et al. 2008). A majority of MRSA are toxin producing strains (TSS-1, SEA,SEB, SED), but how the production of the toxin affects the immunopathogenesis of MRSA-associated diarrhea is unclear (Pinchuk et al. 2010). MRSA has become a major problem in many countries, resulting in increased morbidity, mortality and health costs.

Although the most common site for MRSA colonization is the anterior nares, MRSA can colonize other body locations, such as the hands, axillae, perineum, wounds, catheter sites, throat, genitourinary tracts, and gastrointestinal tract. MRSA colonization can persist from a few days up to several years and results from physical contact with either a colonized patient or MRSA-contaminated objects (such as dressings or medical devices) (Turnidge et al. 2008).

MRSA strains have been isolated from several production animals, including pigs, cattle and chicken. Pigs in particular and the pig farmers and their families, are colonized by MRSA and an association between the emergence of MRSA strains in pigs and the use of antibiotics in pig farming is suggested. During slaughter of MRSA-positive animals, both carcasses and the environment may become contaminated with MRSA (Rusnak et al. 2004).

Strains with increased MIC (minimum inhibitory concentration) to vancomycin have been detected in many countries since first being described in Japan in 1996. These strains are termed VISA (vancomycin-intermediate *S. aureus*) or GISA (glycopeptides-intermediate *S. aureus*), and all VISA strains are MRSA (Turnidge et al. 2008, p. 14).

Clinical Picture

Ingestion

Staphylococcal food poisoning is an intoxication that results from the consumption of foods containing sufficient amounts of preformed enterotoxins. The presentation of each symptom, incubation time, severity, and duration of illness depends on the individual characteristic of the patient. Clinical characteristics of staphylococcal food poisoning comprise acute and rapid onset of symptoms, normally 1–6 hr after the consumption of contaminated food (Rusnak et al. 2004). The initial symptoms are acute salivation, then nausea followed by violent vomiting and abdominal cramps, with or without diarrhea. Other commonly described symptoms are general weakness, headaches, and sometimes a fall in body temperature. In the most severe cases, prostration and low blood pressure are reported (Rusnak et al. 2004, Argudin et al. 2010, Hennekinne et al. 2012). The duration is generally 24 to 48 hr (Hennekinne et al. 2012).

The treatment of healthy people consists of bed rest and maintenance of fluid balance: antibiotics are not useful in this illness. Patients are not contagious and the enterotoxin cannot be transmitted from one person to another (CDC 2010). Sometimes hospitalization for intravenous therapy is required for dehydration, especially when infants, elderly or debilitated people (Lawrynowicz-Paciorek et al. 2007) are affected; however, death is rare and occurs in people most susceptible to dehydration, such as infants, the elderly and people with underlying illnesses (Hennekinne et al. 2012).

The total amount of enterotoxin needed to cause disease is less than 1 μg (Pinchuk et al. 2010). In an outbreak in USA in 1985, due to enterotoxin (SEA)-contaminated chocolate milk, the amount of toxin was reported to be 0.5 ng/mL (Evenson et al. 1988). In a large outbreak in Japan in 2000, the total intake of SEA in low-fat milk per capita was estimated at approximately 20–100 ng (Hennekinne et al. 2012).

Inhalation and ocular exposure

The inhalation of SEB results in fever and respiratory complaints, such as cough, dyspnea, and chest pain, and in severe cases pulmonary edema, shock, and death: gastrointestinal symptoms are also reported (Rusnak et al. 2004). After ocular exposure to SEB, symptoms of conjunctivitis with periocular or facial swelling are reported, with symptoms occurring within 1-6 hr after exposure (Rusnak et al. 2004). However, both the inhalation and ocular route of infection are reported after working with SEB in aerosolized form (Rusnak et al. 2004).

Toxic shock syndrome (TSS)

In the early 1980s, staphylococcal enterotoxin F (SEF) was reclassed as toxic shock syndrome toxin (TSST), when it was discovered the toxin was not an enterotoxin (Montville et al. 2012). TSST causes the disease toxic shock syndrome (TSS), which not is associated with foods. TSS is an acute and potentially fatal illness characterized by high fever, vomiting, diarrhea, renal problems, central system symptoms, diffuse erythematous rash, and desquamation of the skin from hands and feet one to two weeks after onset (Tolan et al. 2012, Montville et al. 2012). Although first described in children, TSS was most frequently associated with women who used tampons during menstruation. TSS emerged in the mid-1970s when superabsorbent tampons were introduced; the superabsorbency led to extended periods of use, which offered extended time for growth of, and toxin production by, *S. aureus* in a moist, warm, nutrient rich environment (Montville et al. 2012). However, since the early 1990s, at least half of the cases have not been associated with menstruation (Tolan et al. 2012). Some enterotoxin-producing strains produce TSST, and some symptoms of toxic shock syndrome appear to be caused by SEA, SEB, and SEC_1 (Jay et al. 2005).

Mechanisms of emesis and diarrhea

In contrast to many other bacterial enterotoxins, specific cells and receptors in the digestive system are not linked to oral intoxication by SE, and it suggested SE stimulate the vagus nerve in the abdominal viscera, which transmits a signal to the vomiting center of the brain. SE are able to penetrate the gut lining and activate local and systematic immune responses, and there it is the release of inflammatory mediators, e.g., histamine and leukotrienes, that causes vomiting. Local immune system activation could also be responsible for the gastrointestinal damage associated with SE ingestion: inflammatory changes are observed in several regions of the gastrointestinal tract, but the most severe lesions appear in the stomach and the upper part of the small intestine (Argudin et al. 2010).

Reservoirs

Humans and animals are the primary reservoirs for staphylococci of humans (Hait 2012). The largest numbers of staphylococci tend to be found near openings on the body surface, such as the anterior nares, axillae, and the inguinal and perineal areas: the number of staphylococci per square centimeter may reach 10^3–10^6 in moist habitats, and 10–10^3 in dry habits (Jay et al. 2005). The frequency of enterotoxigenic staphylococci in the nose, throat and hand lesions was investigated in 86 meat cutters and dressers by Danielsson and Hellberg (1984). Enterotoxin producing staphylococci were present in nasal swabs from 22% clinically well workers and 42% of a group with mild coryza and in throat swabs from 6% clinically well workers and 12% with mild coryza, and four of 16 superficial lesions of the hand harbored enterotoxigenic staphylococci. However, *S. aureus* is also present in food animals and dairy cattle: sheep and goats, particularly if affected by subclinical mastitis, are likely contaminants of milk. *S. aureus* can be isolated from faeces, and sporadically from a wide range of environmental locations, such as soil, marine and fresh water, plant surfaces, dust, and air (Adams and Moss 2007).

Foods Usually Involved

In general, staphylococci may be expected to exist, at least in low numbers, in all food products of animal origin or those handled by humans, unless heat-processing steps are applied to facilitate their destruction (Jay et al. 2005). Any food that provides a convenient medium for staphylococci growth and enterotoxin production may be involved in a staphylococcal food poisoning outbreak. As staphylococci grow over a wide range of temperatures and pH, the bacteria may grow in a wide assortment of foods (Pinchuk et al. 2010), and staphylococcal toxins are heat resistant and cannot be destroyed by cooking.

The foods most frequently involved in food poisoning outbreaks differ from one country to another and this is probably due to differing food habits (Hennekinne et al. 2012).

Foods frequently incriminated in staphylococcal food poisoning include meat and meat products, canned meats, poultry and egg products, unpasteurized milk, cheeses and other dairy products, salads, canned mushrooms, shrimps, seafood, bakery products particularly cream-filled pastries and cakes and sandwich fillings. Salted food products, such as, ham and smoked fish, have also been implicated, according to the capacity of staphylococci to grow at low water activity (Argudin et al. 2010, Le Loir et al. 2003, Loncarevic 2010); however, the foods do not smell bad or look spoiled, as the staphylococcal enterotoxins are odorless, colorless, and tasteless.

Risk Factors and Outbreaks

Five conditions are required to induce staphylococcal food poisoning: a source containing enterotoxin-producing staphylococci; transfer of staphylococci from source to food; food composition with favorable physicochemical characteristics for staphylococci growth and toxin production; favorable temperature and sufficient time for bacterial growth and toxin production; and, ingestion of food containing sufficient amounts of toxin to provoke symptoms (Hennekinne et al. 2012).

Food handlers carrying enterotoxin-producing staphylococci in their noses or on their hands are considered the main source of food contamination. The staphylococci are spread via direct contact through skin fragments or respiratory droplets produced when people cough or sneeze (Montvile et al. 2012, Jay et al. 2005). In foods such as raw meat, sausages, raw milk, and raw milk cheese, contamination from animal origins are more frequent. As animals with subclinical or chronic mastitis often excrete staphylococci in the milk, raw milk and unpasteurized cheeses are risk products. In addition to contamination by food handlers or animals, meat grinders, knives, storage utensils, cutting blocks, and saw blades may also introduce staphylococci into food.

As staphylococci grow over a wide range of temperatures and pH, the bacteria may grow in a wide assortment of foods (Pinchuk et al. 2010). Conditions often associated with outbreaks of staphylococcal illness are inadequate refrigeration, preparing foods too far in advance, poor personal hygiene, inadequate cooking or heating of food, and prolonged use of warming plates when serving foods (Montville et al. 2012). Therefore, a common source of outbreaks is food contaminated with enterotoxin-producing strains being left at temperatures that allow rapid growth of the bacteria (Pinchuk et al. 2010); as staphylococcal toxins are resistant to heat, they cannot be destroyed by cooking.

The best prevention is for foods to be stored at below 6°C if they are to be eaten cold, or kept at 60°C or above if they are to be eaten warm. The number of staphylococcal food poisoning cases reported is underestimated, as symptoms may not be severe and remission usually occurs within 24 to 48 hr, and most patients do not consult a physician. Many outbreaks occur at home and such small outbreaks are not reported to public health officials; thus, most reported cases are from banquets generally involving large numbers of people (Jay et al. 2005). However, staphylococcal food poisoning represents a considerable burden in terms of loss of working days and productivity, and economic losses for the food industries, catering companies and restaurants (Argudine et al. 2010).

Enterotoxin A is the most common toxin in *Staphylococcus*-related food poisoning (Pinchuk et al. 2010). In a study by Danielsson & Hellberg (1984), SEA was the dominant enterotoxin produced by staphylococci isolated from meat workers. Out of 44 enterotoxins-producing strains isolated, 34 produced SEA, 4 strains produced SEB, 4 strains produced type SEC_1, one strain produced both SEA and SEC_1, and one strain produced both SEB and SEC_1.

Some examples of outbreaks

Chocolate milk contaminated with staphylococci was stored at too high temperatures for 4 to 5 hr before pasteurization. The pasteurization killed the staphylococci but had no effect on the SE. This illustrated the importance of eliminating the contamination source during the processing and refrigeration of food and food ingredients whenever possible (Evenson et al. 1988).

One of the largest outbreaks recorded (13,420 cases) occurred in June-July 2000 in Kansai district, Japan. All cases had ingested dairy products, where the main ingredient was powdered skim milk manufactured by a factory in Hokkaido. Operations at the factory had stopped for some hours due to a power supply failure. The pasteurization process, twice at 130°C for 4 or 2 s, killed the staphylococci but did not inactivate the enterotoxin produced during the power failure. The low-fat milk contained 0.38 ng/ml of SEA and the powdered skim milk contained ca 3.7 ng/g. The average amount of SEA consumed per person was estimated as 20–100 ng (Jay et al. 2005, Asao et al. 2003).

After having lunch served at 16 elementary schools in Texas, USA, 1364 children out of 5824 became ill. The food had been prepared in a central kitchen and transported to the schools by truck. Epidemiological studies revealed 95% of the children who became ill had eaten a chicken salad. The afternoon of the day preceding the lunch, frozen chickens were boiled for 3 hrs. After cooking, the chickens were deboned, cooled to room temperature with a fan, ground into small pieces, placed into 12-inch-deep aluminum pans, and stored overnight in a walk-in refrigerator at 5.6–7.2°C (42–45°F). The next morning, the remaining ingredients of the salad were added and the mixture was blended with an electric mixer. The food was placed in thermal containers and transported to the various schools between 09.30 AM to 10.30 AM, where it was kept at room temperature until being served between 11.30 AM and noon. Bacteriological examination of the chicken salad revealed the presence of large numbers of *S. aureus*. Contamination of the chicken probably occurred when it was deboned and the chicken meat was not cooled rapidly enough because it was stored in 12-inch-deep layers. Staphylococcal growth also probably occurred during the period when the food was kept in the warm classrooms. Prevention of this incident would have entailed more rapid cooling of the chicken and adequate refrigeration of the salad from the time of preparation to its consumption (Hait 2012).

Bioterrorist attack

Staphylococcal toxins could be used as a biological agent either by contamination of food/water or by aerosolization and inhalation (CDC 2010). The staphylococcal enterotoxin SEB has been examined as a biological warfare weapon. There was particular interest in weaponizing SEB in the Cold War Era because of its stability and potential simplicity of production and dispersal. SEB could be purified and introduced into water or food systems to affect a large number of people. With SEB as a weapon, the risk of widespread mortality is low; however, it could effectively incapacitate the general population or soldiers on the front line (Pinchuk et al. 2010).

Isolation and Identification

Diagnosis of staphylococcal food poisoning is generally confirmed either by recovery of staphylococci from food remnants or by the detection of SE in food remnants. Staphylococci are heat sensitive, and in some cases, confirmation of staphylococcal food poisoning is difficult, as in heat-treated food matrices, staphylococci may be eliminated without inactivating SE. As SE are not heat sensitive, SE in food remnants need to be analyzed in cases where staphylococci have been eliminated.

Detecting staphylococci in foods

Enumerating staphylococci in foods relies on classic microbial detection and quantification of coagulase positive staphylococci on a selective medium (Le Loir et al. 2003). The most common medium is Baird-Parker (BP) medium, which was developed for the isolation of *S. aureus* (Baird-Parker 1962). The basis of the diagnostic properties of the media is that most *S. aureus* strains are capable of reducing potassium tellurite to metallic tellurium and of clearing egg yolk. On BP medium, a typical *S. aureus* colony is black, shiny, 1.0–1.5 mm in diameter, and often has a narrow, inner white margin caused by precipitation of fatty acids. The colony is surrounded by a clear zone resulting from hydrolysis of the egg-yolk protein lipovitellenin, which extends 2–5 mm into the opaque medium (Baird-Parker 1962). The appearance of colonies on BP agar gives a presumptive identification of *S. aureus* that is often confirmed by tests for the production of coagulase and/or thermostable nuclease. However, not all enterotoxins-producing staphylococci are coagulase positive; neither do all enterotoxin positive strains grow typical colonies on BP agar. In staphylococci isolated from man, only 61% of enterotoxins-producing strains are BP-positive (Danielsson and Hellberg 1977). Conversely, 25% of non-enterotoxins-producing strains present typical colonies on BP agar. The best correlation is between enterotoxin production and coagulase activity, as 82% of the enterotoxins-producing strains synthesize coagulase, as much as 7% of the coagulase negative strains can produce enterotoxin (Danielsson and Hellberg 1977).

BP agar is a selective culturing media containing some ingredients that inhibit competing microorganisms, and if staphylococci cells are injured, they may be inhibited. Therefore, it is useful to use a non-selective agar medium such as blood agar (Blood Agar Base supplemented with 5v/v% defibrinated horse or bovine blood). A non-selective medium such as blood agar can also be used for detecting staphylococci with atypical appearances on BP agar. Sometimes older techniques for demonstrating high numbers of staphylococci in a suspected food item could be useful, such as microscopic investigation of a direct smear from the food stained with methylene blue solution (Merchant and Packer 1967).

Although many attempts have been made to relate enterotoxin production to other biochemical properties of the staphylococcal cell, e.g., production of gelatinase, phosphatase, lysozyme, lecithinase, lipase and DNase production or the fermentation of various carbohydrates, has been unsuccessful. Similarly, attempts to relate enterotoxigenesis with specific bacteriophage types have also been unsuccessful (Jay et al. 2005).

SE detection

For detecting SE in food items suspected of causing poisoning, the toxin must be separated from food constituents, and then, concentrated. Two principles are used for this purpose: the selective adsorption of the enterotoxin from an extract of the food onto ion exchange resins, and the use of physical and chemical procedures for the selective removal of food constituents from the extract, leaving the enterotoxin in solution. These techniques and the concentration of the resulting products have rendered it possible to detect small amounts of enterotoxin in food (Hait 2012).

The common method for detecting SE in food is based on the use of anti-enterotoxin polyclonal or monoclonal antibodies; commercially available kits have been developed according to the different principles, these include enzyme linked immunoassay (EIA), comprising ELISA and enzyme-linked fluorescent assay (ELFA), and reversed passive latex agglutination (RPLA). However, immunoassays can generate false or incomplete diagnoses when used to detect SE in food. For example, false positive tests could occur when food components or some proteins produced by bacteria interfere. As immunological methods are based on anti-enterotoxin antibodies, purified enterotoxin is needed to raise specific antibodies; thus, unknown enterotoxins cannot be detected.

Polymerase chain reaction (PCR) is a commonly used molecular biological method that detects enterotoxins-encoding genes in strains of staphylococci isolated from contaminated foods. However, the method has two limitations: first, staphylococcal strains must be isolated from food, and second, the results give information about the presence or absence of genes encoding the SE, but does not provide information about the expression of these genes in food.

Characterization of Strains

For epidemiological investigations, it is not sufficient to isolate the same bacterium from the incriminated food and from the patient, the isolated strains need to be characterized and compared to be able to establish the route of infection. For this purpose, pulsed field gel electrophoresis (PFGE) is useful as a discriminating and typing method for staphylococcal strains (Kérouanton et al. 2007).

Patients

Individual patients are not usually tested for toxin-producing staphylococci or the actual toxin, as diagnosis of staphylococcal food poisoning is generally based only on the signs and symptoms presented by the patient. The investigation of stool or vomit for enterotoxins-producing staphylococci is usually reserved for outbreaks involving several people (CDC 2010).

Conclusion

To obtain a correct diagnosis in the investigation of food borne outbreaks, a combination of classical microbiological and immunological methods and molecular biological techniques is required.

Risk Assessment and Judgement

As staphylococci are common bacteria in food, a low degree of contamination is tolerated in most foodstuffs. Most people will not become sick through a small number of staphylococci, as there are a minimum number of staphylococci cells required to produce sufficient amounts of enterotoxin in food to cause illness. The acceptable number of staphylococci depends on the type of food and specific regulations within each country. A guideline for food being considered unfit for consumption could be $\geq$ 1000 CFU (colony forming units) per gram food. However, food contaminated with enterotoxins-producing strains should always be considered a source for staphylococcal food poisoning if it is left at temperatures that allow rapid growth of the bacteria.

There is a discussion among food hygienists to replace the quantification of staphylococci in finished food products with controlled analyses during the food processing at times when the staphylococci population is considered highest. This proposal takes into account SE can be produced and remain active in foodstuffs, whereas, the SE-producing staphylococci population has declined and may no longer be detectible when the product is released (Cretenet et al. 2011).

Merely eating a few hundreds, or even thousands of staphylococcal cells newly planted onto a cooked food will be harmless. To cause symptoms of sickness, sufficient amounts of toxin need to form in the food, and the number of organisms in the food must increase to millions per gram of food during unchilled storage.

Prevention of Staphylococcal Food Poisoning

To prevent staphylococcal food poisoning, hands should be washed with plenty of soap and warm water, and preferably rinsed in running water: nails should be kept short, unvarnished, and scrupulously clean. As some skin bacteria, e.g., staphylococci, cling to the skin surface and persist in hair follicles, pores, and crevices in the skin, it is extremely difficult to remove all these organisms through normal hand washing methods. Therefore, risky foods should never be kept longer than two hours in the critical temperature range for enterotoxin production (10–46°C).

Hair should be kept clean and well covered with a cloth, net or paper caps.

The habit of licking fingers to pick up paper or turn pages of a book is not recommended in the kitchen, nor is the use of fingers for tasting food items. Nose picking or fingering the nose may leave staphylococci or other harmful organisms on the fingers and should be followed by vigorously hand washing.

The unguarded cough or sneeze can disperse staphylococci suspended in droplets of moisture from the nose, mouth, or throat that pass infections directly to the food. Susceptible food should not be prepared if there are infected wounds or other skin infections on the hands or wrists or if there is nose or eye infection. Clean, non-suppurative abrasions should be covered with a waterproof dressing and then a disposable glove. Rapid cooling and cold storage is essential for foods not intended to be eaten immediately on arrival from an external source or after cooking. The conditions of storage should ensure foods are kept at temperatures that inhibit the growth and multiplication of staphylococci. Cool and cold storage areas should be readily accessible from the food preparation areas. The best prevention is for foods to be stored at below +6°C if they are to be eaten cold, or kept at 60°C or above if they are to be eaten warm.

References

Adams, M.R. and M.O. Moss. 2007. Food Microbiology, 3rd ed. Chaper 7.14 *Staphylococcus aureus*. RSC Publishing, Cambridge, UK. pp. 252–257.

Argudin, M.A., M.C. Mendoza and M.R. Rodicio. 2010. Food poisoning and *Staphylococcus aureus* enterotoxins. Toxins 2: 1751–1774.

Asao, T., Y. Kumeda, T. Kawai, T. Shibata, H. Oda, K. Haruki, H. Nakazawa and S. Kozaki. 2003. An extensive outbreak of staphylococcal food poisoning due to low-fat milk in Japan: estimation of enterotoxin A in the incriminated milk and powdered skim milk. Epidemiol. Infect. 130: 33–40.

Baird-Parker, A.C. 1962. An improved diagnostic and selective medium for isolating coagulase positive staphylococci. J. Appl. Bact. 25: 12–19.

Bergdoll, M.S., M.J. Surgalla and G.M. Dack. 1959. Staphylococcal enterotoxins. I. Purification. Arch. Biochem. Biophys. 85: 62–69.

Casman, E.P. 1960. Further serological studies of staphylococcal enterotoxin. J. Bacteriol 79: 849–856.

CDC (Centers for Disease Control and Prevention), Atlanta, GA, USA. Page last updated: June 7, 2010. Staphylococcal food poisoning. http://www.cdc.gov/nczved/divisions/dfbmd/diseases/staphylococcal/.

Cretenet, M., E. Sergine and Y. Le Loir. 2011. Unveiling *Staphylococcus aureus* enterotoxin production in dairy products: a review of recent advances to face new challenges. Dairy Sci. & Technol. 91: 127–150.

Danielsson, M.L. and B. Hellberg. 1977. The biochemical activity of enterotoxin and non-enterotoxin producing staphylococci. Acta Vet. Scand. 18: 266–273.

Danielsson, M.L. and B. Hellberg. 1984. Prevalence of enterotoxigenic staphylococci in nose, throat and skin lesions in meat-workers. Acta Vet. Scand. 25: 242–249.

Evenson, M.L., M.W Hinds, R.S. Bernstein and M.S. Bergdoll. 1988. Estimation of human dose of Staphylococcal enterotoxin A from a large outbreak of staphylococcal food poisoning involving chocolate milk. Int. J. Food Microbiol. 7: 311–316.

Euzéby J.P. 2012. List of prokaryotic names with standing in nomenclature—Genus *Staphylooccus*. http://www.bacterio.cict.fr/s/staphylococcus.html.

Hait, J. 2012. *Staphylococcus aureus*. *In:* K.A. Lampel, S. Al-Khaldi and S.M. Cahill (eds.). Bad Bug Book: Foodborne Pathogenic Microorganisms & Natural Toxins Handbook, 2nd ed. Center for Food Safety and Applied Nutrition, of the Food and Drug Administration (FDA), U.S. Department of Health and Human Services. pp. 87–92.

Hennekinne, J.A., M.L, De Buyser and S. Dragacci. 2012. *Staphylococcus aureus* and its food poisoning toxins: characterization and outbreak investigation. FEMS Microbiol Rev. 36: 815–836.

Jay, J.M., M.J. Loessner and D.A Golden. 2005. Modern Food Microbiology. 7th ed. Chapter 23. Staphylococcal gastroenteritis. Food Science Text Series. Springer Science + Business Media, New York, USA. ISBN 0-387-23180-3. pp. 545–566.

Kérouanton, A., J.A. Hennekinne, C. Letertre, L. Petit, O. Chesneau, A. Brisabois and M.L. De Buyser. 2007. Characterization of *Staphylococcus aureus* strains associated with food poisoning outbreaks in France. Int. J. Food Microbiol. 115: 369–375.

Kloos, W.E. and K.H. Schleifer. 1986. Genus IV. *Staphylococcus*. *In*: P.H.A. Sneath, N.S. Mair, M.E. Sharpe and J.G. Holt (eds.). Bergey's Manual of Systematic Bacteriology, Vol 2. Williams & Wilkins, Baltimore, USA. pp. 1013–1035.

Lawrynowicz-Paciorek, M., M. Kochman, K. Piekarska, A. Grochowska and B. Windyga. 2007. The distribution of enterotoxin and enterotoxin-like genes in *Staphylococcus aureus* strains isolated from nasal carriers and food sampes. Int. J. Food Microbiol. 117: 319–323.

Le Loir, Y., F. Baron and M. Gautier. 2003. *Staphylococcus aureus* and food poisoning. Genet. Mol. Res. 2: 63–76.

Loncarevic, S. 2010. *Staphylococcus aureus* food poisoning. Abstract no. 9, 3 pages. The proceeedings of the 2nd Symposium on Food Associated Pathogens. Grythyttan, Sweden. Örebro University. ISBN 978-91-7668-749-9.

Merchant, I.A. and R.A. Packer. 1967. Veterinary Bacteriology and Virology. 7th ed. Chapter 4. Bacteriological techniques and methods p. 65. The Iowa State University Press Ames, Iowa, USA.

Montville, T.J., K.R. Matthews and K.E. Kniel. 2012. Food Microbiology: An Introduction, 3rd ed. Chapter 16. *Staphylococcus aureus*. ASM Press, Washington DC, USA. pp. 231–243.

Pinchuk, I.V., E.J. Beswick and V.E. Reyes. 2010. Staphylococcal enterotoxins. Toxins 2: 2177–2179.

Rusnak, J.M., M. Kortepeter, R. Ulrich, M. Poli and E. Boudreau. 2004. Laboratory Exposures to Staphylococcal enterotoxin B. Emerg. Infect. Dis. 10: 1544–1549.

Tolan, R.W. 2012. Page last updated *Staphylococcus aureus* infection. http://emedicine.medscape.com/article/971358-overview. Steele, RW (ed.).

Turnidge, J., N. Rao, F-Y. Chang, V.G. Fowler, S.M. Kellie, S. Arnold, B.Y. Lee and A. Tristan. 2008. Page last updated *Staphylococcus aureus*. pp. 1, 6, 14, 15, http://www.antimicrobe.org/sample_staphylococcus.asp

Vaughan, V.C. Poisonous or "sick" cheese. 1884. Public Health Pap. Rep. 10: 241–245.

CHAPTER 17

Vibrio as a Food Pathogen

Torkel Wadström[#] and *Åsa Ljungh*[#, *]

The family *Vibrionaceae* now contains > 70 species, of which at least 12 are human pathogens (Table 1). This family earlier comprised also of *Aeromonas* spp. and *Plesiomonas* spp. but, by comparison of DNA-sequences, these now form their own families, *Aeromonadaceae* and *Plesiomonas,* belonging to *Enterobacteriaceae.* All three families, however, have water and water-associated animals and plants as their natural habitat, and may cause food-associated infections.

Looking through the history, outbreaks of diarrhea have not been uncommon and have had serious implications—a very high mortality, particularly among infants who are more vulnerable to the loss of fluid, but also on outcomes of wars. However, it was with John Snow in London 1831-32 that the first connection between a common source of infection, i.e., the Broad Street Pump and a diarrhoeal outbreak became evident. By plotting the location of deaths related to cholera he could show that the majority of cases emanated from persons who got their water from a company that drew its water from the downstream location of the city, instead of the upstream one. Further mapping identified the Broad Street Pump as the source, and when Snow convinced officials not to supply water via this pump, the epidemic was contained.

Section of Medical Microbiology, Department of Laboratory Medicine, Faculty of Medicine, University of Lund, Lund, Sweden.
Emails: Torkel.wadstrom@med.lu.se; Asa.ljungh@med.lu.se
* Corresponding author

Table 1. *Vibrio* species recognized causing human infections.

Species	Gastroenteritis	Wound infections/Septicaemia
V. cholerae O1, O139	+++	(+)
Non-O1 *V. cholerae*	++	(+)
V. mimicus	++	+
V. parahaemolyticus	+++	++
V. fluvialis	++	+
V. hollisae (*Grimontia*)	++	+
V. furnissii	++	(+)
V. vulnificus	+	+++
V. damsela (*Photobacterium*)	(+)	++
V. alginolyticus	++	++
V. harveyi (*V. carchariae*)	(–)	(+)
V. metschnikovii	(+)	(+)
V. anguillarum	+	
V. cincinnatiensis	(+)	(+)
V. owensii	+?	–

Bacteriology and Environmental Aspects

Vibrio spp. are gram-negative, oxidase-positive rods, motile by a unipolar flagellum which survive and thrive in aquatic environments. Some species, but not *V. cholerae*, are halophilic and require 3–6% NaCl in the growth medium. This implicates that faecal samples for isolation of *Vibrionaceae* shall be transported in a special medium, thiosulfate-citrate-bile-salts-sucrose TCBS (Marano et al. 2000). Some species multiply faster at temperatures > 18° C which means that infections are more prevalent in tropical and subtropical areas, and that infections in temperate areas are more prevalent during the summer. An interesting recent study from Chandigarh, India, showed that isolation of *V. cholerae* from freshwater and plankton was dependent on water and air temperatures, chlorophyll, salinity and/ or pH whereas cases of cholera correlated with rainfall and chlorophyll concentration but not with air temperature (Mishra et al. 2011).

In Dhaka, Bangladesh, there is a dual peak of cholera—March–May and September–November. Dhaka is located close to a heavily polluted freshwater river system and flood embankment. Extreme flooding in 2007 caused a severe outbreak of cholera. The infectious clone of *V. cholerae*, isolated from flood waters, was shown to circulate via the faecal-oral route during and between the two seasonal cholera peaks, and also to cause off-season outbreaks of cholera (Alam et al. 2011).

Given aquatic environments as their natural habitat, fish and seafood are prone to bacterial contamination, particularly filter feeders such as mussels and oysters, clams and crabs (Wafaa et al. 2011). This holds for both natural and farm environments, and all seafood types (Baffone et al. 2000). However, *V. vulnificus* was only isolated from crayfish and fish gills (Adebayo-Tayou et al. 2011). The increasing prevalence of infection is probably due to the increase of water temperature.

Several of these species are not only contaminants but are recognized as important pathogens of marine and estuarine animals with substantial losses in commercial production systems and natural waters throughout the world. Species of the so called *V. harveyi*-group are major pathogens for aquatic animals, causing outbreaks of disease in the aquaculture industry. In this group, several species are almost indistinguishable even by 16S rRNA gene analysis (Cano-Gómez et al. 2010). This has certainly lead to misidentifications, also in the case of human food borne infections.

Since the 1980s several new *Vibrio* species have been identified, and some have been reclassified (Table 1). Thus, epidemiological reports on infections caused by non-cholera *Vibrio* conducted earlier than 1980 are misleading regarding prevalences of these.

Overall, increasing prevalences of infections in marine vertebrates, invertebrates and humans have been linked to increasing marine temperatures with reports on foodborne infections in areas where *Vibrio* infections have been rare, such as North America, Europe and Japan (Iwamoto et al. 2010, Eiler et al. 2006). Generally, the infectious dose for gastrointestinal disease is $> 10^6$ but may be as low as 10^2 in patients treated with proton pump inhibitors and immunosuppressed patients.

During suboptimal conditions, such as water temperature below 8°C, *Vibrio* spp. persist by entering a viable but non culturable (VBNC) state (Xu et al. 1982). Colonization of algae and the parasite *Acanthamoeba castellanii* are other mechanisms for survival of *V. cholerae* in waters between epidemics.

Though the number of reported foodborne infections are more commonly caused by *Salmonella* spp. and *Campylobacter* spp., the prevalence of food borne infections caused by *Vibrio* spp. is high, and seems to increase. CDC reported that between 1998 and 2010, the incidence of foodborne *Vibrio* infections increased by at least 115%. Nowadays there are around 80,000 infections annually in the United States. The prevalence is probably higher in several Asian, South American and African countries but official reporting has limitations. However, there is a recent report from Alexandria on *Salmonella* and *Vibrio* spp. in seafood (Wafaa et al. 2011). Alexandria is pretty close to Europe.

Several *Vibrio* spp. produce lysine decarboxylase which enable them to neutralize the acid gastric pH and survive passage through the stomach. It is important to especially warn patients treated with proton pump inhibitors which increases the gastric pH to avoid seafood and potentially undercooked fish in the coastal states of the USA and while travelling to Asia, Africa, and South America. *Vibrio* spp. causing extraintestinal infections like wound infections and septicaemia produce proteases, haemolysins, and other enzymes which facilitate tissue damage and local spread in extraintestinal tissues.

Vibrio cholerae

V. cholerae serotype O1, the cause of cholera, presents in two biovariants, the classical non-haemolytic, and El Tor, the haemolytic. The classical biotype which is thought to have originated in India, has caused 6 pandemics whereas the 7th was caused by El Tor. The El Tor pandemic struck harder than those caused by classical biotypes. Genomic analyses of endemic and pandemic *V. cholerae* strains showed genes unique to all pandemic strains and genes specific for the 7th pandemic El Tor strain as well as to related O139 strains (Dziejman et al. 2002). Some of these specific genes could encode for traits associated with enhanced virulence.

Tsunamis and hurricanes, like Catherine, during the last decade have been followed by increased prevalences of cholera. After the earthquake in Haiti, there was an outbreak of cholera (Hill et al. 2011). *V. cholerae* O1 and genes encoding for cholera toxin were detected in harbor waters and seafood in Port-au-Prince, and later spread to all 10 Haitian departments, again emphasizing the importance of not drinking untreated water and undercooked seafood.

Since 1992, a new serotype, O139, has caused epidemics on the Indian subcontinent, and later spread to South America, probably by consumption of contaminated food, served on aeroplanes. Serotype O139 is thought to emanate from *V. cholerae* El Tor by horizontal gene transfer. Interestingly, the *Vibrio* 7th pandemic island-II is a genomic island present in both El Tor and O139. It shows a strong homology to a 43.4 kb genomic island present in *V. vulnificus* (O´Shea et al. 2004). Serotype El Tor and O139 share biochemical and genetical features in common but differ to that extent that patients who have had an infection with serotype O1 are not protected from infection by serotype O139 (Mishra et al. 2011).

Between 1984 and 2002, *V. cholerae* from another serogroup, O141, has been isolated from the stools of 11 patients with acute or milder diarrhoea, and in one patient also from the blood stream in the US (Crump et al. 2003). The common denominator of the patients was consumption of seafood and/or living in coastal areas. All strains produced cholera toxin.

In most endemic areas, cholera infections peak in March-May. In Dhaka, however, there is another peak in September–November as well (Alam et al. 2011). In a severe outbreak in Dhaka in 2007, subsequent analyses showed that the cause was a hypervirulent clone of El Tor, possessing classical cholera toxin. Strains of this clone were present in the estuarine ecosystem in Bangladesh. Flood waters were shown to transmit the infectious clone and to circulate via the faecal-oral route both during and between the dual cholera peaks in Dhaka (Alam et al. 2011).

Virulence factors

The most recognized virulence factor of *V. cholerae* is the cholera toxin, see below. However, to be able to exert any effect on a mucosal surface, the microbe has to adhere to it. After ingestion, adhesion to the small intestinal epithelium by *V. cholerae* is mediated by a type IV fimbriae, toxin-coregulated pilus, TCP.

Cholera toxin, encoded by a filamentous phage, CTXΦ, consists of a central A unit and 5 B subunits (Sánchez and Holmgren 2008). The B units bind to GM_1-ganglioside which is present in most human cells. After binding, the A unit is internalized by endocytosis and transported to the endoplasmatic reticulum. There, proteolytic cleavage gives A1 and A2. A1 transfers ADP-ribose to a G-protein which activates adenylate cyclase which in turn increases the intracellular cAMP content. Protein kinase A, activated by cAMP, opens ion channels in the intestinal cells. Water and electrolytes, like chloride are secreted, and the absorption of Na_2CO_3 is hampered, giving the watery, rice water-like, diarrhoea of cholera. Cholera toxin thus stimulates a normal biochemical pathway, and is not toxic to any cell. "Toxin" is therefore a misleading term but all have agreed upon naming cholera toxin a cytotonic enterotoxin, in contrast to the cell damaging enterotoxins of, e.g., *Shigella dysenteriae* and *Clostridium perfringens*. Interestingly, TCP is a receptor for the CTXΦ phage.

Cholera toxin also activates phospholipase A2, PLA2, a precursor of prostaglandin, mainly PGE2. The activation is independent of cAMP. The jejunal level of PGE2 has been shown to be correlated to the volume of diarrhea.

Only *V. cholerae* strains carrying these virulence factors can cause human disease. Others may colonize the human gastrointestinal tract without giving any symptoms.

Persons with blood group 01 are more susceptible to cholera infection. The mechanisms behind this have not been clarified.

Other members of *Vibrionaceae*, with the exception of *V. parahaemolyticus* (see below) have not caused epidemics. However, they are also prevalent in brackish water, and cause human gastrointestinal as well as extragastrointestinal disease and cause local outbreaks of gastrointestinal disease (Table 1).

V. mimicus

V. mimicus is similar to *V. cholerae,* hence its name. It is thought to have emanated from a common ancestor, similar to the strain causing the 6th pandemic (Hasan et al. 2010). Few strains produce cholera toxin. The intestinal infection may be quite serious with bloody stools. In a study from the US, 55% of the isolates emanated from hospitalized patients (Shandera et al. 1983). *V. mimicus* has been reported to cause fairly large outbreaks of food borne infections (Chitov et al. 2009). Strains produce a haemolysin (Sultan et al. 2007).

V. parahaemolyticus

V. parahaemolyticus is the most common non-cholera species causing human food borne infections. One serotype of *V. parahaemolyticus,* O3:K6, has caused epidemics in India and spread to other Asian and South American countries (Nair et al. 2007). In 2007 in Chile, the pandemic O3:K6 strain was replaced with a new serotype, O3:K59, in > 25% of the isolates from clinical cases. One of these O3:K59 strains, associated with 11% of the cases, contained genes, identical to those in the pandemic strain, indicating lateral transfer of pathogenicity-related genes to the *V. parahaemolyticus* groups colonizing shellfish (Harth et al. 2009).

V. parahaemolyticus produces thermostable direct haemolysin, TDH, a haemolysin which is enterotoxic but also cardiotoxic. The enterotoxic mechanism is not mediated by cell damage but seems to be ion dependent. Patients present with diarrhoea but may also have cardiac arrhythmia or other symptoms of cardiac involvement (Nishibushi et al. 1992). Strains may also produce toxin-related haemolysin, TRH, and thermo-labile haemolysin, TLH (Honda and Iida1993). In temperate climates, we encounter patients with external otitis, but rarely with gastroenteritis.

V. fluvialis

V. fluvialis, first identified in 1981, is globally distributed in shellfish and estuarine waters, and has been isolated from patients with food poisoning, gastrointestinal and extra-gastrointestinal diseases. It is biochemically similar to *Aeromonas hydrophila* but can be differentiated by the ability of *V. fluvialis* to grow on media containing 6–7% sodium chloride. In a report from South Africa, *V. fluvialis* was isolated from treated wastewater effluent systems which is of major concern (Igbinosa et al. 2010, Igbinosa et al. 2009). Like *V. vulnificus, V. fluvialis* produce a metalloprotease which contributes to the tissue damage. Soft tissue infections may present as necrotizing fasciitis, quite a severe form (Barber and Swygert 2000). They do not produce cholera toxin but probably other, hitherto unidentified toxins in patients often present with bloody diarrhoea.

V. hollisae, Grimontia hollisae

V. hollisae was designated as a separate species, *Grimontia*, in 1982. They are difficult to isolate, and grow poorly on TCBS. Hence, reports on the prevalence of food borne infections caused by *G. hollisae* are likely to be underestimated. However, available reports rank *V. hollisae* as a significant cause of foodborne disease (Elhadi et al. 2004). The strains produce a thermostable haemolysin, similar to that of *V. parahaemolyticus*. The gastroenteritis may be severe and spread to the liver, causing hepatitis (Daumas et al. 2009).

V. furnissii

V. furnissii has been mainly isolated from stool samples from patients with gastroenteritis and healthy persons (Huq et al. 1980). In contrast to infections caused by other *Vibrio* spp., infections have not been linked to consumption of seafood. Retrospectively, it was shown to be implicated in an outbreak on board an aircraft in 1969. The importance of *V. furnissii* as a human enteric pathogen is at present unclear (Dalsgaard et al. 1997).

V. vulnificus

V. vulnificus has a worldwide spread and causes infection through consumption of contaminated seafoods, or sea water contamination of wounds. Though halophilic, *V. vulnificus* was not isolated from oysters and seawater in North Carolina during the severe drought 2007–9, which caused an increase in salinity (Froelich et al. 2012).

Particularly, contamination of wounds may lead to septicaemia with a high mortality rate, first recognized among divers in the American Navy in Chesapeake Bay. *V. vulnificus* has the highest case fatality rate of any food-borne pathogen (Bross et al. 2007). Recently, three biotypes were identified in a study from Israel. Biotype 3 constituted around 20% of the aquaculture isolates but 86% of the clinical isolates, indicating higher virulence of this, very distinct, biotype (Broza et al. 2009). *V. vulnificus* possesses a matrix metalloprotease which activates the bradykinin pathway, leading to increased vascular permability (Miyoshi et al. 1998). This activates several human enzymes, contributing to the typical severe skin damage and even necrotizing fasciitis. However, it is not fundamental for the virulence. Protease-deficient mutants of *V. vulnificus* expressed increased cytolysin activity, and showed similar virulence (Shao and Hor 2000). The presence of a capsular polysaccharide enables the bacterium to circumvent the host´s immune system, and contributes to the tissue damage (Hilton et al. 2006). Persons with cirrhosis are at higher risk of infection, and are at 200 times more likely to die from the infection (Haq and Dayal 2005).

V. damsela, Photobacterium damsela

V. damsela , now classified as *Photobacterium damsela,* rarely cause foodborne gastrointestinal illness but soft tissue infections after exposure to brackish or saltwater. The name reflects its pathogenicity for damsel fish. However, like soft tissue infections caused by *V. vulnificus,* fulminant infections caused by *P. damsela* are frequently fatal, also in immunocompetent hosts (Barber and Swygert 2000). *V. damsela* produces a potent cytolysin (Kreger 1983).

V. alginolyticus

V. alginolyticus causes foodborne as well as wound infections. It was the most commonly isolated *Vibrio* spp. in wound infections in a study from the Ionian Sea (Cavallo and Stabili 2002) and the USA (Bonner et al. 1983). It is widely distributed in the environment, and also causes infections in fish and seafood (Schets et al. 2010, Goulden et al. 2012). In a study from Holland, 80% of isolated *Vibrios* from mussels and oysters in fish shops and noncommercial sampling sites were *V. alginolyticus.* Strains produce proteases and a collagenase (Yishan et al 2011).

V. harveyi

The *V. harveyi* group has been the subject of intense taxonomic work during the last decade. These have shown *V. carchariae* to be a junior synonym of *V. harveyi* (Austin and Zhang 2006). *V. owensii, V. campbelli* and *V. rotiferianus,* all important pathogens for aquatic organisms, also form part of this group (Gomez-Gil et al. 2004). *V. carchariae* has hitherto been isolated only from wound infections caused by shark bites (Pavia et al. 1989), and *V. owensii* from patients with gastroenteritis (Cano-Gómez et al. 2010).

Other *Vibrio* species

V. metschnikovii is commonly isolated from the environment, and has occasionally been isolated from patients with gastroenteritis. Infections caused by *V. cincinnatienses* have rarely been reported at all, possibly due to its recent description as a species.

Treatment

Oral hydration of patients with gastrointestinal infection is mandatory. When oral rehydration solution (ORS) is not available home-available fluid is recommended.

Severely ill patients, needing antibiotic treatment, i.e., patients with extragastrointestinal infections, can be treated with quinolones. The sensitivity to other antibiotics, i.e., ceftazidime, differs. During cholera epidemics, tetracycline is often recommended to reduce further spreading.

Concluding Remarks

With the increasing temperature globally, floodings, and pollution of water circulating to households, we can envisage an increase of the burden of *Vibrio* infections in free-living sea organisms, farmed fish and shellfish, and hence, as human gastrointestinal foodborne infections. Improvement of sanitary systems and minimizing the faecal-oral route of spread are mandatory. Also, the importance of avoidance of ingestion of undercooked fish and seafood has to be made at public awareness programmes. Since several *Vibrio* spp. cause wound infections and other extragastrointestinal infections, it is important for diagnostic laboratories to have diagnostics available also for *Vibrio* spp.

References

Adebayo-Tayo, B.C., I.O. Okonko, C.U. Esen, N.N. Odu, C.C. Onoh and N.J.P. Igwiloh. 2011. Incidence of potentially pathogenic *Vibrio* spp. in fresh seafood from Itu Creek in Uyo, Akwa Ibom State, Nigeria. World Appl. Sci. J. 15: 985–991.

Alam, M., A. Islam, N.A. Bhuiyan, N. Rahim, A. Hossain, G.Y. Khan, D. Ahmed, H. Watanabe, A.S. Akanda, S. Islam, R.B. Sack, A. Huq, R.R. Colwell and A. Cravioto. 2011. Clonal transmission, dual peak, and off-season cholera in Bangladesh. Inf. Ecol. Epidemiol. 1: 7273–7286.

Austin, B. and X.H. Zhang. 2006. *Vibrio harveyi*: a significant pathogen of marine vertebrates and invertebrates. Lett. Appl. Microbiol. 43: 119–124.

Baffone, W., A. Pianetti, F. Bruscolini, E. Barbieri and B. Citterio. 2000. Occurrence and expression of virulence-related properties of *Vibrio species* isolated from widely consumed seafood products. Int. J. Food Microbiol. 54: 9–18.

Barber, G.R. and J.S. Swygert. 2000. Necrotising fasciitis due to *Photobacterium damsela* in a man lashed by a stingray. N. Engl. J. Med. 342: 824.

Bonner, J.R., A.S. Cocker, C.R. Berryman and H.M. Pollock. 1983. Spectrum of Vibrio infections in a Gulf Coast community. Ann. Intern. Med. 99: 464–469.

Bross, M.H., K. Soch, R. Morales and R.B. Mitchell. 2007. *Vibrio vulnificus* infection: Diagnosis and treatment. Am. Family Physician. 76: 539–544.

Broza, Y.Y., Y. Danin-Poleg, L. Lerner, L. Valinsky, M. Broza and Y. Kashi. 2009. Epidemiologic study of *Vibrio vulnificus* infections by using variable number tandem repeats. Emerg. Infect. Dis. 15: 1282–1285.

Cano-Gómez, A., E.F. Goulden, L.Owens and L. Høj. 2010. *Vibrio owensii* sp. Nov., isolated from cultured crustaceans in Australia. FEMS Microbiol. Lett. 302: 175–181.

Cavallo, R.A. and L. Stabili. 2002. Presence of vibrios in seawater and *Mytilus galloprovencialis* (Lam.) from the Mar Piccolo of Taranto (Ionian Sea). Water Res. 36: 3719–3726.

Chitov, T., P. Kirikaew, P. Yungyune, N. Rengprajan and K. Sontitun. 2009. An incidence of large foodborne outbreak associated with *Vibrio mimicus*. Eur. J. Clin. Microbiol. Infect. Dis. 28: 421–424.

Crump, J.A., C.A. Bopp, K.D. Green, K.A. Kubota, R.L. Middendorf, J.G. Wells and E.D. Mintz. 2003. Toxigenic *Vibrio cholerae* serogroup O141-associated cholera-like diarrhea and bloodstream infection in the United States. J. Infect. Dis. 187: 866–868.

Dalsgaard, A., P. Glerup, L.L. Hoybye , A.M. Paarup, R. Meza, M. Bernal, T. Shimada and D.N. Taylor. 1997. *Vibrio furnissii* isolated from humans in Peru: a possible human pathogen? Epidemiol. Infect. 119: 143–149.

Daumas, E.S., S.S. Branger, J.M. Durand, D. Raoult and P.E. Fournier. 2009. *Grimontia hollisae*, a potential agent of gastroenteritis and bacteremia in the Mediterranean area. Eur. J. Clin. Microbiol. 28: 705–707.

Dziejman, M., E. Balon, D. Boyd, C.M. Fraser, J.F. Heidelberg and J.J. Mekalanos. 2002. Comparative genomic analysis of *Vibrio cholerae:* Genes that correlate with cholera endemic and pandemic disease. Proc. Nat. Acad. Sci. USA. 99: 1556–1561.

Eiler, A., M. Johannissón and S. Bertilson. 2006. Environmental influences on *Vibrio* populations in Northern temperate and boreal coastal waters (Baltic and Skagerrak seas). Appl. Environ. Microbiol. 72: 6004–6011.

Elhadi, N., S. Radu, C.H. Chen and M. Nishibushi. 2004. Prevalence of potentially pathogenic *Vibrio* species in the seafood market in Malaysia. J. Food Prot. 67: 1469–1475.

Froelich, B.A., T.C. Williams, R.T. Noble and J.D. Oliver. 2012. Apparent loss of *Vibrio vulnificus* from North Carolina oysters coincides with a drought-induced increase in salinity. Appl. Environ. Microbiol. 79: 3885–3889.

Gomez-Gill, B., S. Soto-Rodríguez, A. Garciá-Gasca, A. Roque, R. Vazquez-Juarez, F.L. Thompson and J. Swings. 2004. Molecular identification of *Vibrio harveyi*-related isolates associated with diseased aquatic organism. Microbiol. 150: 1769–1777.

Goulden, E.F., M.M. R. Hall, D.G. Bourne, L.L. Pereg and I. Hoj. 2012. Pathogenicity and infection cycle of *Vibrio owensii* in larviculture of ornste spiny lobster (*Panulirus ornatus*). Appl. Environ. Microbiol. 78: 2841–2849.

Haq, S.M. and H.H. Dayal. 2005. Chronic liver disease and consumption of raw oysters: a potentially lethal combination—a review of *Vibrio vulnificus* septicemia. Am. J. Gastroenterol. 100: 1195–1199.

Harth, E., L. Matsuda, C. Hernández, M.L. Rioseco, J. Romero, N. Gonzalez-Escalona, J. Martinez-Urtaza and R.T. Espejo. 2009. Epidemiology of *Vibrio parahaemolyticus* outbreak, Southern Chile. Emerg. Infect. Dis. 15: 1269–1279.

Hasan, N.A., C.J. Grim, B.J. Baley, J. Chun, M. Alam, E. Tawiani, M. Hoq, A.C. Munk, E. Saunders, T.S. Brettin, D.C. Bruce, J.F. Challacombe, J.C. Detter, C.S. Han, G. Xie, G.B. Nair and R.R. Colwell. 2010. Comparative genomics of clinical and environmental *Vibrio mimicus*. Proc. Nat. Acad. Sci. USA. 107: 21134–21139.

Hill, V.R., N. Cohen, A.M. Kahler, J.L. Jones, C.A. Bopp, N. Marano, C.L. Tarr, N.M. Garrett, J. Boncy, A. Henry, G.A. Gómez, M. Wellman, M. Curtis, M.M. Freeman, M.Turnsek, R.A. Benner Jr., G. Dahourou, D. Espey, A. DePaola, J.W. Tappero, T. Handzel and R.V. Tauxe. 2011. Toxigenic *Vibrio cholerae* O1 in water and seafood, Haiti. Emerg. Infect. Dis. 17: 2147–2150.

Hilton, T., T. Rosche, B. Froelich, B. Smith and J. Oliver. 2006. Capsular polysaccharide phase variation in *Vibrio vulnificus*. Appl. Environ. Microbiol. 72: 6968–6993.

Honda, T. and T. Iida. 1993. The pathogenicity of *Vibrio parahaemolyticus* and the role of the thermostable direct haemolsyin and related haemolysin. Rev. Med. Microbiol. 4: 106–113.

Huq, M.I., K.M.J. Alam, D.J. Brenner and G.K. Morris.1980. Isolation of a *Vibrio*-like group, EF-6, from patients with diarrhea. J. Clin. Microbiol. 11: 621–624.

Igbinosa, E.O., C.L. Obi and A.I. Okoh. 2009. Occurrence of potentially pathogenic vibrios in the final effluents of a wastewater treatment facility in a rural community of the Eastern Cape Provine of South Africa. Res. Microbiol. 160: 531–537.

Igbinosa, E.O. and A.I. Okoh. 2010. *Vibrio fluvialis:* An unusual enteric pathogen of increasing public health concern. Int. J. Environ. Res. Publ. Health 7: 3628–3643.

Iwamoto, M., T. Ayers, B.E. Mahon and D.L. Swerdlow. 2010. Epidemiology of seafood associated infections in the United States. Clin. Microbiol. Rev. 23: 399–411.

Kreger, A.S. 1983. Cytolytic activity and virulence of *Vibrio damsela,* Infect. Immun. 44: 326–331.

Marano, N.N., N.A. Daniels, A.N. Easton, A. McShan, B. Ray and J.G. Wells. 2000. A survey of stool culturing practices for vibrio species at clinical laboratories in Gulf Coast states. J. Clin. Microbiol. 38: 2267–2270.

Mishra, A., N. Tanenja and M. Sharma. 2011. Environmental and epidemiological surveillance of *Vibrio cholerae* in a cholera-endemic region in India with freshwater environs. J. Appl. Microbiol. 112: 225–237.

Miyoshi, S., H. Nakazawa, K. Kawata, K. Tomochika, K. Tobe and S. Shinoda. 1998. Characterization of the hemorrhagic reaction caused by *Vibrio vulnificus* metalloprotease, a member of the thermolysin family. Infect. Immun. 66: 4851–4855.

Nair, G.B., T. Ramamurthy, S.K. Bhattacharya, B. Dutta, Y. Takeda and D.A. Sack. 2007. Global dissemination of *Vibrio parahaemolyticus* serotype O3:K6 and its serovariants. Clin. Microbiol. Rev. 20: 39–48.

Nishibushi, M., A. Fasano, P.G. Rusell and J.B. Kaper. 1992. Enteroxigenicity of *Vibrio parahaemolyticus* with and without genes encoding thermostable direct hemolysin. Infect. Immun. 60: 3539–3543.

O'Shea, Y.A., S. Finnan, F.J. Reen, J.P. Moreyissey, C. O'Gara and E.F. Boyd. 2004. The *Vibrio* seventh pandemic island-II is a 26.9 kb genomic island present in *Vibrio cholerae* El Tor and O139 isolates that shows homology to a 43.4 kb genomic island present in *V. vulnificus.* Microbiol. 150: 4053–4063.

Pavia, A.T., J.A. Bryan, K.L. Maher, T.R. Hester and J.J. Farmer. 1989. *Vibrio carchariae* infection after a shark bite. Ann. Intern. Med. 111: 85–86.

Sánchez, J. and J. Holmgren. 2008. Cholera toxin structure, gene regulation and pathophysiological and immunological aspects. Cell. Mol. Life Sci. 65: 1347–1360.

Schets, F.M., H.H. van den Berg, S.A. Rutjes and A.M. de Roda Husman. 2010. Pathogenic *Vibrio* species in Dutch shellfish destined for direct human consumption. J. Food Prot. 73: 734–738.

Shandera, W.X., W.M. Johnston, B.R. Davis and P.A. Blake. 1983. Disease from infection with *Vibrio mimicus,* a newly recognized *Vibrio* species. Ann. Intern. Med. 99: 169–171.

Shao, C.P. and L.I. Hor. 2000. Metalloprotease is not essential for *Vibrio vulnificus* infection in mice. Infect. Immun. 68: 3569–3573.

Sultan, Z., T. Mizuno, A. Sakurai, N. Takata, K. Okamoto and S. Mivoshi. 2007. Growth phase dependent activation of the precursor of *Vibrio mimicus* hemolysin (Pro-VMH). J. Health Sci. 53: 430–434.

Wafaa, M.K.B., A.H. Walaa and F.A. Amani. 2011. Detection of *Salmonella* and *Vibrio* species in some seafood in Alexandria. J. Am. Sci. 7: 663–668.

Xu, H.S., N. Roberts, F.L. Singleton, R.W. Attwell, D.J. Grimes and R.R. Colwell. 1982. Survival and viability of nonculturable *Escherichia coli* and *Vibrio cholerae* in estuarine and marine environment. Microb. Ecol. 8: 313–323.

Yishan, L., F. Jiaming, W. Zaohe and J. Jichang. 2011. Genotype analysis of collagenase gene by PCR-SSCP in *Vibrio alginolyticus* and its association with virulence to marine fish. Curr. Microbiol. 62: 1697–1703.

CHAPTER 18

Foodborne Viruses

Magnus Simonsson,[1] Amal Elfaitouri,[2,] Nahla Mohamed,[2,3,*] Jonas Blomberg[2,§] and Christina Öhrmalm[2,*]*

Overview of Foodborne Viral Infections

Viruses that spread via food are in many cases causes of diarrhea, i.e., gastroenteritis. They all spread via the fecal-oral route (Fig. 1 and 2).

Gastroenteritis viruses are basically of six kinds, rotaviruses, astroviruses, noroviruses, sapoviruses, coronaviruses and adenoviruses (Fig. 3). Most types of corona- and adenoviruses cause infection in the respiratory tract but some types also or exclusively give gastroenteritis symptoms. They are all shed through faeces but only the types that cause gastroenteritis have been proven to spread through the faecal-oral route. All mentioned gastroenteritis causing viruses have the potential to spread through food and water. Noroviruses are the most recognized food- and waterborne viruses and are today considered the most common of all food-borne pathogens. Other viruses that are shed through faeces are hepatitis A virus (HAV) and enteroviruses, which belong to the Picornavirus family.

[1] National Food Agency, Box 622, SE - 751 26 Uppsala, Sweden.
Email: magnus.simonsson@slv.se

[2] Clinical Virology, Department of Medical Sciences, Uppsala University, Academic Hospital, S-751 85 Uppsala, Sweden.
Emails: Jonas.Blomberg@medsci.uu.se; Amal.Elfaitouri@medsci.uu.se; Christina.Ohrmalm@medsci.uu.se

[3] Department of Virology, Umeå University, Umeå university Hospital, SE-901 85 Umeå, Sweden.
Email: nahla999@hotmail.com

* The three authors contributed equally.

§ Corresponding author

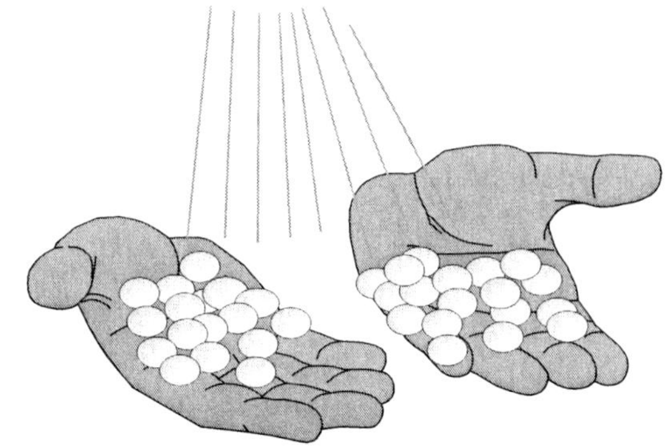

Figure 1. Gastroenteric viruses are often present in copious amounts in faeces (billions of virus particles per ml) and a handwash has to remove most of that. This explains why foodborne virus infections are widespread.

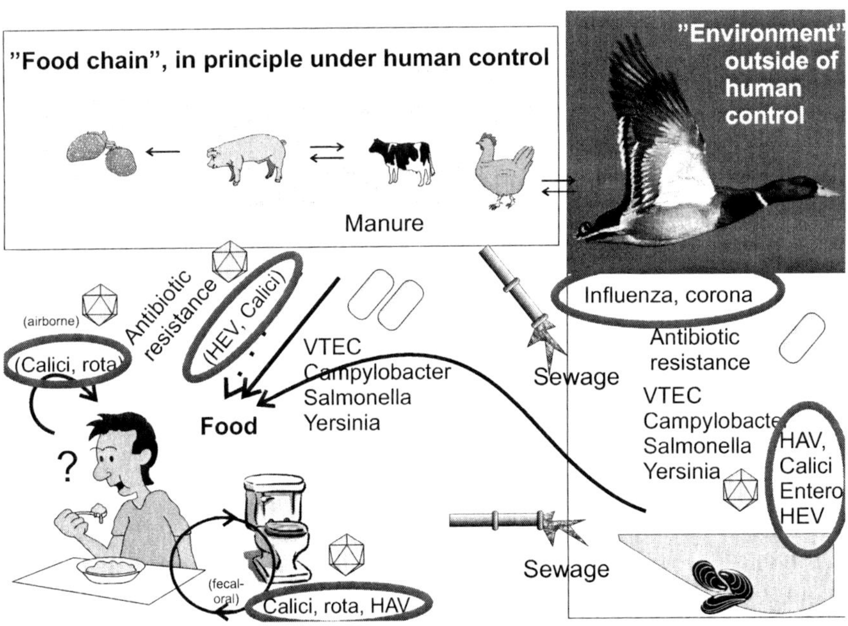

Figure 2. The complex routes of spread of gastroenteric microbes. They (viruses are encircled) are spread in close or wide circles. Narrow circles of spread are from human to human fecal-orally via food or not, but also from human to human airborne. Wider circles spread include food chain and environment. Such a spread occurs within the food chain (animal to animal, animal to berry or green), from food chain to human via manure and food, from wild bird to poultry and vice versa, from food chain to environment via sewage, from human to environment via sewage, within environment via wild animals and from environment (shellfish) to human via food. Gastroenteric bacteria and antibiotic resistance genes are also shown. Surveillance of gastroenteric microbial spread can be performed in diseased humans, foodstuff and drinking water, animals in the food chain, manure, sewage and wild animals (see below).

They are also important food-borne viruses. There is also an increasing concern for food-borne and zoonotic spread of hepatitis E virus (HepEV) in Europe. HepEV has caused large waterborne outbreaks in Asia, northern Africa and Latin America. Besides noroviruses, HAV was identified as the most important food-borne virus by WHO/FAO (see section on HAV). Noroviruses and sapoviruses belong to the calicivirus family and they will be discussed together below.

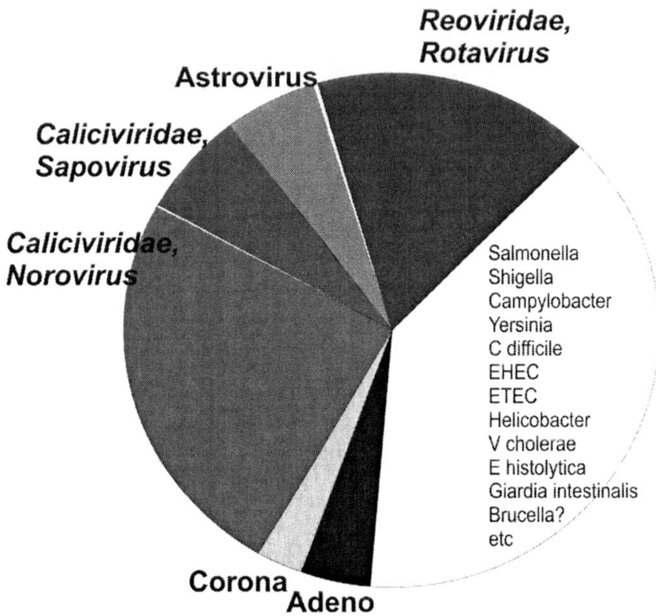

Figure 3. Approximate proportions of viral causes of human gastroenteritis. The figure is based on the Swedish situation, but also has global relevance. Bacteria and protozoa are shown without reference to frequency. Virus names are explained below.

Color image of this figure appears in the color plate section at the end of the book.

How are microbes spread in the environment and to humans?

Microbes in general are spread in the environment in a complex way (Fig. 2). In order to understand how viruses and other microbes get into the food, a general knowledge of how microbes can spread is necessary.

Specifically, how do gastroenteric viruses spread?

Typical routes of viral spread are via skin, air, food, breast milk and sexual contact. In general, most of the spread of gastroenteric viruses takes place via faecal oral spread (Fig. 2). This may be directly due to poor hygiene (did not wash hands after toilet visit) or indirectly due to deficits in sewage and drinking water separation. It can also be due to faecal contamination of food at slaughterhouses or poultry farms. However, the latter route is mostly relevant for gastroenteric bacteria like VTEC, Salmonella and Campylobacter.

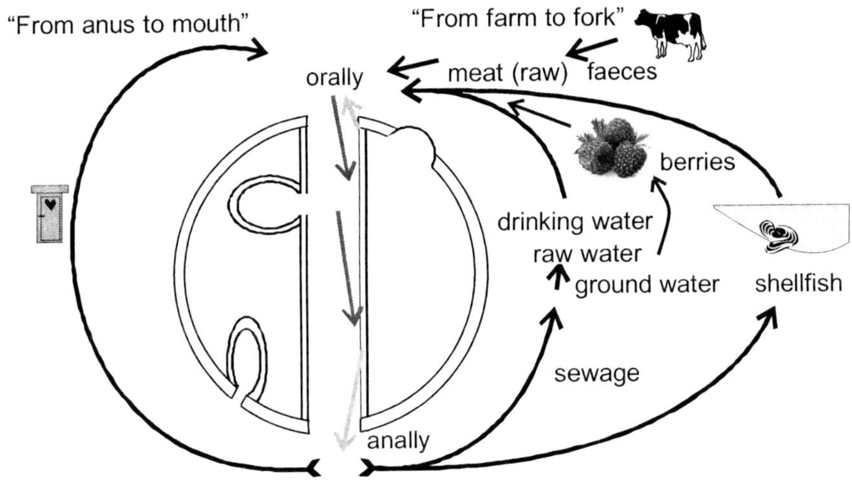

Figure 4. The faecal-oral transmission circles of gastroenteric viruses can be either direct (left) or indirect (right).

An example of the direct route is the cook who does not wash hands properly, and contaminates the food (e.g., salad or princess cake) by using his hands during cooking (Fig. 1).

As viruses do not replicate outside the host the spread of virus depends on its resistance to environmental factors. Most viruses that spread via food and water are non-enveloped viruses with a naked protein capsid. These viruses infect and pass the gastro-intestinal tract and are shed via faeces. They have a natural resistance to low pH and enzymatic digestion. In order to survive in the environment food-borne viruses have also evolved other traits like resistance to UV-light and heat. For instance lightly cooked mussels have been the cause of numerous norovirus infections. Indirect spread of microbes, via sewage, groundwater and drinking water is rather common. The early observation of John Snow in London 1854 is fundamental to our understanding of food- and waterborne microbes. London did not handle faecal waste well at that time. Cholera bacteria from faeces contaminated the groundwater and led to the epidemics. The same problem of insufficient separation between sewage and drinking water or insufficiently purified faecally contaminated raw water still occurs all over the world. Drinking water outbreaks are common and sometimes constitute a mixture of different pathogens (Maunula et al. 2009). Noroviruses that infect all ages and have a very low infectious dose are often the major infection seen in drinking water outbreaks. This includes the outbreak in 2008 in the Swedish municipality Lilla Edet where 2,400 people out of 13,000 inhabitants suffered from gastroenteritis because of contaminated water. The outbreak in the winter resort Sälen in Sweden during 2010, is another example of how norovirus contaminated water can lead to outbreaks of viral gastroenteritis (Hallin 2012).

Figure 5 indicates which of the human viruses has the potential to infect via food. The RNA viruses are more variable than the DNA viruses, which creates diagnostic problems (see below).

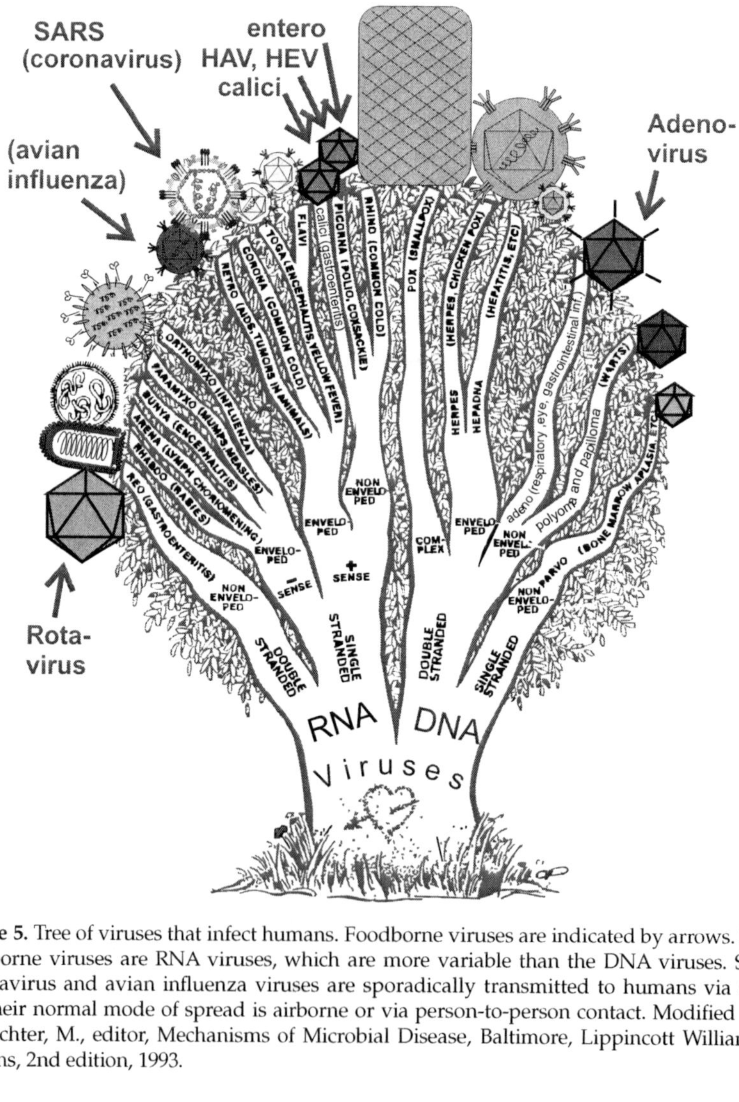

Figure 5. Tree of viruses that infect humans. Foodborne viruses are indicated by arrows. Most foodborne viruses are RNA viruses, which are more variable than the DNA viruses. SARS coronavirus and avian influenza viruses are sporadically transmitted to humans via food, but their normal mode of spread is airborne or via person-to-person contact. Modified from Schaechter, M., editor, Mechanisms of Microbial Disease, Baltimore, Lippincott Williams & Wilkins, 2nd edition, 1993.

Many of the foodborne viruses are RNA viruses which are related to each other (Fig. 6).

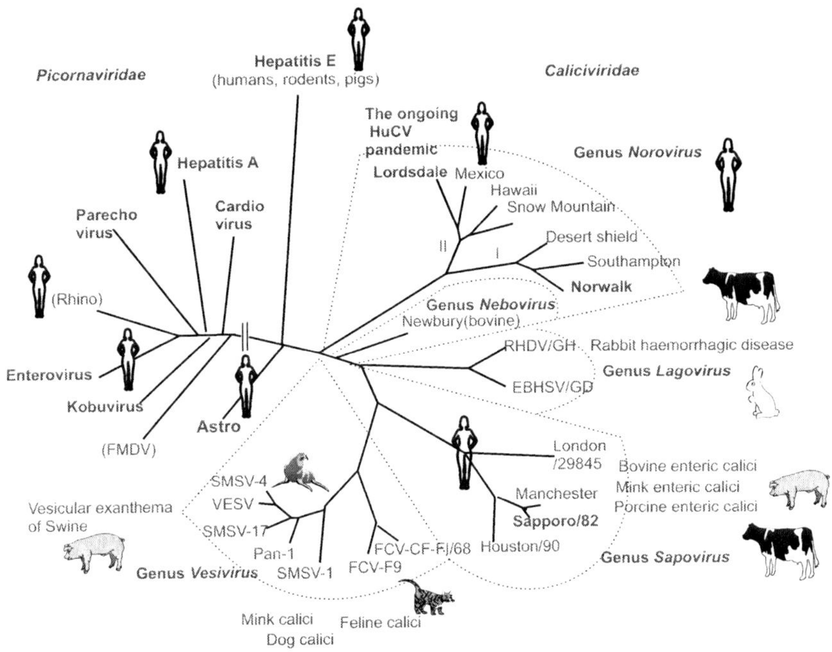

Figure 6. Foodborne human viruses (shown by a human figure) are related to animal viruses. They belong to the virus families *Picorna-* , *Astro-*, *Hepe-* and *Caliciviridae*. *Caliciviridae* contains the genera norovirus, lagovirus, nebovirus, sapovirus and vesivirus, of which the noro and sapoviruses infect humans, and are foodborne. Neboviruses are intermediate between norovirus and lagovirus, and are represented by the bovine virus Newbury agent in the figure. Norovirus genogroups I and II are shown. HAV is Hepatitis A Virus. EBHSV is European Brown Hare Syndrome Virus, FCV is Feline CaliciVirus. VESV is Vesicular Exanthema of Swine Virus, SMSV is San Miguel Sea Lion Virus. FMDV is Foot and Mouth Disease Virus. Animal viruses related to those of humans are also present in the genera cardio-, entero-, kobu- and parechoviruses but are not shown.

Virus Families Involved in Foodborne Infections

Caliciviridae

Noro- and sapoviruses

Morphology and genome

Noro- and sapoviruses are genera in the *Caliciviridae* family and the most common cause of viral gastroenteritis in adults. In electron microscopy (Fig. 7 and 8) calicivirus particles are icosahedral and show cup-shaped pits

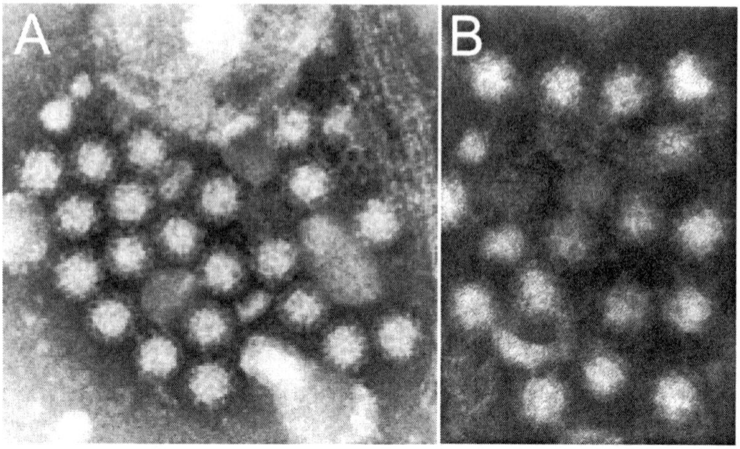

Figure 7. A. Norovirus, and B. Sapovirus in faeces (electron microscopy, courtesy of Kjell-Olof Hedlund, Swedish Institute for Communicable Disease control, SICDC).

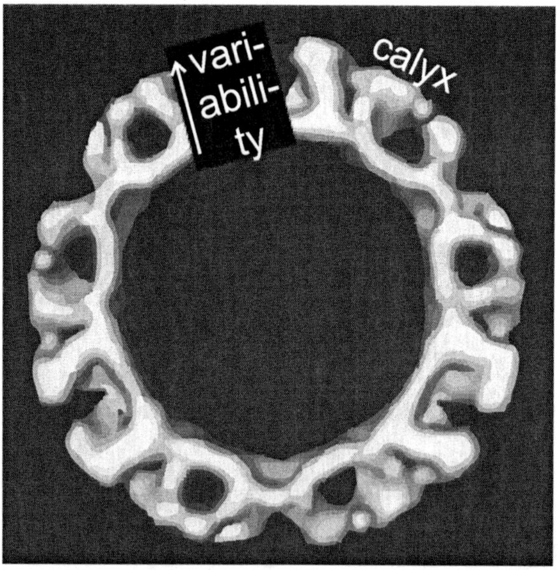

Figure 8. Cross section of a calicivirus, reconstructed from electron micrographs. The receptor binding site has the form of a shallow beaker ("calyx"). Variable structures stick out, more constant ones are hidden inside. However, the bottom of the calyx contains conserved receptor-binding residues.

(calyces), the feature which gave the family its name. The viral genome is a 7600 base pairs polyadenylated single stranded RNA of positive polarity. The capsid is formed by association of the proteins VP1 and VP2. It is rather heat and pH stable.

Taxonomy and basic properties

The caliciviruses have been found in a number of organisms such as humans, cattle, pigs, cats, mice, chickens, reptiles, dolphins and amphibians. The relationships within the noro- and sapoviruses are shown in Fig. 6. The prototypes of norovirus (Norwalk virus) and of sapovirus (Sapporo virus) were identified in 1972 and 1977, respectively. Humans are infected by noro- and sapoviruses.

Human caliciviruses (HuCV) are emerging enteric pathogens that are a common cause of diarrhea in humans worldwide. Caliciviruses are the most variable of all viruses. They come back with new variants regularly.

Human noroviruses belong mainly to genogroup I and II, but genogroup IV infections occur occasionally. Noroviruses belonging to genogroup III are confined to animals, but are similar to human noroviruses, raising the possibility that occasional zoonotic transfers may occur. An especially common subgenogroup is genogroup II.4 (Fig. 9).

The variation within norovirus genogroups is depicted in Fig. 9 (see also Fig. 6). There are so far five norovirus genogroups.

Although sapovirus is less variable it also contains at least five genogroups. Genogroups I, II, IV and V infect humans (with few exceptions) while genogroup III infects pigs (Donaldson et al. 2010). Other sapoviruses, suggested to be named genogroup VI-VIII, have been described among pigs.

Caliciviruses are among the most variable of all viruses. This benefits the virus because it aids escape from herd immunity, and exploitation of new receptor molecules. Point mutations and recombination occur.

Replication, pathogenesis and clinical picture

Caliciviruses infect the gastrointestinal tract through ingestion of contaminated food or direct person-to-person transmission.

Airborne infections also occur via aerosol associated with vomiting. The virions are acid stable so that they can pass through the stomach. The virus primarily replicates in the upper gastrointestinal tract.

The susceptibility to noroviruses is determined by the carbohydrate histo-blood group antigens (HBGA), which occur on small intestine epithelia. Different norovirus strains bind to different HBGA. The placement of fucose residues in the branched HBGA chain is critical (Fig. 10).

HBGAs are complex carbohydrates, oligosaccharides, which are expressed on many cells in the body: red blood cells, and epithelial cells in the respiratory, gastrointestinal and urogenital tracts. The dominant factor behind the susceptibility is the presence of a fucosyl transferase, FUT2, which couples L-fucose in an alpha1-2 linkage at the terminus of the

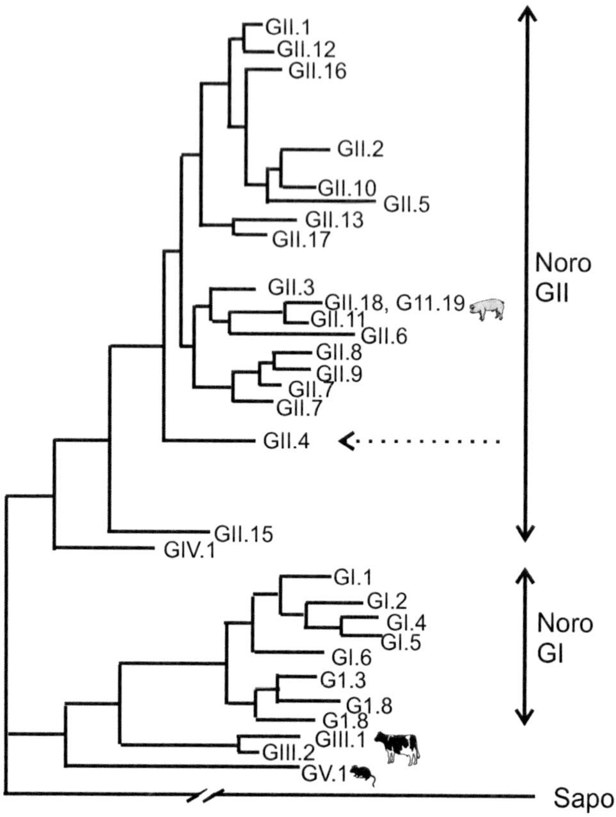

Figure 9. Overview of norovirus variants. Genogroups are sometimes abbreviated "GG", sometimes "G". The currently most expansive genogroup (GII.4) is shown by a dotted arrow. Noroviruses of murine, bovine and swine origins (Wang et al. 2005) are shown. Subgroups GII.11, GII.18 and GII.19 are of swine origin. The rest are of human origin. The tree was rooted on a sapoviral sequence. Modified from (Wang et al. 2005, Donaldson et al. 2010).

HBGA chain in enteric or respiratory mucins. A functional FUT2 confers the "secretor" phenotype. Different norovirus clusters have different host HBGA receptors, but most use this terminal fucose. Around 20% of the population are non-secretors due to lack of FUT2 activity. They are resistant to the major circulating norovirus strains.

Caliciviruses cause an acute gastroenteritis characterized by sudden and often explosive, vomiting, abdominal pain, diarrhea, moderate fever. Other symptoms may be headache, dizziness and muscle ache. Diarrhea occurs in 50–90% of cases. The incubation period is between six hours and two days. Symptoms usually last between two and four days, but in immunosuppressed patients it can last substantially longer. Although

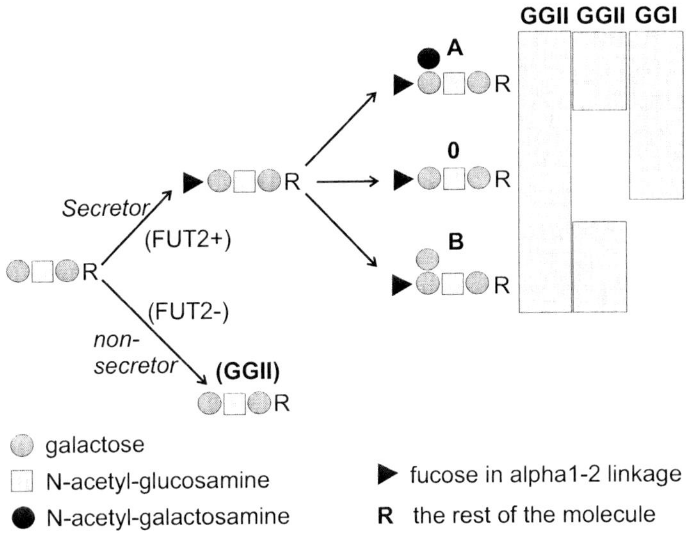

Figure 10. Simplified map of norovirus binding to histo blood group antigens (HBGA) in secreted mucins. Viruses from different outbreaks can infect different persons. This is reflected in the variable pattern of binding of the norovirus capsid protein to HBGA. Particles of the most widely spread genogroup II (GII.4) strains bind to all blood group antigens on secreted mucins. They have a terminal fucose. The influence of nonterminal fucose is disregarded here. Other GGII capsids have a more restricted binding spectrum, and bind to blood group A and/or B HBGA. Genogroup I (GI) strains may bind to blood groups A and 0. Non-secretor binding GII strains are rare variants (Huang et al. 2005, Tan et al. 2008, Huang et al. 2009).

excretion of the virus can last up to eight weeks after loss of symptoms in healthy adults, a symptomless excretion time of two weeks is common. Immunosuppressed individuals may excrete calicivirus in large quantities during months to several years. Symptoms, and their frequency, are: vomiting 50–90%, diarrhea 50–85%, abdominal pain 40–70%, fever 30–40%, headache 30–50% and myalgia 25–35%. The disease is usually harmless but can cause mild or moderate dehydration. Mortality is generally considered to be low at least in developed countries but after investigation of deaths among elderly infected by norovirus in the Netherlands van Asten et al. (van Asten et al. 2011) concludes, "Norovirus should not be regarded as an infection with trivial health risk." An estimate by Patel et al. (Patel et al. 2009) indicates that there are 200,000 deaths worldwide every year caused by noroviruses.

Epidemiology

Human caliciviruses are endemic throughout the world and noroviruses are considered to be the major cause of gastroenteritis. In USA around 23 million individuals are estimated to be infected by noroviruses every year (Mead et al. 1999). In Sweden the incidence amounts to at least 100,000 infections per year. This discrepancy probably illustrates the differences between countries to determine incidence and the difficulty in estimating the frequency of this kind of infections.

There is no solid evidence for human caliciviruses being zoonotic (Bank-Wolf et al. 2010, Summa et al. 2012). If zoonotic transfer occurs it is probably more of a rare event rather than of real epidemiological importance.

The contagiosity of human caliciviruses is very high. Out of the billions of particles per gram of stool or vomitus of an acutely ill person 10–100 virus particles are sufficient to cause disease. The particles are highly stable to pH, detergents, heat and drying. The disease has an incubation time of 6 hours–2 days. Their great variability leads to epidemics which affect all age groups. The variability also explains why it is so difficult to create a calicivirus vaccine. Previous immunity does not give much protection (Donaldson et al. 2010).

Human calicivirus infections occur in the community round the year, but cause regular outbreaks during the winter months in children and adults.The epidemics with direct airborne or contact spread are often caused by Norovirus genogroup II which peak from January to March. They have therefore been given the common name "winter vomiting disease". The most common norovirus genotype, GII.4, was first named "Lordsdale" and accounts for around 80% of all norovirus infections (Patel et al. 2009). This virus gives an especially dramatic disease. Several antigenically distinct variants of GII.4 have been pandemic during the last 20 years. The evolution seems to have been driven by protective herd immunity indicating that protective variant-specific immunity is acquired in many individuals, at least for norovirus GII.4 (Lindesmith et al. 2011). Outbreaks within institutions like hospitals, with direct, airborne or contact, spread from person to person is often caused by GII.4. Person to person spread is a major route of infection for noroviruses. However, around 30% of all norovirus infections could be foodborne (Widdowson et al. 2005, Scallan et al. 2011).

Sapoviruses are a less common cause of gastroenteritis but the frequency seems to increase, which could be due to increased testing for sapovirus in patients. Symptoms are similar to norovirus but infect mainly children. Infections also peak during the winter months. Outbreaks of sapovirus are not common but often food associated when discovered.

Outbreaks of winter vomiting disease often occur in large institutions such as schools, military camps and health care institutions. There are also sporadic cases in the community. The incidence of both hospital and community acquired calicivirus infections has increased during the last 20 years.

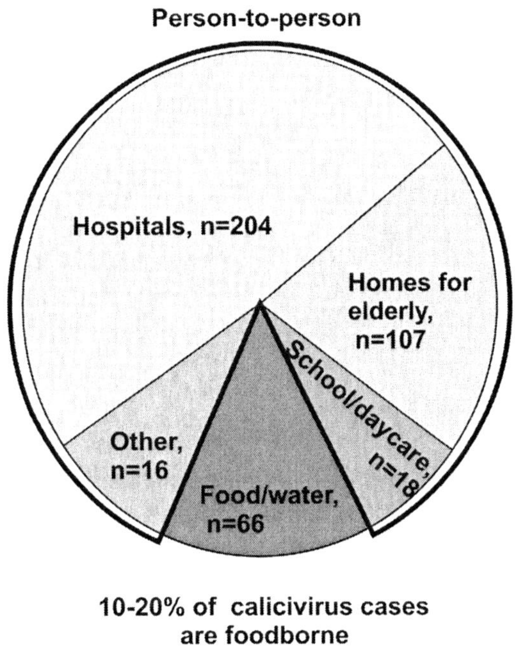

Figure 11. Place and route of infection in 407 outbreaks of gastroenteritis in Sweden 1994–1998 (SICDC). The majority of cases are spread person-to-person.

Color image of this figure appears in the color plate section at the end of the book.

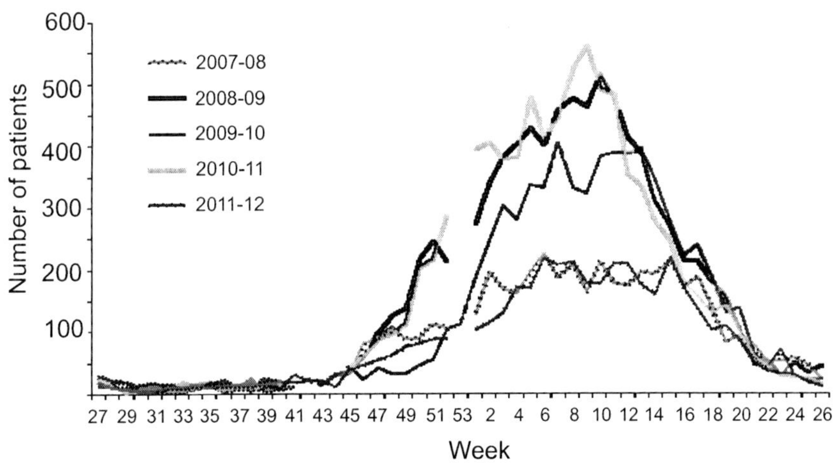

Figure 12. Reported norovirus cases per week, 2007–2012 (SICDC). The majority of cases occur during winter time.

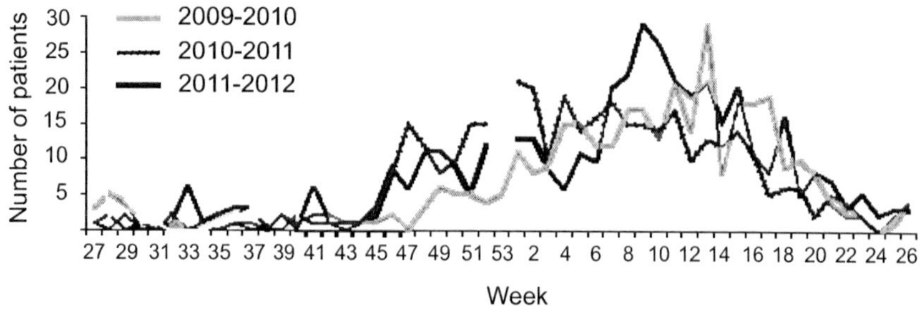

Figure 13. Reported cases of sapovirus infection per week, 2009–2012 (SICDC). The majority of cases occur during winter time.

Foodborne spread

Noroviruses are the most common foodborne viruses and outbreaks are common. These can be direct through food handlers, for example due to poor hygiene during cooking or indirect due to drinking water contamination or contaminated fresh produce like raspberries and bivalve molluscan shellfish. Bivalve molluscan shellfish like oysters and mussels are filter feeders and accumulate virus and bacteria from surrounding water. Water contaminated with human sewage in harvesting areas therefore poses a public health risk. Numerous outbreaks have been recorded after consumption of raw oysters and lightly cooked mussels. Oysters with less than 100 norovirus RNA copies per g of tissue can still cause disease in humans. Especially in the Nordic countries, imported frozen raspberries have been the cause of numerous outbreaks. Today the most common registered food-borne outbreaks in Sweden involve the combination norovirus and raspberries. Norovirus infections peak during winter but norovirus food-borne outbreaks are more evenly distributed throughout the year. It seems that the dominating norovirus genogroup II is less pronounced in food-borne outbreaks and genogroup I is more common. This also seems to be the case for drinking water outbreaks. One speculation is that those viruses are more resistant in the environment.

Even though *sapoviruses* are often found in sewage and molluscan shellfish, these viruses are less commonly involved in food-borne outbreaks than noroviruses. Sapoviruses, like rota- and astroviruses, mostly infect children under the age of five. It is generally assumed that these infections are mostly transmitted by person-to-person contact. The actual transmission via food and water is unknown because disease in a subpopulation like infants would not be recognized as an outbreak. Such sporadic transmissions are more or less impossible to register. The low infectivity dose and the common detection of these viruses in sewage could mean that transmissions other than person-to person can occur but the frequency is unknown.

Diagnosis

Human caliciviruses cannot be grown in cell culture. Therefore, diagnosis and epidemiological surveillance is highly dependent on molecular biological techniques such as PCR and sequencing. Rapid diagnosis is important for adequate handling of outbreaks. Molecular methods can be used for all types of food matrices, including for detection of norovirus in food such as shellfish and drinking water. State of the art today is the use of qPCR, which is rapid, specific, sensitive and can detect less than 10 virus particles. Other methods of diagnosis are electron microscopy and serologic assays. These have a low sensitivity but are sufficient for diagnosis from faecal samples. They are far too insensitive for analyses of food and water matrices. Commercial enzyme-linked immunoassays are available but are also of relatively low sensitivity.

Immunity, prevention and treatment

Noroviruses are variable and there is no immunity that covers many norovirus strains in the population. Immunity to a certain strain may be so weak that reinfection with the same strain can occur. The chances of developing an efficient vaccine are slim but for pandemic GII.4 variants a procedure similar as for influenza virus could be a future possibility (Lindesmith et al. 2011). There are no antiviral drugs to caliciviruses.

For prevention, good hygiene is all important. Food handlers should not work until two days after the disappearance of gastrointestinal symptoms. Even though they may still shed virus the amount of virus is low and its transmission can be curbed by good hygiene. Cascade vomiting can lead to virus spread via aerosol. Dried viruses remain infectious for 1–2 weeks and surfaces should be cleaned thoroughly. Rapid diagnosis, isolation of sick patients, short hospital stays, cleaning and disinfection of the hospital room, as well as stopping admission to new patients, or both is of importance to hinder nosocomial infections. Virus in water can remain infectious much longer, which is a problem when small lakes get contaminated. The viruses can also be transported long distances and contaminate unexpected areas, sometimes used for drinking water production or molluscan shellfish production.

Reoviridae

Rotaviruses

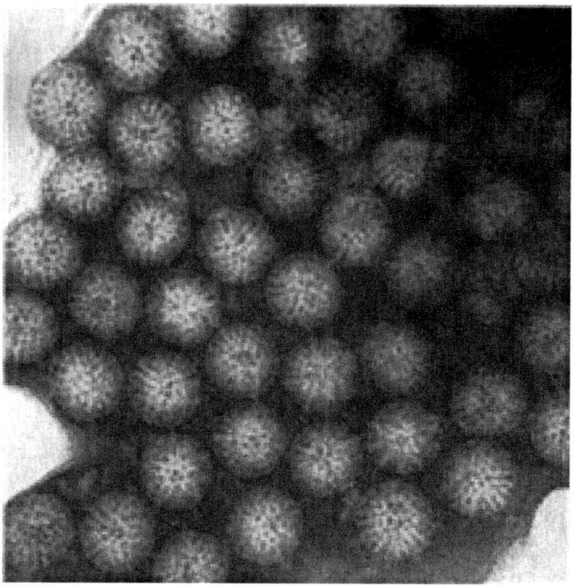

Figure 14. Electron micrograph of Rotavirus. Credit to Kjell-Olof Hedlund, SICDC.

Morphology and genome

Rotaviruses are a genus of the *Reoviridae* family and the most common cause of viral gastroenteritis in infants and young children. The virus structure is unique, with two surface layers. Although the virus particle is icosahedral it looks wheel-like in the electron microscope (Flewett et al. 1974, Flewett et al. 1978). "Rota" means wheel in Latin. Rotaviruses are non-enveloped viruses with a genome composed of 11 segments of double-stranded RNA (dsRNA). The genome encodes for six structural (VP1 to VP4, VP6, and VP7), and six nonstructural proteins (NSP1 to NSP6).

Genogroups and genetic variation

The classification of human rotaviruses is complex. They are classified into seven groups, referred to as A, B, C, D, E, F and G depending on the presence of cross-reactive epitopes primarily located on the internal structural protein VP6 .

Group A rotaviruses are the most frequent pathogens of severe gastroenteritis in humans (Goodgame 1999). Group B rotavirus infections have been associated with outbreaks in China (Qiao et al. 1999, Sen et al. 2001) and India (Chitambar et al. 2011).

There are eighteen serotypes, G-types (based on the glycoprotein (G protein) VP7) and 26 P-types (P means "protease sensitive" a protease-cleaved protein (P protein)), based on the surface protein VP4. Eight serotypes are commonly associated with human diarrhea caused by rotaviruses (G1, G2, G3, G4, G9, P [4], P [6], and P [8]) (Santos et al. 2005). New rotavirus serotypes have arisen in recent years (genotypes 27G and 35P) (Matthijnssens et al. 2011). The significance of these remains unclear, but they could be a problem, because they are not covered by current rotavirus vaccines.

Epidemiology and transmission

Genogroup A is by far the most common among humans. Rotavirus infections at less than six months age and more than five years age tend to be asymptomatic and give degrees of protection against diarrhoeal infection (Velazquez et al. 1996). The infections are the same in developed and developing countries. Globally, rotaviruses cause between 1/2 to one million deaths per year, mostly in low income countries as a result of the severe diarrhoea, dehydration caused by production of a viral toxin (Parashar et al. 2006) and similarly, in Australia (Whitehead et al. 1996). Over 10,000 cases per year occur in Sweden. Rotavirus is often excreted in tens of billions of viral particles per mL in the stool of infected persons. Transmission of as few as 10 infectious particles can result in infection by the faecal-oral route, both through close person-to-person contact and by contaminated food and water (Brassard et al. 2005).

Virus is transmitted from child to child through contact. Several large rotavirus epidemics occurred in China affecting more than a million people (Hung et al. 1984, Su et al. 1986, Fang et al. 1989). Despite the general spread of rotavirus into the environment due to the enormous amount of viruses shed in faeces, there are far less food- and waterborne outbreaks recorded compared to norovirus. The probable main cause for this is the one described for sapovirus above, where mostly children are infected. Rotavirus group A infects nearly all children in high-income countries prior to age three. Adults are more rarely infected. However, immunity is not lifelong and persons above 60 years can be reinfected. Rotavirus B seems to be more associated with food and water transmission, probably due to a higher susceptibility among adults. Waterborne outbreaks are reported from many countries including Sweden. A large water-borne outbreak affected more than twelve thousand people (Lycke et al. 1978, Ijaz et al. 1994). In Japan more than 3000 cases were reported from primary schools. School lunches prepared centrally were suspected but rotavirus could not be found in the food distributed (Konno et al. 1978, Ishimaru et al. 1991, Mikami et al. 2004, Khamrin et al. 2006).

Pathogenesis, replication and clinical picture

Infection begins suddenly with vomiting followed by intense diarrhea and high fever. This often results in severe dehydration. The incubation time is one to three days. The symptoms usually last for four to six days. As in noroviruses, immune deficient individuals can become chronic shedders of rotavirus (Gilger et al. 1992).

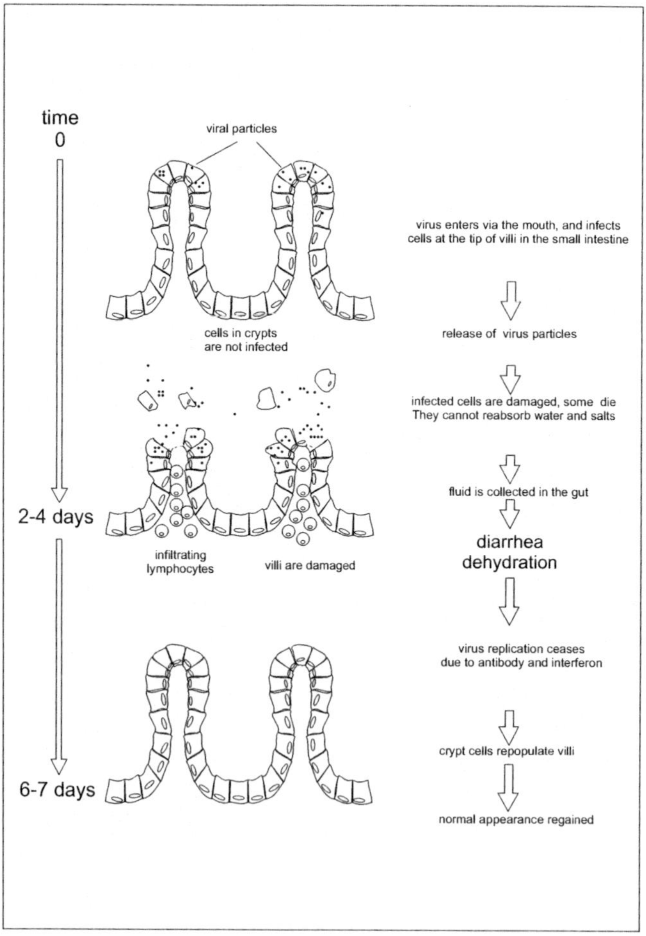

Figure 15. Rotavirus pathogenesis. Rotavirus decreases the resorption of salt and fluid in the intestine, causing a more fluid like faeces, i.e., diarrhea.

The rotavirus protein NSP4 is a toxin which increases gut permeability and reduces reabsorption of sodium ions, leading to diarrhea (Ball et al. 1996).

Prevention and therapy

Rotavirus therapy is aimed at preventing dehydration by oral or parenteral rehydration with oral rehydration solution (ORS) formulae (Desselberger 1999). There are two approved oral attenuated rotavirus vaccines. Both have given protection against severe rotavirus diarrhoea in high-income countries (Finland and the United States). However, the larger number of virus types in low-income countries means that protection may be more limited there. The vaccines are either recombinants between human rotaviruses or between human and bovine rotaviruses.

Picornaviridae

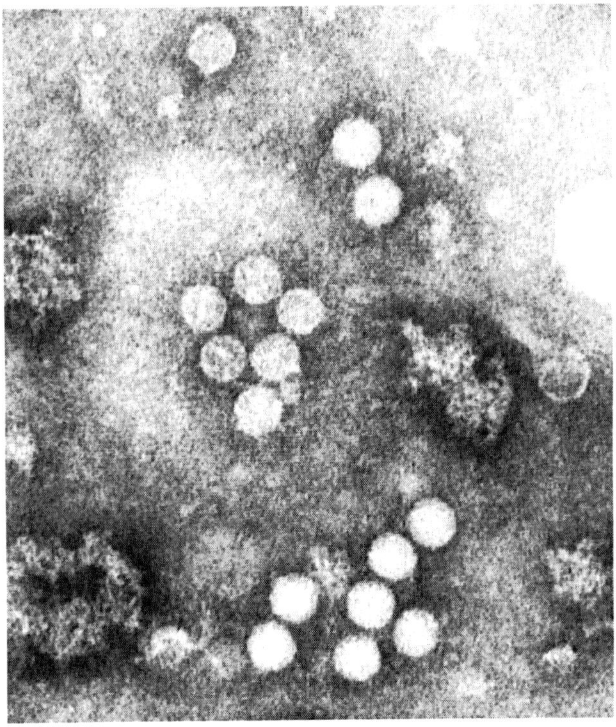

Figure 16. Electron micrograph of polio virus. Credit to Kjell-Olof Hedlund, SICDC.

Taxonomy and basic properties

The *Picornaviridae* family (with wellknown representatives within parenthesis) consists of the genera enterovirus (including Rhino-, Polio-, ECHO- and Coxsackie-viruses), parechovirus, kobuvirus (Aichivirus), hepatovirus (Hepatitis A virus), cardiovirus (Saffoldvirus) and aftovirus (Foot and mouth disease virus). Picornaviruses are 20–30 nm diameter icosahedral non-enveloped RNA viruses (Fig. 16). They are therefore resistant to many solvent detergents and disinfectants. However, hypochlorite, phenol, glutaraldehyde and formaldehyde inactivate picornaviruses.

Enteroviruses have been studied most. They are divided into Rhinovirus A-C and Human Enterovirus (HepEV) A-D. Poliovirus belongs to HepEV-B.

Although they are RNA viruses, picornaviruses are relatively invariant. An example is poliovirus. It contains three serotypes but still the poliovirus vaccine covers all three.

Replication, pathogenesis and clinical picture

Picornavirus infections are often asymptomatic, but can occasionally be severe. Rhinoviruses primarily replicate in upper respiratory epithelium, hepatoviruses in liver and gut, while enteroviruses, parechoviruses and kobuviruses primarily replicate in gut but also other tissues. Thus, faecal-oral transmission occurs for most picornaviruses. Rhinoviruses spread primarily via respiratory secretions on fingers.

The incubation period of picornaviruses varies between one and two weeks but can range from one-two days (rhinovirus) to 30 days (hepatovirus).

Diagnosis of picornavirus infection

The major diagnostic techniques are PCR and serology. Virus isolation may be useful for research, but is seldom used in routine diagnostic work.

Immunity, prevention and treatment of picornavirus infection

Picornaviruses generally give rise to type-specific immunity after infection. However, reinfection with the same rhinovirus strain can occur. The only vaccines available are against hepatitis A and poliovirus. Passive immunization with gammaglobulin is possible for most picornaviruses. It can also be used for treatment in severe cases, e.g., neonatal enterovirus infection.

More on specific picornaviruses

Rhinovirus

Rhinoviruses are a common cause of "common cold", a mild upper respiratory infection. They can occasionally be foodborne.

Aftovirus

Foot-and mouth disease virus (FMDV) can be present in meat from slaughtered infected livestock, and be a rare foodborne infection of humans. However FMDV cases in humans are mild. The greatest problem with FMDV is its effects on the animals.

Hepatovirus

Hepatitis A virus causes a relatively benign hepatitis. There are around 100 cases per year in Sweden. Asymptomatic infection is common among children but could be more sever in adults. Infection leads to lifelong immunity. In countries with high prevalence almost all adults are immune. In countries with a somewhat higher prevalence, like south of Europe, the risk for infections and food- and water-borne transmission to adults is higher. There is only one serotype of HAV. The HAV vaccination gives essentially 100% protection to all HAV strains. In clinical disease, after an incubation time of three to four weeks, nausea and vomiting appear. They are followed by a hepatitis which lasts 2–4 weeks. Nearly all patients recover without sequelae. The long incubation period makes it difficult to discover food- and water-borne outbreaks but after norovirus WHO/FAO consider HAV to be the most important among food-borne viruses (Scallan et al. 2011). HAV has been implicated in many food-associated outbreaks. The first shellfish associated outbreak with 600 cases of hepatitis was recorded in Sweden 1955-56 (Roos 1956). In China in 1988 300,000 people were infected with HAV after consumption of lightly cooked clams harvested in a sewage contaminated area (Halliday et al. 1991). Water-borne outbreaks also occur (Linglof et al. 1981, Scallan et al. 2011).

Enteroviruses

These infections are often asymptomatic, but can also provide more or less severe symptoms from various organs including the respiratory tract, heart, muscle, brain, eyes and skin. can cause conjunctivitis, herpangina, hand-foot-and-mouth disease (in humans; do not confuse with foot-and-mouth disease caused by aftovirus), myalgia and pleurodynia ("Bornholm disease"), perimyocarditis, aseptic meningitis, meningoencephalitis and, rarely, palsy. Newborns are especially susceptible and develop a sepsis-like disease, often with meningoencephalitis.

Enterovirus epidemiology. Enterovirus infections preferentially occur during late summer and autumn. Summer colds are often caused by rhinoviruses. However, foodborne spread is often not seasonal. A new human enterovirus type, enterovirus 71 (which belongs to HepEV-A), seemingly new to humans, was found 1972 (Blomberg et al. 1974). It rapidly spread over the world, causing foot-and-mouth-disease but sometimes more severe disease like meningoencephalitis with a mortality of 1–2%. Its spread is mostly by human-human contact, but may also foodborne. Only a few (around 1/1000) of the infected individuals are symptomatic. But many in the population are infected, and shed virus. This leads to a very much larger shedding of enterovirus than would be expected. Enterovirus is often detected in sewage and molluscan shellfish but enteroviruses are seldom involved in food- and water-borne outbreaks (Lopez-Sabater et al. 1997). The risk for transmission through food and water could although be underestimated as enterovirus infections in the population more or less always appear as sporadic cases.

Foodborne spread of picornaviruses. A major route for spread of picornaviruses is through fecal-oral transmission. Common transmission chains involve cakes, water and mussels, especially for Hepatitis A virus. Older and immunosuppressed individuals can develop carriership with chronic secretion of picornavirus in faeces. This poses a risk for outbreaks in institutions for the elderly, and in hospitals.

Newly discovered picornaviruses

The recent discovery of a large number of new picornaviruses will require an intense effort to understand their epidemiology and pathogenesis.

Parechoviruses. They cause a broad spectrum of disease, from gastroenteritis, respiratory infection, palsies, encephalitis, aseptic meningitis, myocarditis, neonatal sepsis-like illness and rash with hemorrhage. Their spread via food has not been studied much.

Kobuviruses. These seem to be rather common causes of gastroenteritis, with faecal-oral spread. In a number of cases, spread via food has been documented. For example, Aichi virus has been isolated from shellfish in association with gastroenteritis outbreaks in Japan and France (Sdiri-Loulizi et al. 2010). Similar viruses have been isolated from cattle and swine. Thus some kobuvirus outbreaks in humans may have a zoonotic origin. Much remains to be investigated regarding the epidemiology and pathogenesis of kobuviruses.

Cardioviruses. In mice, they cause neurological, gastroenterological and myocardial disease. This group of viruses was recently discovered in humans. They have been associated with myocarditis in humans, acute and chronic encephalitis, meningitis, sudden infant death, upper respiratory infection and gastroenteritis. Their spread via food has not been studied much.

Hepeviridae

Hepatitis E virus

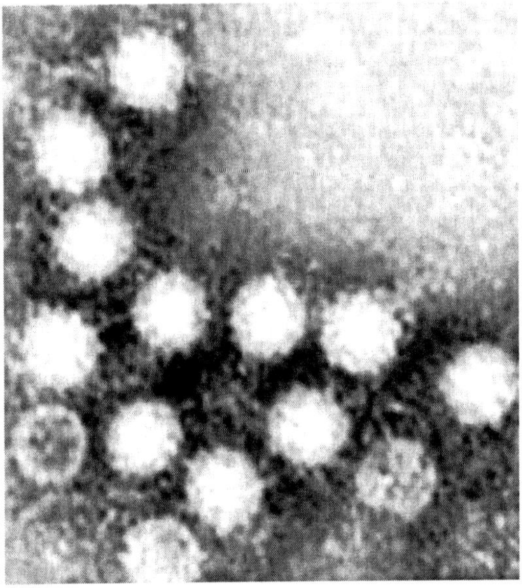

Figure 17. Electron micrograph of Hepatitis E virus. Credit to Kjell-Olof Hedlund, SICDC.

Taxonomy and basic characteristics

Hepatitis E virus (HepEV) is the single species in the *Hepeviridae* family. They are nonenveloped icosahedral RNA viruses, with a diameter of 30 nm (Fig. 17).

There are four genotypes of HepEV. Type 1 and 2 infects only humans and is endemic in Asia, Africa and Latin America. Type 3 and 4 also infect animals and has also been isolated from Europe, USA and Australia. Most patients diagnosed with hepatitis E in Europe have been infected by type 1 in Asia or Africa. However there are also endemic HepEV type 3 cases (usually in men over 60), which often seems to be of zoonotic origin, from pigs. Genotype 3 is present in the pig population world-wide while genotype 4 is present in the pig population of East and South-East Asia.

Replication, pathogenesis and clinical picture

HepEV infects gut epithelium, spreads via lymphatics into blood, then to the liver, where the main replication takes place. Transmission is generally fecal-oral.

HepEV infection is often asymptomatic or subclinical. If clinical, it has a somewhat longer incubation time than hepatitis A, about four weeks. During the first ten days the patient has general symptoms with malaise, abdominal pain, nausea and fever. Then follow hepatitis symptoms with jaundice. Hepatitis E is similar to, but generally more severe than Hepatitis A. Especially susceptible are pregnant women, where mortality can reach 20% in the second half of gestation. Other risk groups are children under the age of two, and immunocompromised patients. The latter risk the development of a chronic infection.

Epidemiology

HepEV causes large waterborne outbreak in Asia, Africa and Latin America, but also infects via food or other contact with domestic swine, wild boar and deer. HepEV is responsible for about half of all cases of acute hepatitis in countries in Asia, Africa, and Central and South America where it is endemic. It is spread faecal-oral route and has since the 1950s caused large waterborne outbreaks in these countries. HepEV also infects domestic swine, wild boars, deer and other mammals. HepEV type 3 infection is increasing. It infects humans through food or contact with infected animals in Europe and the United States, where hepatitis E has not previously been known to occur.

Foodborne spread

Waterborne outbreaks with more than 10,000 cases, have been discovered in India, Russia, Tunisia and Somalia. Eating of seafood contaminated by sewage and undercooked pig liver, pig or wild boar and deer meat can also lead to HepEV infection in humans.

Diagnosis

Both PCR and serology are available.

Immunity, prevention and treatment

Even though immunity develops after HepEV infection, there are no commonly available vaccines. Intravenous gamma globulin has been tried in severe cases.

Other viruses

Astrovirus

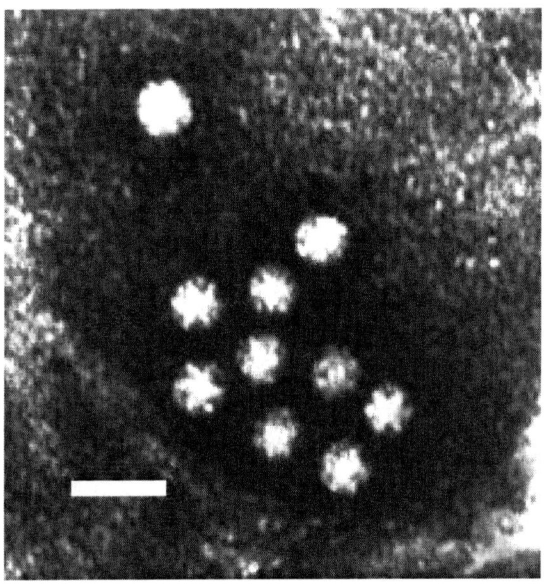

Figure 18. Human astrovirus. Credit to F.P. Williams, U.S. EPA.

Taxonomy and basic characteristics

Astroviruses have now been found in numerous mammalian animal species. Astroviruses are nonenveloped 28–35 nm diameter, icosahedral viruses that have a characteristic five- or sixpointed star-like surface structure when viewed by electron microscopy.

Replication, pathogenesis and clinical picture

Astroviruses infect epithelium in the the small intestine and cause a deranged water and salt balance there. Human astroviruses occurs in 5–10%

of gastroenteritis in young children. Humans of all ages can be infected, but children, elderly and those that are immunocompromised are most prone.

Epidemiology

The majority of children have been infected by the age of 5. There are around 1000 diagnosed cases per year in Sweden, causing 5–10% of childhood gastroenteritis cases. It is most common during winter. The reason for this pattern is unknown, but spread may be promoted by connections between sewage and drinking water which may occur during heavy rainfall.

Foodborne spread

The main transmission for Astroviruses is person-to-person, mainly among children, and there is little evidence for food and water transmission. Nevertheless, contaminated shellfish and water has been implicated in outbreaks (Oishi et al. 1994, Utagawa et al. 1994, Glass et al. 1996). Food supplied to 14 different schools in Japan affected thousands of children and adults. Environmental spread of astroviruses is well known and has been recorded in shellfish. The spread through water and food could therefore be of greater importance than indicated by the relatively few recorded outbreaks.

Diagnosis

(PCR and serology are available)

Immunity, prevention and treatment

There is no vaccine or treatment available but antibodies seem to provide protection through adult life, until the antibody titre begins to decline later in life.

Enteric adenoviruses

Taxonomy and basic characteristics

The adenovirus family is large and present in most vertebrates. Adenoviruses are relatively complex large (70–90 nm) nonenveloped icosahedral DNA viruses. They are resistant to lipid solvents and detergents, but are inactivated by formaldehyde and hypochlorite. Essentially all human adenoviruses are shed via faeces. However, it is only human adenovirus genotypes 40 and 41 ("enteric" adenoviruses) which cause gastroenteritis.

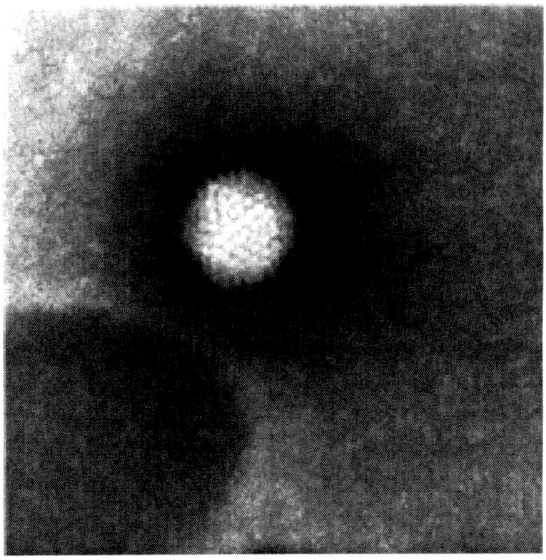

Figure 19. Electron micrograph of human enteric adenovirus from stool. Credit to Kjell-Olof Hedlund, SICDC.

Replication, pathogenesis and clinical picture

Most adenoviruses infect gut and respiratory epithelium, then spread to the rest of the body. Enteric adenoviruses are confined to gastrointestinal epithelium. Adenovirus infection is often asymptomatic or subclinical. If clinical, types 40 and 41 cause a gastroenteritis which is similar to rotaviral diarrhea. Like rotaviruses, they occur primarily during the first years of life. However, their incidence is around one tenth of that of rotaviruses. The disease is similar to that of rotavirus gastroenteritis. Especially susceptible are immunocompromised (especially bone marrow transplanted and AIDS) patients. They can develop pneumonia, encephalitis and hepatitis, with a high mortality. The infection can also become chronic, with sustained excretion of virus in faeces.

Epidemiology

Enteric adenoviruses occur endemically all over the world. In Sweden there are around 2000 cases per year.

Foodborne spread

Gastroenteric adenoviruses have been found in raw water and shellfish. It is commonly found in sewage. As a DNA virus is easier to detect by molecular methods in environmental samples adenoviruses has been proposed to act as a indicators for enteric viruses in the environment and in food matrices. Despite the general occurrence of adenovirus in water and shellfish there have not been any recorded adenovirus outbreaks associated with food or water transmission.

Diagnosis

PCR and serology are the most common methods.

Immunity, prevention and treatment

There are no commonly available vaccines. Intravenous gammaglobulin can be tried in severe cases. The nucleoside analog Cidofovir can also be used against severe adenovirus infections.

Tick-borne encephalitis virus

Tick-borne encephalitis virus, an enveloped RNA virus belonging to the family flaviviridae, is generally not considered a food-borne virus as the main transmission is by tick bites. However, it is now clear that a less common mode of transmission through consumption of goat, sheep or cow's milk occurs occasionally. It is estimated that in Eastern and Central Europe this route is responsible for 5% of all cases (Balogh et al. 2010).

Technologies for Detection of Foodborne Viruses

A majority of the many hundreds of millions of food- and waterborne disease cases, which occur globally every year, are due to viruses. A surveillance system for foodborne viruses should be highly sensitive, inexpensive, rapid and comprehensive. This surveillance poses problems, which surpass the demands for diagnostic tests in humans. The surveillance technique must be part of an institutional surveillance system and sample the food chain at several strategic stages. For optimal results, the surveillance should be instigated at strategic stages in the food chain (Figure 2). The surveillance system must find its place together with established methods. In contrast to bacterial detection methods, where bacterial culture still plays a large role, most foodborne viruses cannot be grown in cell culture and therefore cannot be detected by virus isolation on cell culture. Virology has therefore become much more committed to nucleic acid based detection methods like PCR than bacteriology has. The real-time (RT)-PCR methods are sensitive and quick and can detect the virus in a few hours.

There is presently a method standard available for analysis of viruses in foodstuff. The standard is in the format of a technical specification and is named ISO TS 15216. Preparatory, enrichment and analytical procedures are also described in (Schwartzbrod et al. 1989, Gajardo et al. 1991, Pinto et al. 2007, Sanchez et al. 2007, Wyn-Jones 2007, Rodriguez-Lazaro et al. 2012).

There are some basic challenges during development of a nucleic-acid based surveillance tool for foodborne viruses; 1. The sample volume is often great. The sample may have to be concentrated by ultrafiltration, precipitation or two-phase extraction. Nucleic acid extraction may have to be performed from large volumes; 2. Inhibitors of nucleic acid detection may have to be removed; 3. Large amounts of non-target nucleic acid may make detection of target nucleic acid difficult; 4. The sequence variation of the target may require variation-tolerant detection methods.

The sample has to be chosen carefully with a great knowledge of where in the sample the virus occurs, e.g., on the surface or inside the vegetable, or within the clam. The virus could be eluted with a buffer from the surface of the food, but this will increase the sample volume. The virus can also be homogenized with the food into a solution. The question of sample volume is especially relevant for analysis of raw and treated sewage water where tens of liters may have to be analyzed. River and drinking water may require even greater volumes.

To recover and concentrate the virus several successive techniques can be used: precipitation with PEG (polyethylene glycol), flocculation, ultrafiltration, ultracentrifugation and adsorption plus elution from filters. After enrichment the nucleic acid is extracted, amplified, detected and identified with, e.g., real-time PCR. In some cases, inhibitors can follow through the different steps of enrichment and extraction, which will affect the detection of the nucleic acid. A comprehensive approach could therefore be to combine target nucleic acid enrichment with ISO standardized sample preparation protocols (Fig. 21). It has however not yet been used in practice.

The amplification system can be singleplex, duplex or multiplex, i.e., amplification and detection of one specific target or several possible targets (pathogens) is made simultaneously in the same analysis. Many different primer and probe techniques can be used. The primers can be target-specific or consist of a random oligonucleotide sequence. These could be combined with a generic sequence into one oligonucleotide, act separately in consecutive order, or be linked as in a padlock probe. The variation in the pathogens genomes could require long primers and probes and wobbling between different bases in one or several nucleotide positions in a target specific primer, creating a degenerated primer. The primer can also contain nucleotide analogues, like deoxyInosines, that hybridize to all four types of nucleotide bases, to cover up for variation in the target. An illustration of the profound variation in some foodborne RNA viruses is given in Fig. 20, where the variation in norovirus genome is depicted. The great variation is a serious complication for nucleic acid based detection methods. It is common that 10–40% of noroviral outbreaks cannot be detected using such methods. Special techniques for managing variation are needed.

Diagnosis and surveillance of gastroenteric agents

Different techniques can be used to detect and identify the amplified pathogen nucleic acid, like size separating gel electrophoresis, real-time PCR, melting point analysis, microarray (solid phase or bead liquid array), or sequencing. Many singleplex and multiplex gastroenteritis methods have been developed using different strategies of amplification and read-out platforms (Palacios et al. 2007, Pham et al. 2010, van Maarseveen et al. 2010, Chen et al. 2011, Khamrin et al. 2011, Liu et al. 2011, Öhrmalm et al. 2012).

Real-time PCR is a common and remarkably sensitive technique. Despite a sensitivity of 1–10 copies per reaction it has a relatively low tendency to contaminate, because it is not necessary to open the PCR tubes after completion of the analysis. A very high sensitivity of the detection method is needed, since presence of only a few virus particles per 100 mL of foodstuff are enough to cause disease. Frozen raspberries have caused gastroenteritis outbreaks repeatedly due to contamination with Nororvirus. If 100 mL of raspberries containing 10 Norovirus RNA molecules is extracted to 1 mL of extract, and a maximum of 50% of the Norovirus RNA is recovered from the extract, the PCR must have a sensitivity of 1–10 target molecules to give a positive result. This demanding task calls for extensive optimization of the entire procedure (Vennema et al. 2002, Vinje et al. 2003, van den Berg et al. 2005, Boxman et al. 2007, de Wit et al. 2007, Kostela et al. 2008, Siebenga et al. 2008, Boxman et al. 2009a, Boxman et al. 2009b, Teunis et al. 2009, ter Waarbeek et al. 2010, Verhoef et al. 2010, Amdiouni et al. 2012).

Despite its great sensitivity, RT-PCR can normally detect only a few different target sequences per reaction. It will therefore probably be complemented gradually by multiplex techniques, where most of the causes of gastroenteritis are detected. High multiplexity can be gained by hybridising the amplicons to a microarray with millions of probes (Wang et al. 2002, Palacios et al. 2007, Chen et al. 2011), but the method has limited sensitivity and is time consuming due to slow diffusion times. Detecting and identifying the target in a liquid bead array, as in the Luminex system (Luminex corporation, Austin Tx) (Liu et al. 2011, Öhrmalm et al. 2012) is quicker due to shorter hybridization time because bead-bound probes have a higher likelihood of a collision with a target. VOCMA (Variation

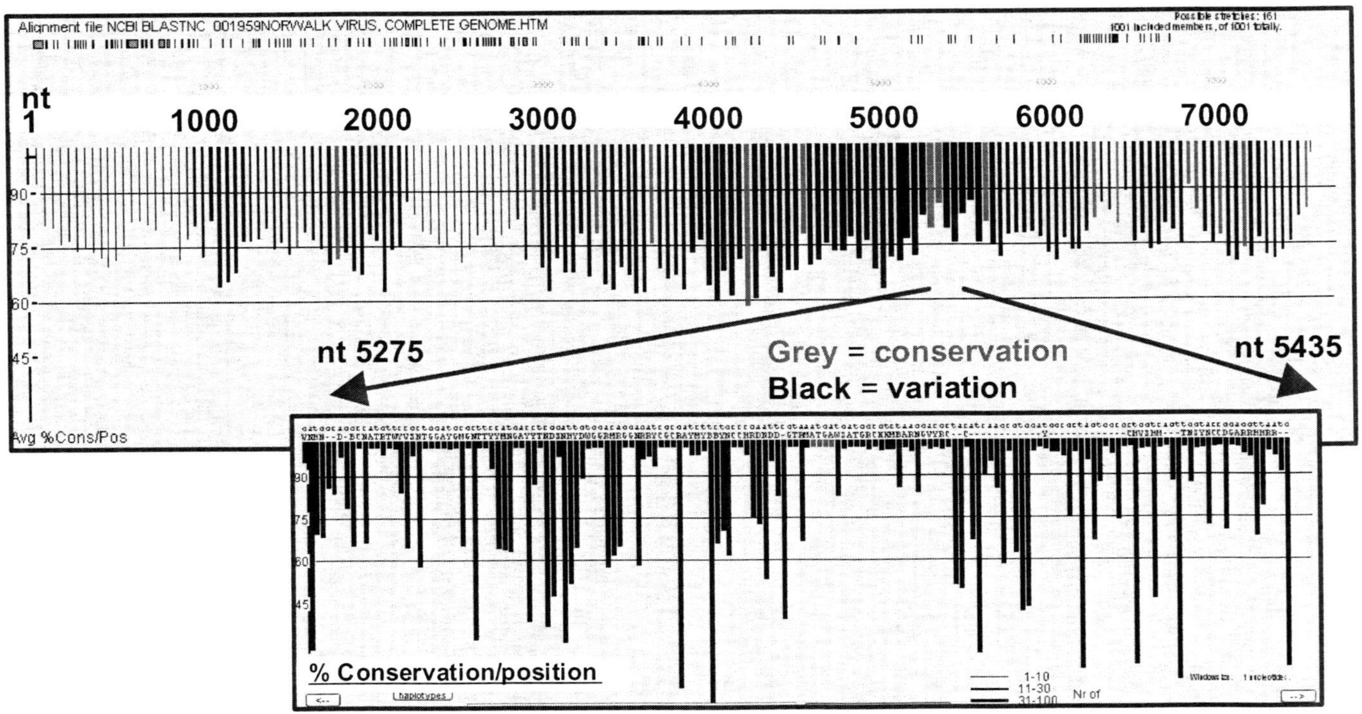

Figure 20. The variability of the Noroviral RNA genome, as depicted by the computer program ConSort (J.B., unpublished, and (Öhrmalm et al. 2010, Öhrmalm et al. 2012)). The variation diagram was based on 1472 noroviral genomes. Only a few of the 8000 nucleotides (nt) are invariant.

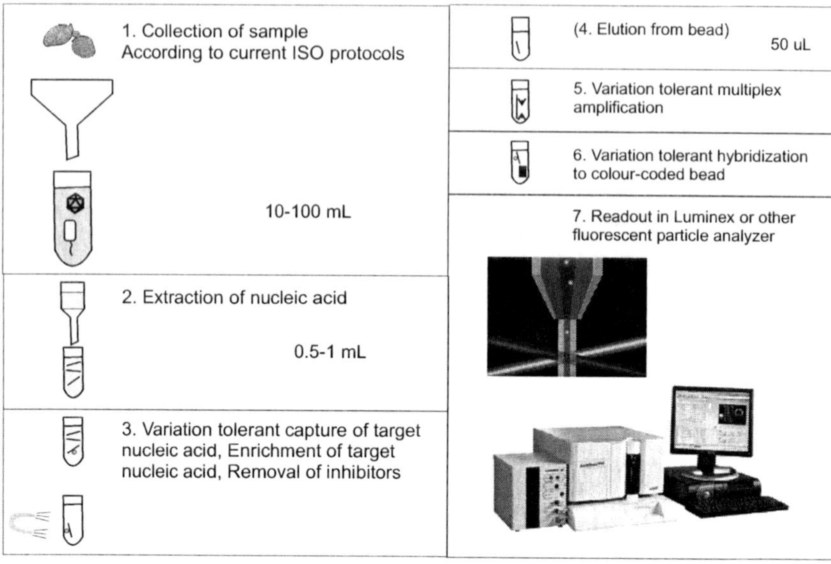

Figure 21. Outline of analysis of the stages of analysis for foodborne pathogens, exemplified by VOCMA (Variation tOlerant Capture Multiplex Assay (Öhrmalm et al. 2009, Öhrmalm et al. 2012)). VOCMA allows enrichment from inhibitors and large sample volumes, and can tolerate variation in target nucleic acid. A combination of enrichment both pre and post nucleic acid extraction is necessary.

tolerant Capture Multiplex Assay) is an example of such a technique (Fig. 21) (Öhrmalm et al. 2009, Öhrmalm et al. 2012). It can enrich target nucleic acid, and is tolerant to the highly variable targets. It is multiplex, and covers the most common gastroenteric viruses. In preliminary tests, bacteria and protozoa which can cause gastroenteritis were also included. It has potential for both clinical diagnosis, and for surveillance of food, sewage and strategic steps in the food chain. The xTAG® Gastrointestinal Pathogen Panel (xTAG GPP) from Luminex detects 15 major gastrointestinal pathogens with the xTAG technology. One of the primers has a unique tag, which allows the biotinylated product to be hybridized to an anti-tag coupled to the colour-coded bead. The Seeplex® Diarrhea ACE Detection (Seegene (Seoul, Korea) detects 15 pathogens, divided into 3 panels, with DPO™ (Dual Priming Oligonucleotide) technology and size separating capillary electrophoresis.

It is safe to conclude that methods for detection of foodborne viruses will develop much in the coming years, and will allow a much improved level of surveillance for these pathogens.

Conclusions and Trends

Foodborne viruses are common, and cause much disease. To control them calls for efficient and precisely directed surveillance methods. Ideally, diagnosis and surveillance should be done with broadly targeted multiplex methods. It is only if we know where in the food chain that the viral contamination occurs that an efficient surveillance can be achieved.

References

Amdiouni, H., L. Maunula, K. Hajjami, A. Faouzi, A. Soukri and J. Nourlil. 2012. Recovery comparison of two virus concentration methods from wastewater using cell culture and real-time PCR. Curr. Microbiol. 65: 432–437.

Ball, J.M., P. Tian, C.Q. Zeng, A.P. Morris and M.K. Estes. 1996. Age-dependent diarrhea induced by a rotaviral nonstructural glycoprotein. Science. 272(5258): 101–104.

Balogh, Z., E. Ferenczi, K. Szeles, F. Stefanoff, W. Gut, K.N. Szomor, M. Takacs and G. Berencsi . 2010. Tick-borne encephalitis outbreak in Hungary due to consumption of raw goat milk. J. Virol. Meth. 163: 481–485.

Bank-Wolf, B.R., M. König and H.J. Thiel. 2010. Zoonotic aspects of infections with noroviruses and sapoviruses. Vet. Microbiol. 140: 204–212.

Blomberg, J., E. Lycke, K. Ahlfors, T. Johnsson, S. Wolontis and G. von Zeipel. 1974. Letter: New enterovirus type associated with epidemic of aseptic meningitis and-or hand, foot, and mouth disease. Lancet. 2(7872): 112.

Boxman, I., R. Dijkman, L. Verhoef, A. Maat, G. van Dijk, H. Vennema and M. Koopmans. 2009a. Norovirus on swabs taken from hands illustrate route of transmission: a case study. J. Food Protect. 72: 1753–1755.

Boxman, I.L., R. Dijkman, N.A. te Loeke, G. Hägele, J.J. Tilburg, H. Vennema and M. Koopmans. 2009b. Environmental swabs as a tool in norovirus outbreak investigation, including outbreaks on cruise ships. J. Food Protect. 72: 111–119.

Boxman, I.L., J.J. Tilburg, N.A. te Loeke, H. Vennema, E. de Boer and M. Koopmans. 2007. An efficient and rapid method for recovery of norovirus from food associated with outbreaks of gastroenteritis. J. Food Protect. 70: 504–508.

Brassard, J., K. Seyer, A. Houde, C. Simard and Y.L. Trottier. 2005. Concentration and detection of hepatitis A virus and rotavirus in spring water samples by reverse transcription-PCR. J. Virol. Meth. 123: 163–169.

Chen, E.C., S.A. Miller, J.L. DeRisi and C.Y. Chiu. 2011. Using a pan-viral microarray assay (Virochip) to screen clinical samples for viral pathogens. J. Vis. Exp. 50: 2536. Published online 2011 April 27. doi: 10.3791/2536.

Chitambar, S.D., A. Lahon, V.S. Tatte, N.H. Maniya, G.U. Tambe, K.I. Khatri, H.S. Desai, M.R. Ugare, S.V. Kulkarni and A.F. Waghmare. 2011. Occurrence of group B rotavirus infections in the outbreaks of acute gastroenteritis from western India. Indian J. Med. Res. 134: 399–400.

de Wit, M.A., M.A. Widdowson, H. Vennema, E. de Bruin, T. Fernandes and M. Koopmans. 2007. Large outbreak of norovirus: the baker who should have known better. J. Infect. 55: 188–193.

Desselberger, U. 1999. Rotavirus infections: guidelines for treatment and prevention. Drugs 58: 447–452.

Donaldson, E.F., L.C. Lindesmith, A.D. Lobue and R.S. Baric. 2010. Viral shape-shifting: norovirus evasion of the human immune system. Nat. Rev. Microbiol. 8: 231–241.

Fang, Z.Y., Q. Ye, M.S. Ho, H. Dong, S. Qing, M.E. Penaranda, T. Hung, L. Wen and R.I. Glass RI. 1989. Investigation of an outbreak of adult diarrhea rotavirus in China. J. Infect. Dis. 160: 948–953.

Flewett, T.H., A.S. Bryden, H. Davies, G.N. Woode, J.C. Bridger and J.M. Derrick. 1974. Relation between viruses from acute gastroenteritis of children and newborn calves. Lancet. 2(7872): 61–63.

Flewett, T.H. and G.N. Woode. 1978 .The rotaviruses. Arch. Virol. 57: 1–23.

Gajardo, R., J.M. Diez, J. Jofre and A. Bosch. 1991. Adsorption-elution with negatively and positively-charged glass powder for the concentration of hepatitis A virus from water. J. Virol. Meth. 31: 345–351.

Gilger, M.A., D.O. Matson, M,E. Conner, H.M. Rosenblatt, M.J. Finegold and M.K. Estes. 1992. Extraintestinal rotavirus infections in children with immunodeficiency. J. Pediatr. 120: 912–917.

Glass, R.I., J. Noel, D. Mitchell, J.E. Herrmann, N.R Blacklow, L.K. Pickering, P. Dennehy, G. Ruiz-Palacios, M.L. de Guerrero and S.S. Monroe. 1996. The changing epidemiology of astrovirus-associated gastroenteritis: a review. Arch. Virol. Suppl. 12: 287–300.

Goodgame, R.W. 1999. Viral infections of the gastrointestinal tract. Curr. Gastroenterol. Rep. 1: 292–300.

Halliday, M.L., L.Y. Kang, T.K. Zhou, M.D. Hu, Q.C. Pan, T.Y. Fu, Y.S. Huang and S.L. Hu. 1991. An epidemic of hepatitis A attributable to the ingestion of raw clams in Shanghai, China. J. Infect. Dis. 164: 852–859.

Hallin, E. 2012. Norovirus in water—A literature review. Svenskt Vatten Utveckling, Svenskt vatten AB. 10: 52.

Huang, P., T. Farkas, W. Zhong, M. Tan, S. Thornton, A.L. Morrow and X. Jiang. 2005. Norovirus and histo-blood group antigens: demonstration of a wide spectrum of strain specificities and classification of two major binding groups among multiple binding patterns. J. Virol. 79: 6714–6722.

Huang, P., A.L. Morrow and X. Jiang. 2009. The carbohydrate moiety and high molecular weight carrier of histo-blood group antigens are both required for norovirus-receptor recognition. Glycoconj. J. 26: 1085–1096.

Hung, T., G.M. Chen, C.G. Wang, H.L. Yao, Z.Y. Fang, T.X. Chao, Z.Y. Chou, W. Ye, X.J. Chang and S.S. Den. 1984. Waterborne outbreak of rotavirus diarrhoea in adults in China caused by a novel rotavirus. Lancet. 1(8387): 1139–1142.

Ijaz, M.K., S. Alharbi, S.A. Uduman, Y. Cheema, M.M. Sheek-Hussen, A.R. Alkhair, A.G. Shalabi, S.S. Ijaz, S.A. Bin-Othman, S.A. Sattar and L.F. Liddle. 1994. Seasonality and prevalence of rotavirus in Al-Ain, United Arab Emirates. Clin. Diagn. Virol. 2: 323–329.

Ishimaru, Y., S. Nakano, H. Nakano, M. Oseto and Y. Yamashita. 1991. Epidemiology of group C rotavirus gastroenteritis in Matsuyama, Japan. Acta Paediatr. Jpn. 33: 50–56.

Khamrin, P., M. Okame, A. Thongprachum, N. Nantachit, S. Nishimura, S. Okitsu, N. Maneekarn and H. Ushijima. 2011. A single-tube multiplex PCR for rapid detection in feces of 10 viruses causing diarrhea. J. Virol. Meth. 173: 390–393.

Khamrin, P., S. Peerakome, L. Wongsawasdi, S. Tonusin, P. Sornchai, V. Maneerat, C. Khamwan, F. Yagyu, S. Okitsu, H. Ushijima and N. Maneekarn. 2006. Emergence of human G9 rotavirus with an exceptionally high frequency in children admitted to hospital with diarrhea in Chiang Mai, Thailand. J. Med. Virol. 78: 273–280.

Konno, T., H. Suzuki, A. Imai, T. Kutsuzawa, N. Ishida, N. Katsushima, M. Sakamoto, S. Kitaoka, R. Tsuboi and M. Adachi. 1978. A long-term survey of rotavirus infection in Japanese children with acute gastroenteritis. J. Infect. Dis. 138: 569–576.

Kostela, J., M. Ayers, J. Nishikawa, L. McIntyre, M. Petric and R. Tellier. 2008. Amplification by long RT-PCR of near full-length norovirus genomes. J. Virol. Meth. 149: 226–230.

Lindesmith, L.C., E.F. Donaldson and R.S. Baric. 2011. Norovirus GII.4 strain antigenic variation. J. Virol. 85: 231–242.

Linglöf, T., U.B. Lindström, T. Johnsson, L. Magnius , F. Nordbring and K. Roos. 1981. [An epidemic of water-borne hepatitis A, in Swedish]. Läkartidningen. 78(36): 3032–3034.

Liu, J., G. Kibiki, V. Maro, A. Maro, H. Kumburu, N. Swai, M. Taniuchi, J. Gratz, D. Toney, G. Kang and E. Houpt. 2011. Multiplex reverse transcription PCR Luminex assay for detection and quantitation of viral agents of gastroenteritis. J. Clin. Virol. 50: 308–313.

Lopez-Sabater, E.I., M.Y. Deng and D.O. Cliver. 1997. Magnetic immunoseparation PCR assay (MIPA) for detection of hepatitis A virus (HAV) in American oyster (*Crassostrea virginica*). Lett. Appl. Microbiol. 24: 101–104.

Lycke, E., J. Blomberg, G. Berg, A. Eriksson and L. Madsen. 1978. Epidemic acute diarrhoea in adults associated with infantile gastroenteritis virus. Lancet 2(8098): 1056–1057.

Matthijnssens, J., M. Ciarlet, S.M. McDonald, H. Attoui, K. Bányai, J.R. Brister, J. Buesa, M.D. Esona, M.K. Estes, J.R. Gentsch, M. Iturriza-Gómara, R. Johne, C.D. Kirkwood, V. Martella, P.P.C. Mertens, O. Nakagomi, V. Parreño, M. Rahman, F.M. Ruggeri, L.J. Saif, N. Santos, A. Steyer, K. Taniguchi, J.T. Patton, U. Desselberger and M. Van Ranst. 2011. Uniformity of rotavirus strain nomenclature proposed by the Rotavirus Classification Working Group (RCWG). Arch. Virol. 156: 1397–1413.

Maunula, L., M. Roivainen, M. Keränen, S. Mäkela, K. Söderberg, M. Summa, C.H. von Bonsdorff, M. Lappalainen, T. Korhonen, M. Kuusi and T. Niskanen. 2009. Detection of human norovirus from frozen raspberries in a cluster of gastroenteritis outbreaks. Euro surveillance : bulletin europeen sur les maladies transmissibles = European communicable disease bulletin 14(49).

Mead, P.S., L. Slutsker, V. Dietz, L.F. McCaig, J.S. Bresee, C. Shapiro, P.M. Griffin and R.V. Tauxe. 1999. Food-related illness and death in the United States. Emerg. Infect. Dis. 5: 607–625.

Mikami, T., T. Nakagomi, R. Tsutsui, K. Ishikawa, Y. Onodera, K. Arisawa and O. Nakagomi. 2004. An outbreak of gastroenteritis during school trip caused by serotype G2 group A rotavirus. J. Med. Virol. 73: 460–464.

Oishi, I., K. Yamazaki, T. Kimoto, Y. Minekawa, E. Utagawa, S. Yamazaki, S. Inouye, G.S. Grohmann, S.S. Monroe, S.E. Stine, C. Carcamo, T. Ando and R.I. Glass. 1994. A large outbreak of acute gastroenteritis associated with astrovirus among students and teachers in Osaka, Japan. J. Infect. Dis. 170: 439–443.

Palacios, G., P. L. Quan, O.J. Jabado, S. Conlan, D.L. Hirschberg, Y. Liu, J. Zhai, N. Renwick, J. Hui, H. Hegyi, A. Grolla, J.E. Strong, J.S. Towner, T.W. Geisbert, P.B. Jahrling, C. Büchen-Osmond, H. Ellerbrok, M.P. Sanchez-Seco, Y. Lussier, P. Formenty, S.T. Nichol, H. Feldmann, T. Briese and W.I. Lipkin. 2007. Panmicrobial oligonucleotide array for diagnosis of infectious diseases. Emerg. Infect. Dis. 13: 73–81.

Parashar, U.D., C.J. Gibson, J.S. Bresee and R.I. Glass. 2006. Rotavirus and severe childhood diarrhea. Emerg. Infect. Dis. 12: 304–306.

Patel, M.M., A.J. Hall, J. Vinjé and U.D. Parashar. 2009. Noroviruses: a comprehensive review. J. Clin. Virol. 44: 1–8.

Pham, N.T., Q.D. Trinh, W. Chan-It, P. Khamrin, H. Shimizu, S. Okitsu, M. Mizuguchi and H. Ushijima. 2010. A novel RT-multiplex PCR for detection of Aichi virus, human parechovirus, enteroviruses and human bocavirus among infants and children with acute gastroenteritis. J. Virol. Meth. 169: 193–197.

Pintó, R.M., D. Alegre, A. Domínguez, W.M. El-Senousy, G. Sánchez, C. Villena, M.I. Costafreda, L. Aragonès and A. Bosch. 2007. Hepatitis A virus in urban sewage from two Mediterranean countries. Epidemiol. Infec. 135: 270–273.

Qiao, H., M. Nilsson, E.R. Abreu, K.O. Hedlund, K. Johansen, G. Zaori and L. Svensson. 1999. Viral diarrhea in children in Beijing, China. J. Med. Virol. 57: 390–396.

Rodriguez-Lazaro, D., N. Cook, F.M. Ruggeri, J. Sellwood, A. Nasser, M.S. Nascimento, M. D'Agostino, R. Santos, J.C. Saiz, A. Rzeżutka, A. Bosch, R. Gironés, A. Carducci, M. Muscillo, K. Kovač, M. Diez-Valcarce, A. Vantarakis, C.H. von Bonsdorff, A.M. de Roda Husman, M. Hernández and W.H. van der Poel. 2012. Virus hazards from food, water and other contaminated environments. FEMS Microbial. Rev. 36: 786–814.

Roos, B. 1956. [Hepatitis epidemic transmitted by oysters. In Swedish]. Svenska Läkartidningen 53(16): 989–1003.

Sanchez, G., A. Bosch and R.M. Pintó. 2007. Hepatitis A virus detection in food: current and future prospects. Lett. Appl. Microbiol. 45: 1–5.

Santos, N. and Y. Hoshino. 2005. Global distribution of rotavirus serotypes/genotypes and its implication for the development and implementation of an effective rotavirus vaccine. Rev. Med. Virol. 15: 29–56.

Scallan, E., R.M. Hoekstra, F.J. Angulo, R.V. Tauxe, M.A. Widdowson, S.L. Roy, J.L. Jones and P.M. Griffin. 2011. Foodborne illness acquired in the United States—major pathogens. Emerg. Infect. Dis. 17: 7–15.

Schwartzbrod, L., A. Bosch, F. Lucena, R. Girones, C. Beril and J. Jofre. 1989. Recovery of solid-associated virus: evaluation of different procedures. Zentralbl. Hyg. Umweltmed. (Int. J. Hyg. Environ. Med.) 188: 559–565.

Sdiri-Loulizi, K., M. Hassine, Z. Aouni, H. Gharbi-Khelifi, N. Sakly, S. Chouchane, M.N. Guédiche, P. Pothier, M. Aouni and K. Ambert-Balay. 2010. First molecular detection of Aichi virus in sewage and shellfish samples in the Monastir region of Tunisia. Arch. Virol. 155: 1509–1513.

Sen, A., N. Kobayashi, S. Das, T. Krishnan, S.K. Bhattacharya and T.N. Naik. 2001. The evolution of human group B rotaviruses. Lancet. 357(9251): 198–199.

Siebenga, J., A. Kroneman, H. Vennema, E. Duizer, M. Koopmans and Food-borne Viruses in Europe network. 2008. Food-borne viruses in Europe network report: the norovirus GII.4 2006b (for US named Minerva-like, for Japan Kobe034-like, for UK V6) variant now dominant in early seasonal surveillance. Euro. Surveill. 13(2). pii: 8009.

Su, C.Q., Y.L. Wu, H.K. Shen, D.B. Wang, Y.H. Chen, D.M. Wu, L.N. He and Z.L. Yang. 1986. An outbreak of epidemic diarrhoea in adults caused by a new rotavirus in Anhui Province of China in the summer of 1983. J. Med. Virol. 19: 167–173.

Summa, M., C.H. von Bonsdorff and L. Maunula. 2012. Pet dogs—a transmission route for human noroviruses? J. Clin. Virol. 53: 244–247.

Tan, M., M. Xia, S. Cao, P. Huang, T. Farkas, J. Meller, R.S. Hegde, X. Li, Z. Rao and X. Jiang. 2008. Elucidation of strain-specific interaction of a GII-4 norovirus with HBGA receptors by site-directed mutagenesis study. Virology. 379: 324–334.

ter Waarbeek, H.L., N.H. Dukers-Muijrers, H. Vennema and C.J. Hoebe. 2010. Waterborne gastroenteritis outbreak at a scouting camp caused by two norovirus genogroups: GI and GII. J. Clin. Virol. 47: 268–272.

Teunis, P.F., S.A. Rutjes, T. Westrell and A.M. de Roda Husman. 2009. Characterization of drinking water treatment for virus risk assessment. Water Res. 43: 395–404.

Utagawa, E.T., S. Nishizawa, S. Sekine, Y. Hayashi, Y. Ishihara, I. Oishi, A. Iwasaki, I. Yamashita, K. Miyamura and S. Yamazaki. 1994. Astrovirus as a cause of gastroenteritis in Japan. J. Clin. Microbiol. 32: 1841–1845.

van Asten, L., J. Siebenga, C. van den Wijngaard, R. Verheij, H. van Vliet, M. Kretzschmar, H. Boshuizen, W. van Pelt and M. Koopmans. 2011. Unspecified gastroenteritis illness and deaths in the elderly associated with norovirus epidemics. Epidemiol. 22: 336–343.

van den Berg, H., W. Lodder, W. van der Poel, H. Vennema and A.M. de Roda Husman. 2005. Genetic diversity of noroviruses in raw and treated sewage water. Res. Microbiol. 156: 532–540.

van Maarseveen, N.M., E. Wessels, C.S. de Brouwer, A.C. Vossen and E.C. Claas. 2010. Diagnosis of viral gastroenteritis by simultaneous detection of Adenovirus group F, Astrovirus, Rotavirus group A, Norovirus genogroups I and II and Sapovirus in two internally controlled multiplex real-time PCR assays. J. Clin. Virol. 49: 205–210.

Velazquez, F.R., D.O. Matson, J.J. Calva, L. Guerrero, A.L. Morrow, S. Carter-Campbell, R.I. Glass, M.K. Estes, L.K. Pickering and G.M. Ruiz-Palacios. 1996. Rotavirus infections in infants as protection against subsequent infections. N. Engl. J. Med. 335: 1022–1028.

Vennema, H., E. de Bruin and M. Koopmans. 2002. Rational optimization of generic primers used for Norwalk-like virus detection by reverse transcriptase polymerase chain reaction. J. Clin. Virol. 25: 233–235.

Verhoef, L., H. Vennema, W. van Pelt, D. Lees, H. Boshuizen, K. Henshilwood, M. Koopmans and Food-Borne Viruses in Europe Network. 2010. Use of norovirus genotype profiles to differentiate origins of foodborne outbreaks. Emerg. Infect. Dis. 16: 617–624.

Wang, D., L. Coscoy, M. Zylberberg, P.C. Avila, H.A. Boushey, D. Ganem and J.L. DeRisi. 2002. Microarray-based detection and genotyping of viral pathogens. Proc. Natl. Acad. Sci. USA 99(24): 15687–15692.

Wang, Q.H., M.G. Han, S. Cheetham, M. Souza, J.A. Funk and L.J. Saif. 2005. Porcine noroviruses related to human noroviruses. Emerg. Infect. Dis. 11: 1874–1881.

Whitehead, F.J., R.T. Couper, L. Moore, A.J. Bourne and R.W. Byard. 1996. Dehydration deaths in infants and young children. Am. J. Forensic. Med. Pathol. 17: 73–78.

Widdowson, M.A., A. Sulka, S.N. Bulens, R.S. Beard, S.S. Chaves, R. Hammond, E.D.P. Salehi, E. Swanson, J. Totaro, R. Woron, P.S. Mead, J.S. Bresee, S.S. Monroe and R.I. Glass. 2005. Norovirus and foodborne disease, United States, 1991–2000. Emerg. Infect. Dis. 11: 95–102.

Vinje, J., H. Vennema, L. Maunula , C.H. von Bonsdorff, M. Hoehne, E. Schreier, A. Richards, J. Green, D. Brown, S.S. Beard, S.S. Monroe, E. de Bruin, L. Svensson and M.P. Koopmans. 2003. International collaborative study to compare reverse transcriptase PCR assays for detection and genotyping of noroviruses. J. Clin. Microbiol. 41: 1423–1433.

Wyn-Jones, P. 2007. The Detection of Waterborne Viruses. *In:* A. Bosch (ed.). Perspectives in Medical Virology, Vol. 17. Human Viruses in Water. Amsterdam, Elsevier Science pp. 177–204.

Öhrmalm, C., R. Eriksson et al. 2009. Multiplex nucleic acid detection. U. S. P. Office. United States of America. 1000012534.

Öhrmalm, C., R. Eriksson, M. Jobs, M. Simonson, M. Strømme, K. Bondeson, B. Herrmann, A. Melhus and J. Blomberg. 2012. Variation-tolerant capture and multiplex detection of nucleic acids: application to detection of microbes. J. Clin. Microbiol. 50: 3208–3215.

Öhrmalm, C., M. Jobs, R. Eriksson, S. Golbob, A. Elfaitouri, F. Benachenhou, M. Strømme and J. Blomberg. 2010. Hybridization properties of long nucleic acid probes for detection of variable target sequences, and development of a hybridization prediction algorithm. Nucleic Acids Res. 38: e195. DOI: 10.1093/nar/gkq777.

CHAPTER 19

Enteropathogenic *Yersinia* in Foods

Riikka Laukkanen-Ninios,[#,*] *Maria Fredriksson-Ahomaa*[#] and *Hannu Korkeala*[#]

Introduction

Enteropathogenic *Yersinia* comprises of pathogenic *Yersinia enterocolitica* and *Yersinia pseudotuberculosis*. Enteropathogenic *Yersinia* causes enteric yersiniosis, the third most frequently reported bacterial zoonosis in the EU after *Campylobacter* and *Salmonella* (EFSA and ECDC 2011) and foods are considered to be the main transmission route of enteropathogenic *Yersinia* to humans. The pathogenicity of *Yersinia* varies from nonpathogenic to highly pathogenic and correct identification and subtyping. Detection of virulence markers are therefore needed to determine clinical significance of isolated strains.

Classification

Yersinia enterocolitica

Y. enterocolitica is a diverse species that is subdivided into two subspecies *Y. enterocolitica* ssp. *enterocolitica* and *Y. enterocolitica* ssp. *palearctica* (Neubauer

[#] University of Helsinki, Faculty of Veterinary Medicine, Department of Food Hygiene and Environmental Health, P.O. Box 66 (Agnes Sjöberginkatu 2), 00014 University of Helsinki. Emails: riikka.laukkanen-ninios@helsinki.fi; maria.fredriksson-ahomaa@helsinki.fi; hannu.korkeala@helsinki.fi

[*] Corresponding author

et al. 2000) and further subtyped by biotyping. Currently, *Y. enterocolitica* is grouped into six biotypes (1A, 1B, 2–5) (Wauters et al. 1987, Wauters et al. 1988) (Table 1). Biotype 1B belongs to *Y. enterocolitica* ssp. *enterocolitica* and serotypes 1A, 2–5 belong to *Y. enterocolitica* ssp. *palearctica* (Neubauer et al. 2000). The biotypes are also associated with the pathogenicity of *Y. enterocolitica* strains. Biotype 1B has been associated with high pathogenicity and biotypes 2–5 with low pathogenicity (Carniel 2001). Strains belonging to biotype 1A are usually regarded as nonpathogenic, although recently there has been a discussion of the pathogenicity of this group and it is suggested that some of these strains could be opportunistic pathogens (Batzilla et al. 2011, Bhagat and Virdi 2011).

Multiple serotypes of *Y. enterocolitica* strains have been described (Wauters et al. 1991) and O:3, O:5,27, O:8, and O:9 are the most common serotypes associated with pathogenicity. Serotype O:8 belongs to biotype 1B, whereas O:3, O:5,27 and O:9 are associated with biotypes 2–5 and lower pathogenicity. Slide agglutination using commercial antisera O:3, O:5, O:8, O:9 and O:27 is still the most commonly used method for serotyping but there are also PCR assays available for identification of serotypes O:3 (Weynants et al. 1996, Thisted Lambertz and Danielsson-Tham 2005) and O:9 (Jacobsen et al. 2005).

Table 1. Biotyping scheme of *Yersinia enterocolitica* strains (Wauters et al. 1987).

Test	Biotype					
	1A	1B	2	3	4	5
Lipase	+	+	–	–	–	–
Esculin/salicin	+/–	–	–	–	–	–
Indole	+	+	(+)[a]	–	–	–
Xylose	+	+	+	+	–	V[b]
Trehalose/NO_3	+	+	+	+	+	–
Pyrazinamidase	+	–	–	–	–	–
Voges Proskauer	+	+	+	+	+	(+)

[a] () delayed reaction
[b] V variable reactions

Yersinia pseudotuberculosis

Unlike *Y. enterocolitica*, the genetic diversity of *Y. pseudotuberculosis* is limited (Laukkanen-Ninios et al. 2011) and there is little variation in biochemical reactions among *Y. pseudotuberculosis* strains (Brenner et al. 1976). A simple biotyping scheme, using raffinose, melibiose, and citrate has been suggested for *Y. pseudotuberculosis* (Tsubokura and Aleksić 1995), but without a clear clinical relevance it is used infrequently. A serotyping scheme consisting of O:1–15 with three subtypes (a–c) in O:1 and O:2 and two subtypes (a, b) in O:4 and O:5 (Tsubokura and Aleksić 1995, Bogdanovich et al. 2003) is used. Commercially, antisera are available for serotypes O:1–O:6, excluding the subtypes. In addition, a PCR method has been developed for the O genotypes O:1–O:15 and subtypes (Bogdanovich et al. 2003).

Table 2. Biotyping scheme of *Yersinia pseudotuberculosis* strains according to (Tsubokura and Aleksić 1995).

Test	Biotype			
	1	2	3	4
Melibiose	+	–	–	+
Citrate	–	–	+	–
Raffinose	–	–	–	+

Major Virulence Factors

All enteropathogenic *Yersinia* strains need a 70-kb virulence plasmid called pYV (plasmid for Yersinia virulence) coding for the type III secretion system andits effector *Yersinia* outer proteins (Yops) needed for the bacteria to survive and replicate within lymphoid tissues of their animal or human hosts (Viboud and Bliska 2005). Virulence plasmid also codes YadA, major adhesin in the host (Leo and Skurnik 2011). Other adhesins, Inv and Ail, are chromosomally encoded and common target for PCR protocols for the detection of enteropathogenic *Yersinia* (Fredriksson-Ahomaa et al. 2006a, Fredriksson-Ahomaa 2009, Leo and Skurnik 2011).The adhesins mediate epithelial cell binding and invasion, but YadA and Ail also induce serum resistance (Leo and Skurnik 2011). *Y. enterocolitica* biotype 1A lack pYV and most of the chromosomal virulence factors, although some of the strains can carry some chromosomal virulence-associated genes (Bhagat and Virdi 2011).

Y. enterocolitica biotype 1B (O:8) and *Y. pseudotuberculosis* serotypes O:1 carry the high-pathogenicity island (HPI) in the chromosome coding for the yersinia bactinsiderophore, which enhances systemic dissemination of the bacteria in the host (Carniel 2002). *Y. pseudotuberculosis* O:3 harbor HPI with truncated left end coding for non-functional yersiniabactin (Fukushima et al. 2001b, Carniel 2002).

The superantigen *Yersinia pseudotuberculosis*-derived mitogen (YPM) has been associated with patients with systemic symptoms (Abe et al. 1997). In Siberia, a large plasmid, pVM82, has also been associated with *Y. pseudotuberculosis* strains causing Far East scarlet-like fever (Carniel 2002).

Distribution in Animals

Animals have been suspected as reservoirs for enteropathogenic *Yersinia*, since *Yersinia* have been recovered from diverse animal sources ranging from farm animals, domestic pets and experimental animals to wild and captive animals (Fredriksson-Ahomaa et al. 2006a, Fredriksson-Ahomaa 2009). Pigs are considered as a major reservoir of pathogenic *Y. enterocolitica* and the pathogen can be frequently isolated from the faeces and tonsils of asymptomatic pigs throughout the world (Table 3). Pathogenic *Y. enterocolitica* can also occasionally be isolated in apparently healthy wild animals such as birds, rodents, and hares as well as pets such as cats and dogs, although, most of the *Y. enterocolitica* strains isolated from wild and domestic animals other than pigs are biotype 1A (Tennant et al. 2003, Fredriksson-Ahomaa et al. 2006a).

Y. pseudotuberculosis has been isolated from many apparently healthy wild animals such as birds, rodents and other small mammals, hares and rabbits, raccoon dogs, martens, deer, and wild boars. In addition, it has been isolated from domestic animals, such as farmed cattle, deer, and pigs, as well as dogs and cats (Fredriksson-Ahomaa 2009). Unlike with *Y. enterocolitica*, no major reservoir of *Y. pseudotuberculosis* has been recognized, but wild animals, particularly small mammals are suspected.

Distribution in pigs

Pathogenic *Y. enterocolitica* is widespread in pigs and the percentage of *Y. enterocolitica*-positive farms is high in most countries. *Y. enterocolitica* 4/O:3 and *Y. pseudotuberculosis* O:3 are the most prevalent subtypes in pigs in Europe and USA (Table 3). In addition to bioserotype 4/O:3, *Y. enterocolitica* 2/O:5, 27 has also frequently been isolated from pigs in the US and Canada (Bhaduri and Wesley 2006, Poljak et al. 2010). In England, *Y. enterocolitica* 2/O:9 and 2/O:5, 27 (Ortiz Martínez et al. 2010) and in China, *Y. enterocolitica* 2/O:9 and 3/O:3 (Wang et al. 2009) are the most common bioserotypes. *Y. pseudotuberculosis* has been isolated sporadically from pigs in Asia, America, and Europe, although the prevalence of *Y. pseudotuberculosis*-positive pigs is markedly lower than that of *Y. enterocolitica*. Enteropathogenic *Yersinia* is usually detected in fattening pigs but less often in sows or piglets (Niskanen et al. 2002, Korte et al. 2004, Gürtler et al. 2005, Bowman et al. 2007, Niskanen et al. 2008).

Table 3. Recent reported isolation rates of pathogenic *Yersinia enterocolitica* (YE) and *Yersinia pseudotuberculosis* (YP) in pigs at slaughterhouse.

Country	No. of pigs	Species	No. of positive pigs (%)		Biotype/serotype	Reference
			Tonsils	Feces		
Belgium	201	YE	89 (44)		3/O:9 (8)[a], 4/O:3 (81)	(Ortiz Martínez et al. 2011)
		YP	5 (2)		1/O:1 (1), 1/O:2 (1), 2/O:3 (3)	
Canada	395	YE	7 (2)		4/O:3 (5), 2/O:5, 27 (2)	(O'Sullivan et al. 2011)
China	2252	YP		3 (0.1)		(Fukushima et al. 2001a)
Estonia	151	YE	135 (89)		4/O:3	(Ortiz Martínez et al. 2009)
		YP	2 (1)		2/O:3	
Finland	301	YE	177 (59)	90 (30)	4/O:3	(Laukkanen et al. 2010b)
		YP	8 (3)	13 (4)	O:3	
Germany	164	YE	101 (62)	17 (10)	4/O:3	(Bucher et al. 2008)
Greece	455	YE	58 (13)		4/O:3	(Kechagia et al. 2007)
		YP	3 (0.7)			
Italy	428	YE	137 (32)		2/O:5 (1), 4/O:3 (136)	(Ortiz Martínez et al. 2011)
		YP	5 (1)		1/O:1 (3), 2/O:2 (1)	
Latvia	404	YE	143 (35)		4/O:3	(Terentjeva and Berzins 2010)
		YP	12 (3)		ND	

Latvia	109	YE	70 (64)		4/O:3	(Ortiz Martínez et al. 2009)
	109	YP	5 (5)		2/O:3	
Russia	197	YE	66 (34)		4/O:3	(Ortiz Martínez et al. 2009)
		YP	13 (7)		2/O:3	
Spain	200	YE	185 (93)		4/O:3	(Ortiz Martínez et al. 2011)
Switzerland	212	YE	72 (34)		2/O:5,27 (6), 2/O:9 (1), 4/O:3 (69)	(Fredriksson-Ahomaa et al. 2007)
UK	630	YE	278 (44)		2/O:5 (97), 2/O:9 (124), 4/O:3 (39)	(Ortiz Martínez et al. 2010)
		YP	114 (18)		1/O:1 (32), 1/O:4 (29), 2/O:3 (41)	

[a] Number of positive samples

Distribution in Environment

Enteropathogenic *Yersinia* are detected in the environment (soil and water) infrequently, whereas apathogenic *Yersinia* are common (Tennant et al. 2003). However, pathogenic *Y. enterocolitica* has been detected in surface waters and more often in cold than warm water (Sandery et al. 1996, Cheyne et al. 2010) and *Y. pseudotuberculosis* has been isolated from the environment in Japan, but less frequently anywhere else (Fukushima et al. 1995). Although environment does not appear to be a reservoir for enteropathogenic *Yersinia*, contaminated water can serve as a vehicle for pathogenic *Yersinia* infecting people.

Characterization of Strains by Genotyping Methods

Many genotyping methods have been used to subtype enteropathogenic *Yersinia* (Fredriksson-Ahomaa et al. 2006a). Pulsed-field gel electrophoresis (PFGE) has been widely used molecular typing method for many bacteria, and it is a commonly used technique in epidemiological studies of *Y. enterocolitica* and *Y. pseudotuberculosis*. It allows subtyping of strains belonging to the same bioserotype (Iteman et al. 1995, Iteman et al. 1996), although, sometimes 2 or 3 enzymes are needed to discriminate strains belonging to same bioserotype (Fredriksson-Ahomaa et al. 1999a). The complexity of the profiles can make comparison of large numbers of restriction patterns of enteropathogenic *Yersinia* strains difficult. Furthermore, the method is very laborious and time consuming. Amplified fragment length polymorphism (AFLP), a fairly recently adopted PCR-based typing method, which allows differentiation of *Y. enterocolitica* strains within serotype-related clusters (Fearnley et al. 2005, Kuehni-Boghenbor et al. 2006).

Multilocus variable-number tandem-repeat analysis (MLVA) is a PCR-based typing method which is based on variable-number tandem repeats (VNTR) present in most bacterial genomes including *Yersinia* (de Benito et al. 2004). This method is increasingly used for typing pathogenic bacteria. The MLVA is easy to perform and it seems to be more discriminatory than other genotyping methods, e.g., PFGE for characterisation of *Y. enterocolitica* strains (Sihvonen et al. 2011). However, only a slight correlation of the MLVA genotypes and the geographic distribution of the isolates or genotypes and serogroups is observed (Gierczynski et al. 2007). MLVA seems at this moment to be best suited for screening potential outbreaks and local epidemiological investigations. Multilocus sequence typing (MLST) has been used to assess the genetic relationship, phylogenetic traits and clustering of the *Yersinia* spp. (Kotetishvili et al. 2005). In addition, two MLST schemes for *Y. pseudotuberculosis* have been developed to study population structure of the species (Ch'ng et al. 2011, Laukkanen-Ninios et al. 2011). Due to low genetic diversity of *Y. pseudotuberculosis,* MLST can be applied in global epidemiology and population structure, but not in local epidemiological investigations.

Yersiniosis in Humans

Enteropathogenic *Yersinia* usually causes self-limiting gastroenteritis. Symptoms include fever, abdominal pain, and diarrhea, sometimes also nausea and vomiting (Smego et al. 1999). The abdominal pain in the right lower abdomen especially due to *Y. pseudotuberculosis* infectious and might be mistaken for appendicitis (pseudo appendicitis), leading to unnecessary surgery (Nuorti et al. 2004). The incubation period of *Y. enterocolitica* usually ranges from 1 to 11 days with symptoms typically persisting for 5 to 14 days and occasionally lasting for several months (Cover and Aber 1989). Based on an outbreak investigation, the median incubation period for *Y. pseudotuberculosis* infection is 8 days (4–18 days) and the length of illness varies 5–36 days (Jalava et al. 2006). The minimum infectious dose for *Yersinia* infection in humans is unknown. Gastroenteritis caused by enteropathogenic *Yersinia* is normally self-limiting and does not require antimicrobial treatment.

The most frequent post-infections sequelae are reactive arthritis and erythema nodosum (Hannu et al. 2003, Jalava et al. 2006, Rosner et al. 2010). The joint infection, most commonly in the knees, ankles or wrists, usually develop within one week to one month after the initial infection and generally resolves after one to six months (Granfors et al. 1989, Hannu et al. 2003). Patients possessing the histocompatibility gene HLA-B27 are at particular risk of developing prolonged and more severe *Yersinia*-related reactive arthritis, urethritis, and conjunctivitis, although the reasons underlying this predisposition are incompletely known (Leirisalo-Repo 2005).

In addition to enteric form, *Y. pseudotuberculosis* infection has shown more severe symptoms in Japan and Russia than in Europe. The symptoms include erythematous skin rash, conjunctivitis, skin desquamation, strawberry tongue, and toxic shock syndrome and the infection has been known in Russia as Far East scarlet-like fever and in Japan Izumi fever and associated with Kawasaki Disease (Vincent et al. 2007).

In patients with underlying disease or immunosuppression, yersiniosis can result in septicemia (Smego et al. 1999). Septicemia cases caused by *Yersinia*-contaminated blood transfusions have also been reported (Guinet et al. 2011). The mortality of patients with bacteremia caused by *Y. pseudotuberculosis* is 36 percent (Kaasch et al. 2012) and the overall fatality rate among the *Y. enterocolitica* post-transfusion sepsis case reports published after 1991 is 46 percent (Guinet et al. 2011). Third generation cephalosporins trimethoprim-sulfamethoxazole, tetracycline, and fluoroquinolones are used in severe *Y. enterocolitica* infections (Guinet et al. 2011). *Y. enterocolitica* strains produce β-lactamases, which render them resistant to penicillins and first-generation cephalosporins *in vitro* (Pham et al. 2000) whereas *Y. pseudotuberculosis* lacks β-lactamase activity and is hence susceptible to penicillins (Kanazawa and Kuramata 1974).

Incidence of Yersiniosis

Yersiniosis cases have been reported in most parts of the world, but the reported incidence varies drastically. In the EU, the reported mean incidence of 1.7 per 100,000 with the incidence varying 0.02–14.4 between member states (EFSA and ECDC 2011). In North-America, the reported incidence of yersiniosis has been 0.3 per 100,000 in USA and 1.5 per 100,000 in Canada (Public Health Agency of Canada 2010, CDC 2011b). Of the yersiniosis cases reported, the majority is caused by *Y. enterocolitica* (EFSA 2006, Thisted Lambertz 2007, Long et al. 2010) and *Y. enterocolitica* 4/O:3 is the most prevalent pathogenic *Y. enterocolitica* bioserotype in both Europe and North America (EFSA 2006, Public Health Agency of Canada 2010). Circa one percent of the cases are caused by *Y. pseudotuberculosis* in EU and USA (EFSA 2005, EFSA 2006, Long et al. 2010). Most of the yersiniosis cases caused by pathogenic *Y. enterocolitica* are reported in children (EFSA 2007, Huovinen et al. 2010, Long et al. 2010). In 1996–2005, the case-fatality rate of yersiniosis in USA was 0.7 percent from all laboratory-confirmed infections and 6 percent from sterile site infections (Barton Behravesh et al. 2011).

Transmission to Humans

Most of the yersiniosis cases reported, especially those caused by *Y. enterocolitica*, are sporadic. The source of *Y. enterocolitica* infections in humans has been identified as pork, raw or undercooked in particular, by case-control studies and comparing *Y. enterocolitica* genotypes from pigs and humans (Tauxe et al. 1987, Ostroff et al. 1994, Jones 2003, Fredriksson-Ahomaa et al. 2006b, Grahek-Ogden et al. 2007, Boqvist et al. 2009, Huovinen et al. 2010). Particularly, small children can also contact *Y. enterocolitica* infection from pets, or from the contaminated environment, e.g., via pacifiers (Fredriksson-Ahomaa et al. 2001a, Jones 2003, Boqvist et al. 2009, Wang et al. 2010). During the past decade, some outbreaks of *Y. enterocolitica* have been reported in Europe, Asia and North America and the source of infection has been pork products (Norway, USA), dairy products (USA), lunch salad containing vegetables, ham, mayonnaise and unknown source (Japan), and bagged salad mix (suspected, Norway) (Jones 2003, Sakai et al. 2005, Grahek-Ogden et al. 2007, Tafjord Heier et al. 2007, Moriki et al. 2010, CDC 2011a, MacDonald et al. 2011).

Y. pseudotuberculosis outbreaks have been reported mainly in Finland, Japan, and Russia (Tsubokura et al. 1989, Nuorti et al. 2004, Anonymous 2008, Rimhanen-Finne et al. 2009). The source of infection has been vegetables linked with school or other institutional kitchens especially in recent years in Finland, but also water and milk have been indicated (Fredriksson-Ahomaa 2009). Unlike with *Y. enterocolitica*, the role of pork as a vehicle of *Y. pseudotuberculosis* is uncertain since *Y. pseudotuberculosis* cases related to pork have not been reported. However, the sources of sporadic *Y. pseudotuberculosis* infections are unknown and *Y. pseudotuberculosis* has been isolated from carcasses and pork (Fukushima 1985, Laukkanen et al. 2008), indicating a possible route from pigs to humans.

Pork Production—Transmission and Prevention

Enteropathogenic *Yersinia* are transmitted from pig farms to slaughterhouse and from pigs to carcasses and pluck sets, although cross-contamination at the slaughterhouse also occurs (Laukkanen et al. 2008, Laukkanen et al. 2009). High prevalence of enteropathogenic *Yersinia* in pigs predisposes to carcass contamination at the slaughterhouse (Laukkanen et al. 2008, Laukkanen et al. 2009). Therefore, lowering the prevalence of enteropathogenic *Yersinia* in pigs would lower the prevalence of these pathogens in pork and prevent human yersiniosis cases.

On farm

The epidemiology of enteropathogenic *Yersinia* is still somewhat unclear, no unambiguous difference in the prevalence of enteropathogenic *Yersinia* in pigs between production types and capacities has been reported. Pathogenic *Y. enterocolitica* may be more prevalent in conventional and pathogenic *Y. pseudotuberculosis* in organic production, whereas fattening vs. farrow-to-finish production or production capacity do not appear to affect the prevalence significantly (Nowak et al. 2006, Laukkanen et al. 2008, Laukkanen et al. 2009, Laukkanen et al. 2010b, Ortiz Martínez et al. 2010, Virtanen et al. 2011). Repeated sampling on a farm shows that enteropathogenic *Yersinia* persist on the farm (Pilon et al. 2000, Niskanen et al. 2008, Poljak et al. 2010), but the contamination sources are still unclear. Both environmental and pig-to-pig transmission of enteropathogenic *Yersinia* on farms has been suggested (Fukushima et al. 1983, Pilon et al. 2000, Laukkanen et al. 2008, Laukkanen et al. 2009). In herds with low prevalence of *Y. enterocolitica* in pigs, the transmission of infection by the environment may be low and pig-to-pig transmission more likely (Pilon et al. 2000). In a Finnish study, the use of municipal water instead of own well was a protective factor against carriage and faecal shedding of *Y. enterocolitica* and feed containing a prebiotic component may reduce the carriage (Virtanen et al. 2011).

It is possible to establish and maintain a closed breeding pyramid free of pathogenic *Y. enterocolitica* (Nesbakken et al. 2007). However, since the prevalence of *Y. enterocolitica* and *Y. pseudotuberculosis*-positive farms in Europe varies between 36–100 percent and 13–80 percent, respectively (Ortiz Martínez et al. 2009, Ortiz Martínez et al. 2010, Ortiz Martínez et al. 2011), stamping out enteropathogenic *Yersinia* on farms or separation of carrier herds during transportation, lairage and slaughter is difficult. More information on the epidemiology of enteropathogenic *Yersinia* on farms is needed for cost-effective on-farm prevention methods.

At the slaughterhouse

At the slaughterhouse, intestinal content and particularly tonsils of the slaughtered pigs are the most important contamination sources of carcasses because of the high prevalence of enteropathogenic *Yersinia* in these sites (Table 3). Bagging of the rectum reduces the carcass contamination from feces significantly, but does not prevent contamination completely, since tonsillar carriage of enteropathogenic *Yersinia* is very high (Andersen 1988, Nesbakken et al. 1994, Laukkanen et al. 2010b). Changes in removal of the head have been suggested to reduce contamination of carcasses and pluck sets (Christensen and Lüthje 1994, Fredriksson-Ahomaa et al. 2000). Meat inspection procedures, such as incision of the lymph nodes, have also been suggested to spread pathogenic *Y. enterocolitica* on the carcasses, since enteropathogenic *Yersinia* can colonize lymphatic tissue and are isolated, e.g., from submaxillary lymph nodes (Nesbakken et al. 2003).

Slaughterhouse environment and slaughter equipment can also contaminate pig carcasses and pluck sets. Same genotypes of pathogenic *Y. enterocolitica* can be isolated from carcasses, offals and the slaughterhouse environment and equipment (Fredriksson-Ahomaa et al. 2000). In addition, the same genotype of pathogenic *Y. enterocolitica* and *Y. pseudotuberculosis* was isolated from carcass-positive pigs and their subsequent carcasses at the slaughterhouse in 80 percent (70/88) and 100 percent (16/16) of the cases, respectively and from pluck set-positive pigs and their subsequent pluck sets in 70 percent (23/33) and 67 percent (4/6) (Laukkanen et al. 2008, Laukkanen et al. 2009, Laukkanen et al. 2010b).

Since *Y. enterocolitica* can frequently be isolated from the head area, cutting of the head meat should be carried out on a separate work table and in a separate room (Borch et al. 1996). Based on the previously published results (Andersen 1988, Christensen and Lüthje 1994, Nesbakken et al. 1994), the combined prevalence of *Listeria monocytogenes* and enteropathogenic *Yersinia* in Finnish pork was estimated to decrease from one to eleven percent to zero to two percent if the head was removed intact and the rectum sealed off (Ranta et al. 2010).

At retail and pork products

Enteropathogenic *Yersinia* have been detected from pork products and edible offals frequently (*Y. enterocolitica* more frequently than *Y. pseudotuberculosis*) (Fukushima 1985) (Table 4). Pathogenic *Y. enterocolitica* has also been isolated in butcher shop environment enabling cross-contamination at the retail level, if good manufacturing practices are not followed (Fredriksson-Ahomaa et al. 2004). Enteropathogenic *Yersinia* is destroyed during normal heating. In food preparation, undercooked and raw pork products and cross-contamination from raw to cooked products need to be avoided.

Fresh Produce—Transmission and Prevention

Y. pseudotuberculosis O:1 outbreaks in Finland have been in the spring or early summer and caused by carrots from the previous season (Jalava et al. 2006, Kangas et al. 2008, Rimhanen-Finne et al. 2009). In one outbreak caused by *Y. pseudotuberculosis* O:3 (Nuorti et al. 2004) iceberg lettuce was indicated in addition to carrots. Fresh produce can be contaminated with pathogenic bacteria in production phases from growing to preparation for consumption (Beuchat and Ryu 1997). Wild animals, such as deer and small mammals have been suspected as a source of *Y. pseudotuberculosis* contamination for the vegetables, but the route of contamination is still uncertain (Nuorti et al. 2004, Jalava et al. 2006, Kangas et al. 2008). Untreated irrigation water contaminated with deer faeces was suspected as the contamination source in one outbreak (Nuorti et al. 2004) and *Y. pseudotuberculosis* O:1 was isolated from a pooled sample of common shrew intestines on one carrot farm in another outbreak (Kangas et al. 2008).

Enteropathogenic *Yersinia* is usually undetected in fresh vegetables. In two studies on retail fresh produce, pathogenic *Y. enterocolitica* was detected in three percent of lettuce samples with PCR, but not with isolation methods (Fredriksson-Ahomaa et al. 2001b, Johannessen et al. 2002). *Y. pseudotuberculosis* has been isolated from lettuce on one occasion in India (Pingulkar et al. 2001). However, *Y. pseudotuberculosis* has not been isolated from fresh carrots in Europe (Thisted-Lambertz et al. 2008, Pitkälä et al. 2009). *Y. pseudotuberculosis* has been isolated in outbreak investigations from wall or floor surface samples and from the carrot-peeling

Table 4. Prevalence of pathogenic *Yersinia enterocolitica* and *Yersinia pseudotuberculosis* in raw (not heat treated) foods.

Sample	Species	No. of samples	Detection		No. of positive samples (%)	Reference
			Isolation (%)	PCR (%)		
Pig tongues	YE	51	40 (78)	47 (92)	50 (98)	(Fredriksson-Ahomaa et al. 1999b)
	YE	99	79 (80)	82 (83)	99 (100)	(Fredriksson-Ahomaa et al. 2001b)
	YE	129	34 (26)	58 (45)	58 (45)	(Messelhäusser et al. 2011)
Pig offal	YE	100	51 (51)		51 (51)	(Bucher et al. 2008)
	YE	34	17 (50)	21 (62)	21 (62)	(Fredriksson-Ahomaa et al. 2000)
	YE	354	33 (9)		33 (9)	(Laukkanen et al. 2009)
	YP	354	6 (2)		6 (2)	(Laukkanen et al. 2008)
Broiler carcass	YE	98	0 (0)	9 (9)	9 (9)	(Lindblad et al. 2006)
Pig carcass	YE	267	0 (0)	73 (27)	73 (27)	(Boyapalle et al. 2001)
	YE	518	0 (0)	81 (16)	81 (16)	(Lindblad et al. 2007)
	YE	301	65 (22)		65 (22)	(Laukkanen et al. 2010b)
	YP	301	6 (2)		6 (2)	
Chitterlings	YE	350	28 (8)	278 (79)		(Boyapalle et al. 2001)
Ground pork	YE	350	0 (0)	133 (38)		(Boyapalle et al. 2001)
	YE	255	4 (2)	63 (25)	63 (25)	(Fredriksson-Ahomaa et al. 1999b)
	YE	100	5 (5)	35 (35)		(Thisted Lambertz et al. 2007)
	YE	102	1 (1)	5 (5)	5 (5)	(Messelhäusser et al. 2011)
Pork	YE	300	6 (2)	50 (17)	54 (18)	(Johannessen et al. 2000)
	YE	97	0 (0)	11 (11)	11 (11)	(Thisted Lambertz et al. 2007)
	YE	34	6 (18)	11 (32)	11 (32)	(Hudson et al. 2008)
	YE	125	2 (2)		2 (2)	(Bonardi et al. 2010)
	YE	153	7 (5)	10 (7)	10 (7)	(Messelhäusser et al. 2011)

Game	YE	60		23 (38)	23 (38)	(Bucher et al. 2008)
	YE	51	0 (0)	3 (6)	3 (6)	(Messelhäusser et al. 2011)
Poultry meat	YE	102	0 (0)	1 (1)	1 (1)	(Bucher et al. 2008)
	YE	615	2 (0.3)		2 (0.3)	(Fukushima et al. 1997)
	YE	80	1 (1)		1 (1)	(Bonardi et al. 2010)
Fish	YE	200	0 (0)	0 (0)	0 (0)	(Fredriksson-Ahomaa and Korkeala 2003)
	YP	50	2 (4)		2 (4)	(Okwori et al. 2009)
Lettuce	YE	200	0 (0)	6 (3)	6 (3)	(Johannessen et al. 2002)
	YE	101	0 (0)	3 (3)	3 (3)	(Fredriksson-Ahomaa and Korkeala 2003)
	YP	4	+[a]			(Pingulkar et al. 2001)
Milk	YE	61	0 (0)		0 (0)	(Messelhäusser et al. 2011)
Fermented milk	YE	250	3 (1)		3 (1)	(Okwori et al. 2009)
	YP	250	4 (2)		4 (2)	

[a] detected, but the number of positive samples is not given

line as well as from spoiled carrots, and from fluid draining from spoiled carrots (Jalava et al. 2006, Kangas et al. 2008, Rimhanen-Finne et al. 2009). Since spoilage of the carrots and long storage seem to be related to the outbreaks caused by *Y. pseudotuberculosis*, Finnish Food Safety Authority has given recommendations on the storage and processing of the carrots and monitoring of *Y. pseudotuberculosis* in processing plants in order to prevent *Y. pseudotuberculosis* outbreaks. However, more information on the epidemiology of these *Y. pseudotuberculosis* outbreaks is needed.

Other Sources of Yersiniosis—Milk and Water

Milk has been a source of *Y. enterocolitica* outbreaks and less rarely *Y. pseudotuberculosis* particularly in North America (Greenwood and Hooper 1990, Nowgesic et al. 1999, Ackers et al. 2000, CDC 2011a). However, enteropathogenic *Yersinia* have seldom been isolated from milk (Hamama et al. 1992, Jayarao et al. 2006) and pasteurization kills the bacteria. In a recent study, 62.5°C for 15 s was enough to produce maximum reduction (> 6.8 mean $\log_{10}$ reduction) in milk inoculated with 10^7cfu per ml of *Y. enterocolitica* (Pearce et al. 2012). The *Y. enterocolitica* outbreaks associated with milk or dairy products have been involved with post pasteurization contamination (Greenwood and Hooper 1990, Ackers et al. 2000, CDC 2011a).

Water has been a suspected cause of *Y. pseudotuberculosis* infections particularly in Japan (Tsubokura et al. 1989), where the faeces of wild rodents have also been suggested to contaminate water leading to human infection (Fukushima et al. 1988). In Korea, untreated mountain spring water was linked to *Y. pseudotuberculosis* infection (Han et al. 2003).

Detection Methods

Conventional isolation methods are commonly used in the detection of enteropathogenic *Yersinia,* although the isolation of enteropathogenic *Yersinia, Y. pseudotuberculosis* in particular, is difficult and time consuming. The isolation methods are based on cold (4°C) enrichment taking advantage of the psycrotrophic nature of enteropathogenic *Yersinia* or on selective enrichment designed for *Yersinia* (Fredriksson-Ahomaa and Korkeala 2003, Fredriksson-Ahomaa 2009, Laukkanen et al. 2010a). Phosphate buffered saline (PBS) or PBS supplemented with sugars and bile salts are frequently used in the cold enrichment. Selective enrichments such as irgasan-ticarcillin-potassium chlorite (ITC) have 1–2 days' incubation times in 25–30°C, whereas samples are cold enriched up to 3 weeks. Cefsulodin-irgasan-novobiosin (CIN) is the most used selective agar plate (Fredriksson-Ahomaa and Korkeala 2003, Fredriksson-Ahomaa 2009).

Different PCR methods targeting either chromosomally located virulence genes, pYV or both have been designed for the detection of enteropathogenic *Yersinia* (Fredriksson-Ahomaa and Korkeala 2003, Fredriksson-Ahomaa 2009). Since enteropathogenic *Yersinia* are usually isolated after enrichment, the quantification of the pathogens is challenging. Most probable number of the pathogens have been counted using serial dilutions to estimate the contamination of pork with *Y. enterocolitica* (Hudson et al. 2008, Bonardi et al. 2010).

Concluding Remarks

Enteropathogenic *Yersinia* are important foodborne pathogens. However, the transmission sources of *Y. enterocolitica* and *Y. pseudotuberculosis* seem to differ. Pigs and pork are important sources of *Y. enterocolitica* infections in humans, whereas vegetables are associated with *Y. pseudotuberculosis* infections. Although the reported incidence of yersiniosis has been declining in the recent years along with many other pathogens, yersiniosis is still 3rd most reported zoonosis in EU and *Y. pseudotuberculosis* and fresh produce have become relevant food safety issues emphasizing the importance of effective control measures in the food production chain from farm to fork.

References

Abe, J., M. Onimaru, S. Matsumoto, S. Noma, K. Baba, Y. Ito, T. Kohsaka and T. Takeda. 1997. Clinical role for a superantigen in *Yersinia pseudotuberculosis* infection. J. Clin. Invest. 99: 1823–1830.

Ackers, M.L., S. Schoenfeld, J. Markman, M.G. Smith, M.A. Nicholson, W. DeWitt, D.N. Cameron, P.M. Griffin and L. Slutsker. 2000. An outbreak of *Yersinia enterocolitica* O:8 infections associated with pasteurized milk. J. Infect. Dis. 181: 1834–1837.

Andersen, J.K. 1988. Contamination of freshly slaughtered pig carcasses with human pathogenic *Yersinia enterocolitica*. Int. J. Food Microbiol. 7: 193–202.

Anonymous, 2008: Yersiniosis—Russia (Krasnoyarsk). Promed-mail 18 July 2008, accession no 20080718.2184. Available at: http://www.promedmail.org(Accessed on November 4th 2009).

Barton Behravesh, C., T.F. Jones, D.J. Vugia, C. Long, R. Marcus, K. Smith, S. Thomas, S. Zansky, K.E. Fullerton, O.L. Henao, E. Scallan and FoodNet Working Group. 2011. Deaths associated with bacterial pathogens transmitted commonly through food: foodborne diseases active surveillance network (FoodNet), 1996–2005. J. Infect. Dis. 204: 263–267.

Batzilla, J., J. Heesemann and A. Rakin. 2011. The pathogenic potential of *Yersinia enterocolitica* 1A. Int. J. Med. Microbiol. 301: 556–561.

Beuchat, L.R. and J.H. Ryu. 1997. Produce handling and processing practices. Emerg. Infect. Dis. 3: 459–465.

Bhaduri, S. and I. Wesley. 2006. Isolation and characterization of *Yersinia enterocolitica* from swine faeces recovered during the National Animal Health Monitoring System Swine 2000 study. J. Food Prot. 69: 2107–2112.

Bhagat, N. and J.S. Virdi. 2011. The enigma of *Yersinia enterocolitica* biovar 1A. Crit. Rev. Microbiol. 37: 25–39.

Bogdanovich, T., E. Carniel, H. Fukushima and M. Skurnik. 2003. Use of O-antigen gene cluster-specific PCRs for the identification and O-genotyping of *Yersinia pseudotuberculosis* and *Yersinia pestis*. J. Clin. Microbiol. 41: 5103–5112.

Bonardi, S., A. Paris, L. Bassi, F. Salmi, C. Bacci, E. Riboldi, E. Boni, M. D'Incau, S. Tagliabue and F. Brindani. 2010. Detection, semiquantitative enumeration, and antimicrobial susceptibility of *Yersinia enterocolitica* in pork and chicken meats in Italy. J. Food Prot. 73: 1785–1792.

Boqvist, S., H. Pettersson, A. Svensson and Y. Andersson. 2009. Sources of sporadic *Yersinia enterocolitica* infection in children in Sweden, 2004: a case-control study. Epidemiol. Infect. 137: 897–905.

Borch, E., T. Nesbakken and H. Christensen. 1996. Hazard identification in swine slaughter with respect to foodborne bacteria. Int. J. Food Microbiol. 30: 9–25.

Bowman, A.S., C. Glendening, T.E. Wittum, J.T. LeJeune, R.W. Stich and J.A. Funk. 2007. Prevalence of *Yersinia enterocolitica* in different phases of production on swine farms. J. Food Prot. 70: 11–16.

Boyapalle, S., I.V. Wesley, H.S. Hurd and P.G. Reddy. 2001. Comparison of culture, multiplex, and 5' nuclease polymerase chain reaction assays for the rapid detection of *Yersinia enterocolitica* in swine and pork products. J. Food Prot. 64: 1352–1361.

Brenner, D.J., A.G. Steigerwalt, D.P. Falcao, R.E. Weaver and G.R. Fanning. 1976. Characterization of *Yersinia enterocolitica* and *Yersinia pseudotuberculosis* by deoxyribonucleic acid hybridization and by biochemical reactions. Int. J. Syst. Bacteriol. 26: 180–194.

Bucher, M., C. Meyer, B. Grotzbach, S. Wacheck, A. Stolle and M. Fredriksson-Ahomaa. 2008. Epidemiological data on pathogenic *Yersinia enterocolitica* in Southern Germany during 2000–2006. Foodborne Pathog. Dis. 5: 273–280.

Carniel, E. 2001. The *Yersinia* high-pathogenicity island: an iron-uptake island. Microbes Infect. 3: 561–569.

Carniel, E. 2002. Plasmids and pathogenicity islands of *Yersinia*. Curr. Top. Microbiol. Immunol. 264: 89–108.

CDC. 2011a. Notes from the field: *Yersinia enterocolitica* infections associated with pasteurized milk—southwestern Pennsylvania. MMWR Morb. Mortal. Wkly. Rep. 60: 1428.

CDC. 2011b. Vital signs: incidence and trends of infection with pathogens transmitted commonly through food—foodborne diseases active surveillance network, 10 U.S. sites, 1996–2010. MMWR Morb. Mortal. Wkly. Rep. 60: 749–755.

Cheyne, B.M., M.I. Van Dyke, W.B. Anderson and P.M. Huck. 2010. The detection of *Yersinia enterocolitica* in surface water by quantitative PCR amplification of the *ail* and *yadA* genes. J. Water. Health. 8: 487–499.

Ch'ng, S.L., S. Octavia, Q. Xia, A. Duong, M.M. Tanaka, H. Fukushima and R. Lan. 2011. Population structure and evolution of pathogenicity of *Yersinia pseudotuberculosis*. Appl. Environ. Microbiol. 77: 768–775.

Christensen, H. and H. Lüthje. 1994. Reduced spread of pathogens as a result of changed pluck removal technique. Proceedings of the 40th International Congress of Meat Science and Technology S-III. 06.

Cover, T.L. and R. C. Aber. 1989. *Yersinia enterocolitica*. N. Engl. J. Med. 321: 16–24.

de Benito, I., M.E. Cano, J. Aguero and J.M. Garcia Lobo. 2004. A polymorphic tandem repeat potentially useful for typing in the chromosome of *Yersinia enterocolitica*. Microbiology 150: 199–204.

EFSA. 2005. The community summary report on trends and sources of zoonoses, zoonotic agents, antimicrobial resistance and foodborne outbreaks in the European Union in 2004. The EFSA Journal 310: 155–162.

EFSA. 2006. The community summary report on trends and sources of zoonoses, zoonotic agents, antimicrobial resistance and foodborne outbreaks in the European Union in 2005. The EFSA Journal 94: 159–288.

EFSA. 2007. The community summary report on trends and sources of zoonoses, zoonotic agents, antimicrobial resistance and foodborne outbreaks in the European Union in 2006. The EFSA Journal 130: 189–196.

EFSA and ECDC. 2011. The European Union summary report on trends and sources of zoonoses, zoonotic agents and food-borne outbreaks in 2009. The EFSA Journal 2090: 370pp.

Fearnley, C., S.L. On, B. Kokotovic, G. Manning, T. Cheasty and D.G. Newell. 2005. Application of fluorescent amplified fragment length polymorphism for comparison of human and animal isolates of *Yersinia enterocolitica*. Appl. Environ. Microbiol. 71: 4960–4965.

Fredriksson-Ahomaa, M., T. Autio and H. Korkeala. 1999a. Efficient subtyping of *Yersinia enterocolitica* bioserotype 4/O:3 with pulsed-field gel electrophoresis. Lett. Appl. Microbiol. 29: 308–312.

Fredriksson-Ahomaa, M., S. Hielm and H. Korkeala. 1999b. High prevalence of *yadA*-positive *Yersinia enterocolitica* in pig tongues and minced meat at the retail level in Finland. J. Food Prot. 62: 123–127.

Fredriksson-Ahomaa, M., T. Korte and H. Korkeala. 2000. Contamination of carcasses, offals, and the environment with *yadA*-positive *Yersinia enterocolitica* in a pig slaughterhouse. J. Food Prot. 63: 31–35.

Fredriksson-Ahomaa, M., T. Korte and H. Korkeala. 2001a. Transmission of *Yersinia enterocolitica* 4/O:3 to pets via contaminated pork. Lett. Appl. Microbiol. 32: 375–378.

Fredriksson-Ahomaa, M., U. Lyhs, T. Korte and H. Korkeala. 2001b. Prevalence of pathogenic *Yersinia enterocolitica* in food samples at retail level in Finland. Arch. Lebensmittelhyg. 52: 66–68.

Fredriksson-Ahomaa, M. and H. Korkeala. 2003. Low occurrence of pathogenic *Yersinia enterocolitica* in clinical, food, and environmental samples: a methodological problem. Clin. Microbiol. Rev. 16: 220–229.

Fredriksson-Ahomaa, M., U. Koch, C. Klemm, M. Bucher and A. Stolle. 2004. Different genotypes of *Yersinia enterocolitica* 4/O:3 strains widely distributed in butcher shops in the Munich area. Int. J. Food Microbiol. 95: 89–94.

Fredriksson-Ahomaa, M., A. Stolle and H. Korkeala. 2006a. Molecular epidemiology of *Yersinia enterocolitica* infections. FEMS Immunol. Med. Microbiol. 47: 315–329.

Fredriksson-Ahomaa, M., A. Stolle, A. Siitonen and H. Korkeala. 2006b. Sporadic human *Yersinia enterocolitica* infections caused by bioserotype 4/O:3 originate mainly from pigs. J. Med. Microbiol. 55: 747–749.

Fredriksson-Ahomaa, M., A. Stolle and R. Stephan. 2007. Prevalence of pathogenic *Yersinia enterocolitica* in pigs slaughtered at a Swiss abattoir. Int. J. Food Microbiol. 119: 207–212.

Fredriksson-Ahomaa, M. 2009. Epidemiology of human *Yersinia pseudotuberculosis* infection. Arch. Lebensmittelhyg. 60: 82–87.

Fukushima, H., R. Nakamura, Y. Ito, K. Saito, M. Tsubokura and K. Otsuki. 1983. Ecological studies of *Yersinia enterocolitica*. I. Dissemination of *Y. enterocolitica* in pigs. Vet. Microbiol. 8: 469–483.

Fukushima, H. 1985. Direct isolation of *Yersinia enterocolitica* and *Yersinia pseudotuberculosis* from meat. Appl. Environ. Microbiol. 50: 710–712.

Fukushima, H., M. Gomyoda, K. Shiozawa, S. Kaneko and M. Tsubokura. 1988. *Yersinia pseudotuberculosis* infection contracted through water contaminated by a wild animal. J. Clin. Microbiol. 26: 584–585.

Fukushima, H., M. Gomyoda, M. Tsubokura and S. Aleksić. 1995. Isolation of *Yersinia pseudotuberculosis* from river waters in Japan and Germany using direct KOH and HeLa cell treatments. Zentralbl. Bakteriol. 282: 40–49.

Fukushima, H., K. Hoshina, H. Itogawa and M. Gomyoda. 1997. Introduction into Japan of pathogenic *Yersinia* through imported pork, beef and fowl. Int. J. Food Microbiol. 35: 205–212.

Fukushima, H., Q. Hao, K. Wu, X. Hu, J. Chen, Z. Guo, H. Dai, C. Qin, S. Lu and M. Gomyoda. 2001a. *Yersinia enterocolitica* O9 as a possible barrier against *Y. pestis* in natural plague foci in Ningxia, China. Curr. Microbiol. 42: 1–7.

Fukushima, H., Y. Matsuda, R. Seki, M. Tsubokura, N. Takeda, F.N. Shubin, I. K. Paik and X.B. Zheng. 2001b. Geographical heterogeneity between Far Eastern and Western countries in prevalence of the virulence plasmid, the superantigen *Yersinia pseudotuberculosis*-derived mitogen, and the high-pathogenicity island among *Yersinia pseudotuberculosis* strains. J. Clin. Microbiol. 39: 3541–3547.

Gierczynski, R., A. Golubov, H. Neubauer, J.N. Pham and A. Rakin. 2007. Development of multiple-locus variable-number tandem-repeat analysis for *Yersinia enterocolitica* subsp. *palearctica* and its application to bioserogroup 4/O3 subtyping. J. Clin. Microbiol. 45: 2508–2515.

Grahek-Ogden, D., B. Schimmer, K.S. Cudjoe, K. Nygård and G. Kapperud. 2007. Outbreak of *Yersinia enterocolitica* serogroup O:9 infection and processed pork, Norway. Emerg. Infect. Dis. 13: 754–756.

Granfors, K., S. Jalkanen, R. von Essen, R. Lahesmaa-Rantala, O. Isomaki, K. Pekkola-Heino, R. Merilahti-Palo, R. Saario, H. Isomaki and A. Toivanen. 1989. *Yersinia* antigens in synovial-fluid cells from patients with reactive arthritis. N. Engl. J. Med. 320: 216–221.

Greenwood, M.H. and W.L. Hooper. 1990. Excretion of *Yersinia* spp. associated with consumption of pasteurized milk. Epidemiol. Infect. 104: 345–350.

Guinet, F., E. Carniel and A. Leclercq. 2011. Transfusion-transmitted *Yersinia enterocolitica* sepsis. Clin. Infect. Dis. 53: 583–591.

Gürtler, M., T. Alter, S. Kasimir, M. Linnebur and K. Fehlhaber. 2005. Prevalence of *Yersinia enterocolitica* in fattening pigs. J. Food Prot. 68: 850–854.

Hamama, A., A. el Marrakchi and F. el Othmani. 1992. Occurrence of *Yersinia enterocolitica* in milk and dairy products in Morocco. Int. J. Food Microbiol. 16: 69–77.

Han, T.H., I.K. Paik and S.J. Kim. 2003. Molecular relatedness between isolates *Yersinia pseudotuberculosis* from a patient and an isolate from mountain spring water. J. Korean Med. Sci. 18: 425–428.

Hannu, T., L. Mattila, J.P. Nuorti, P. Ruutu, J. Mikkola, A. Siitonen and M. Leirisalo-Repo. 2003. Reactive arthritis after an outbreak of *Yersinia pseudotuberculosis* serotype O:3 infection. Ann. Rheum. Dis. 62: 866–869.

Hudson, J.A., N.J. King, A.J. Cornelius, T. Bigwood, K. Thom and S. Monson. 2008. Detection, isolation and enumeration of *Yersinia enterocolitica* from raw pork. Int. J. Food Microbiol. 123: 25–31.

Huovinen, E., L.M. Sihvonen, M.J. Virtanen, K. Haukka, A. Siitonen and M. Kuusi. 2010. Symptoms and sources of *Yersinia enterocolitica*-infection: a case-control study. BMC Infect. Dis. 10: 122.

Iteman, I., H. Najdenski and E. Carniel. 1995. High genomic polymorphism in *Yersinia pseudotuberculosis*. Contrib. Microbiol. Immunol. 13: 106–111.

Iteman, I., A. Guiyoule and E. Carniel. 1996. Comparison of three molecular methods for typing and subtyping pathogenic *Yersinia enterocolitica* strains. J. Med. Microbiol. 45: 48–56.

Jacobsen, N.R., T. Bogdanovich, M. Skurnik, P.S. Lübeck, P. Ahrens and J. Hoorfar. 2005. A real-time PCR assay for the specific identification of serotype O:9 of *Yersinia enterocolitica*. J. Microbiol. Methods 63: 151–156.

Jalava, K., M. Hakkinen, M. Valkonen, U.M. Nakari, T. Palo, S. Hallanvuo, J. Ollgren, A. Siitonen and J.P. Nuorti. 2006. An outbreak of gastrointestinal illness and erythema nodosum from grated carrots contaminated with *Yersinia pseudotuberculosis*. J. Infect. Dis. 194: 1209–1216.

Jayarao, B.M., S.C. Donaldson, B.A. Straley, A.A. Sawant, N.V. Hegde and J.L. Brown. 2006. A survey of foodborne pathogens in bulk tank milk and raw milk consumption among farm families in pennsylvania. J. Dairy Sci. 89: 2451–2458.

Johannessen, G.S., G. Kapperud and H. Kruse. 2000. Occurrence of pathogenic *Yersinia enterocolitica* in Norwegian pork products determined by a PCR method and a traditional culturing method. Int. J. Food Microbiol. 54: 75–80.

Johannessen, G.S., S. Loncarevic and H. Kruse. 2002. Bacteriological analysis of fresh produce in Norway. Int. J. Food Microbiol. 77: 199–204.

Jones, T.F. 2003. From pig to pacifier: chitterling-associated yersiniosis outbreak among black infants. Emerg. Infect. Dis. 9: 1007–1009.

Kaasch, A.J., J. Dinter, T. Goeser, G. Plum and H. Seifert. 2012. *Yersinia pseudotuberculosis* bloodstream infection and septic arthritis: case report and review of the literature. Infection 40: 185–190.

Kanazawa, Y. and T. Kuramata. 1974. Difference in susceptibility to benzylpenicillin between *Yersinia enterocolitica* and *Yersinia pseudotuberculosis*. Jpn. J. Microbiol. 18: 483–485.

Kangas, S., J. Takkinen, M. Hakkinen, U.M. Nakari, T. Johansson, H. Henttonen, L. Virtaluoto, A. Siitonen, J. Ollgren and M. Kuusi. 2008. *Yersinia pseudotuberculosis* O:1 traced to raw carrots, Finland. Emerg. Infect. Dis. 14: 1959–1961.

Kechagia, N., C. Nicolaou, V. Ioannidou, E. Kourti, A. Ioannidis, N.J. Legakis and S. Chatzipanagiotou. 2007. Detection of chromosomal and plasmid—encoded virulence determinants in *Yersinia enterocolitica* and other *Yersinia* spp. isolated from food animals in Greece. Int. J. Food Microbiol. 118: 326–331.

Korte, T., M. Fredriksson-Ahomaa, T. Niskanen and H. Korkeala. 2004. Low prevalence of *yadA*-positive *Yersinia enterocolitica* in sows. Foodborne Pathog. Dis. 1: 45–52.

Kotetishvili, M., A. Kreger, G. Wauters, J.G. Morris Jr., A. Sulakvelidze and O.C. Stine. 2005. Multilocus sequence typing for studying genetic relationships among *Yersinia* species. J. Clin. Microbiol. 43: 2674–2684.

Kuehni-Boghenbor, K., S.L. On, B. Kokotovic, A. Baumgartner, T.M. Wassenaar, M. Wittwer, B. Bissig-Choisat and J. Frey. 2006. Genotyping of human and porcine *Yersinia enterocolitica, Yersinia intermedia,* and *Yersinia bercovieri* strains from Switzerland by amplified fragment length polymorphism analysis. Appl. Environ. Microbiol. 72: 4061–4066.

Laukkanen, R., P. Ortiz Martínez, K. Siekkinen, J. Ranta, R. Maijala and H. Korkeala. 2008. Transmission of *Yersinia pseudotuberculosis* in the pork production chain from farm to slaughterhouse. Appl. Environ. Microbiol. 74: 5444–5450.

Laukkanen, R., P. Ortiz Martínez, K. Siekkinen, J. Ranta, R. Maijala and H. Korkeala. 2009. Contamination of carcasses with human pathogenic *Yersinia enterocolitica* 4/O:3 originates from pigs infected on farms. Foodborne Pathog. Dis. 6: 681–688.

Laukkanen, R., M. Hakkinen, J. Lundén, M. Fredriksson-Ahomaa, T. Johansson and H. Korkeala. 2010a. Evaluation of isolation methods for pathogenic *Yersinia enterocolitica* from pig intestinal content. J. Appl. Microbiol. 108: 956–964.

Laukkanen, R., J. Ranta, X. Dong, M. Hakkinen, P. Ortiz Martínez, J. Lundén, T. Johansson and H. Korkeala. 2010b. Reduction of enteropathogenic *Yersinia* in the pig slaughterhouse by using bagging of the rectum. J. Food Prot. 73: 2161–2168.

Laukkanen-Ninios, R., X. Didelot, K.A. Jolley, G. Morelli, V. Sangal, P. Kristo, C. Brehony, P.F.M. Imori, H. Fukushima, A. Siitonen, G. Tseneva, E. Voskressenskaya, J.P. Falcao, H. Korkeala, M.C.J. Maiden, C. Mazzoni, E. Carniel, M. Skurnik and M. Achtman. 2011. Population structure of *Yersinia pseudotuberculosis* complex according to multilocus sequence typing. Environ. Microbiol. 13: 3114–3127.

Leirisalo-Repo, M. 2005. Reactive arthritis. Scand. J. Rheumatol. 34: 251–259.

Leo, J.C. and M. Skurnik. 2011. Adhesins of human pathogens from the genus *Yersinia*. Adv. Exp. Med. Biol. 715: 1–15.

Lindblad, M., H. Lindmark, S.T. Lambertz and R. Lindqvist. 2006. Microbiological baseline study of broiler chickens at Swedish slaughterhouses. J. Food Prot. 69: 2875–2882.

Lindblad, M., H. Lindmark, S.T. Lambertz and R. Lindqvist. 2007. Microbiological baseline study of swine carcasses at Swedish slaughterhouses. J. Food Prot. 70: 1790–1797.

Long, C.,T.F., Jones, D.J. Vugia, J. Scheftel, N. Strockbine, P. Ryan, B. Shiferaw, R.V. Tauxe and L.H. Gould. 2010. *Yersinia pseudotuberculosis* and *Y. enterocolitica* Infections, FoodNet, 1996–2007. Emerg. Infect. Dis. 16: 566–567.

MacDonald, E., B. Heier, T. Stalheim, K. Cudjoe, T. Skjerdal, A. Wester, B. Lindstedt and L. Vold. 2011. *Yersinia enterocolitica* O:9 infections associated with bagged salad mix in Norway, February to April 2011. Euro Surveill. 16: 19866.

Messelhäusser, U., P. Kämpf, J. Colditz, H. Bauer, H. Schreiner, C. Höller and U. Busch. 2011. Qualitative and quantitative detection of human pathogenic *Yersinia enterocolitica* in different food matrices at retail level in Bavaria. Foodborne Pathog. Dis. 8: 39–44.

Moriki, S., A. Nobata, H. Shibata, A. Nagai, N. Minami, T. Taketani and H. Fukushima. 2010. Familial outbreak of *Yersinia enterocolitica* serotype O9 biotype 2. J. Infect. Chemother. 16: 56–58.

Nesbakken, T., E. Nerbrink, O.J. Røtterud and E. Borch. 1994. Reduction of *Yersinia enterocolitica* and *Listeria* spp. on pig carcasses by enclosure of the rectum during slaughter. Int. J. Food Microbiol. 23: 197–208.

Nesbakken, T., K. Eckner, H.K. Hoidal and O.J. Røtterud. 2003. Occurrence of *Yersinia enterocolitica* and *Campylobacter* spp. in slaughter pigs and consequences for meat inspection, slaughtering, and dressing procedures. Int. J. Food Microbiol. 80: 231–240.

Nesbakken, T., T. Iversen and B. Lium. 2007. Pig herds free from human pathogenic *Yersinia enterocolitica*. Emerg. Infect. Dis. 13: 1860–1864.

Neubauer, H., S. Aleksić, A. Hensel, E. Finke and H. Meyer. 2000. *Yersinia enterocolitica* 16S rRNA gene types belong to the same genospecies but form three homology groups. Int. J. Med. Microbiol. 290: 61–64.

Niskanen, T., M. Fredriksson-Ahomaa and H. Korkeala. 2002. *Yersinia pseudotuberculosis* with limited genetic diversity is a common finding in tonsils of fattening pigs. J. Food Prot. 65: 540–545.

Niskanen, T., R. Laukkanen, M. Fredriksson-Ahomaa and H. Korkeala. 2008. Distribution of *virF/lcrF*-positive *Yersinia pseudotuberculosis* serotype O:3 at farm level. Zoonoses Public Health 55: 214–221.

Nowak, B., T.V. Mueffling, K. Caspari and J. Hartung. 2006. Validation of a method for the detection of virulent *Yersinia enterocolitica* and their distribution in slaughter pigs from conventional and alternative housing systems. Vet. Microbiol. 117: 219–228.

Nowgesic, E., M. Fyfe, J. Hockin, A. King, H. Ng, A. Paccagnella, A. Trinidad, L. Wilcott, R. Smith, A. Denney, L. Struck, G. Embree, K. Higo, J.I. Chan, P. Markey, S. Martin and D. Bush. 1999. Outbreak of *Yersinia pseudotuberculosis* in British Columbia—November 1998. Can. Commun. Dis. Rep. 25: 97–100.

Nuorti, J.P., T. Niskanen, S. Hallanvuo, J. Mikkola, E. Kela, M. Hatakka, M. Fredriksson-Ahomaa, O. Lyytikäinen, A. Siitonen, H. Korkeala and P. Ruutu. 2004. A widespread outbreak of *Yersinia pseudotuberculosis* O:3 infection from iceberg lettuce. J. Infect. Dis. 189: 766–774.

Okwori, A.E.J., P. Ortiz Martínez, M. Fredriksson-Ahomaa, S.E. Agina and H. Korkeala. 2009. Pathogenic *Yersinia enterocolitica* 2/O:9 and *Yersinia pseudotuberculosis* 1/O:1 strains isolated from human and non-human sources in the Plateau State of Nigeria. Food Microbiol. 26: 872–875.

Ortiz Martínez, P., M. Fredriksson-Ahomaa, Y. Sokolova, M. Roasto, A. Berzins and H. Korkeala. 2009. Prevalence of enteropathogenic *Yersinia* in Estonian, Latvian, and Russian (Leningrad region) pigs. Foodborne Pathog. Dis. 6: 719–724.

Ortiz Martínez, P., S. Mylona, I. Drake, M. Fredriksson-Ahomaa, H. Korkeala and J.E. Corry. 2010. Wide variety of bioserotypes of enteropathogenic *Yersinia* in tonsils of English pigs at slaughter. Int. J. Food Microbiol. 139: 64–69.

Ortiz Martínez, P., M. Fredriksson-Ahomaa, A. Pallotti, R. Rosmini, K. Houf and H. Korkeala. 2011. Variation in the prevalence of enteropathogenic *Yersinia* in slaughter pigs from Belgium, Italy, and Spain. Foodborne Pathog. Dis. 8: 445–450.

Ostroff, S.M., G. Kapperud, L.C. Hutwagner, T. Nesbakken, N.H. Bean, J. Lassen and R.V. Tauxe. 1994. Sources of sporadic *Yersinia enterocolitica* infections in Norway: a prospective case-control study. Epidemiol. Infect. 112: 133–141.

O'Sullivan, T., R. Friendship, T. Blackwell, D. Pearl, B. McEwen, S. Carman, D. Slavic and C. Dewey. 2011. Microbiological identification and analysis of swine tonsils collected from carcasses at slaughter Can. J. Vet. Res. 75: 106–111.

Pearce, L.E., B. W. Smythe, R.A. Crawford, E. Oakley, S.C. Hathaway and J.M. Shepherd. 2012. Pasteurization of milk: The heat inactivation kinetics of milk-borne dairy pathogens under commercial-type conditions of turbulent flow. J. Dairy Sci. 95: 20–35.

Pham, J.N., S.M. Bell, L. Martin and E. Carniel. 2000. The beta-lactamases and beta-lactam antibiotic susceptibility of *Yersinia enterocolitica*. J. Antimicrob. Chemother. 46: 951–957.

Pilon, J., R. Higgins and S. Quessy. 2000. Epidemiological study of *Yersinia enterocolitica* in swine herds in Quebec. Can. Vet. J. 41: 383–387.

Pingulkar, K., A. Kamat and D. Bongirwar. 2001. Microbiological quality of fresh leafy vegetables, salad components and ready-to-eat salads: an evidence of inhibition of *Listeria monocytogenes* in tomatoes. Int. J. Food Sci. Nutr. 52: 15–23.

Pitkälä, A., T. Virtanen, S. Joutsen, A. Leimi and P. Tuominen. 2009. Risk profile of *Yersinia enterocolitica* and *Yersinia pseudotuberculosis* [*Yersinia enterocolitica* ja *Yersinia pseudotuberculosis* suomalaisissa elintarvikkeissa-riskiprofiili]. Evira Research Reports 2/2009: 1–73.

Poljak, Z., C.E. Dewey, S.W. Martin, T. Rosendal, J. Christensen, B. Ciebin and R.M. Friendship. 2010. Prevalence of *Yersinia enterocolitica* shedding and bioserotype distribution in Ontario finisher pig herds in 2001, 2002, and 2004. Prev. Vet. Med. 93: 110–120.

Public Health Agency of Canada. 2010. C-EnterNet short report 2009. 1–12.

Ranta, J., K. Siekkinen, L. Nuotio, R. Laukkanen, S. Hellström, H. Korkeala and R. Maijala. 2010. Causal hidden variable model of pathogenic contamination from pig to pork. Stat. Model. 10: 69–87.

Rimhanen-Finne, R., T. Niskanen, S. Hallanvuo, P. Makary, K. Haukka, S. Pajunen, A. Siitonen, R. Ristolainen, H. Poyry, J. Ollgren and M. Kuusi. 2009. *Yersinia pseudotuberculosis* causing a large outbreak associated with carrots in Finland, 2006. Epidemiol. Infect. 137: 342–347.

Rosner, B.M., K. Stark and D. Werber. 2010. Epidemiology of reported *Yersinia enterocolitica* infections in Germany, 2001–2008. BMC Public Health 10: 337.

Sakai, T., A. Nakayama, M. Hashida, Y. Yamamoto, H. Takebe and S. Imai. 2005. Outbreak of food poisoning by *Yersinia enterocolitica* serotype O8 in Nara prefecture: the first case report in Japan. Jpn. J. Infect. Dis. 58: 257–258.

Sandery, M., T. Stinear and C. Kaucner. 1996. Detection of pathogenic *Yersinia enterocolitica* in environmental waters by PCR. J. Appl. Bacteriol. 80: 327–332.

Sihvonen, L.M., S. Toivonen, K. Haukka, M. Kuusi, M. Skurnik and A. Siitonen. 2011. Multilocus variable-number tandem-repeat analysis, pulsed-field gel electrophoresis, and antimicrobial susceptibility patterns in discrimination of sporadic and outbreak-related strains of *Yersinia enterocolitica* BMC Microbiol. 11: 42.

Smego, R.A., J. Frean and H.J. Koornhof. 1999. Yersiniosis I: microbiological and clinicoepidemiological aspects of plague and non-plague *Yersinia* infections. Eur. J. Clin. Microbiol. Infect. Dis. 18: 1–15.

Tafjord Heier, B., K. Nygård, J. Lassen. 2007: Yersiniose i Norge 2006. Available at: www.fhi.no/artikler/?id=65105(Accessed on July 13th 2009).

Tauxe, R.V., J. Vandepitte, G. Wauters, S.M. Martin, V. Goossens, P. De Mol, R. Van Noyen and G. Thiers. 1987. *Yersinia enterocolitica* infections and pork: the missing link. Lancet 1: 1129–1132.

Tennant, S.M., T. H. Grant and R.M. Robins-Browne. 2003. Pathogenicity of *Yersinia enterocolitica* biotype 1A. FEMS Immunol. Med. Microbiol. 38: 127–137.

Terentjeva, M. and A. Berzins. 2010. Prevalence and antimicrobial resistance of *Yersinia enterocolitica* and *Yersinia pseudotuberculosis* in slaughter pigs in Latvia. J. Food Prot. 73: 1335–1338.

Thisted Lambertz, S. 2007. Riskprofil—*Yersinia enterocolitica*. Livsmedelsverkets Rapport 8: 1–39.

Thisted Lambertz, S., K. Granath, M. Fredriksson-Ahomaa, K.E. Johansson and M.L. Danielsson-Tham. 2007. Evaluation of a combined culture and PCR method (NMKL-163A) for detection of presumptive pathogenic *Yersinia enterocolitica* in pork products. J. Food Prot. 70: 335–340.

Thisted Lambertz, S. and M.L. Danielsson-Tham. 2005. Identification and characterization of pathogenic *Yersinia enterocolitica* isolates by PCR and pulsed-field gel electrophoresis. Appl. Environ. Microbiol. 71: 3674–3681.

Thisted-Lambertz, S., C. Nilsson and S. Hallanvuo. 2008. TaqMan-based real-time PCR method for detection of *Yersinia pseudotuberculosis* in food. Appl. Environ. Microbiol. 74: 6465–6469.

Tsubokura, M., K. Otsuki, K. Sato, M. Tanaka, T. Hongo, H. Fukushima, T. Maruyama and M. Inoue. 1989. Special features of distribution of *Yersinia pseudotuberculosis* in Japan. J. Clin. Microbiol. 27: 790–791.

Tsubokura, M. and S. Aleksić. 1995. A simplified antigenic scheme for serotyping of *Yersinia pseudotuberculosis*: phenotypic characterization of reference strains and preparation of O and H factor sera. Contrib. Microbiol. Immunol. 13: 99–105.

Viboud, G. and J. Bliska. 2005. *Yersinia* outer proteins: Role in modulation of host cell signaling responses and pathogenesis. Annu. Rev. Microbiol. 59: 69–89.

Vincent, P., E. Salo, M. Skurnik, H. Fukushima and M. Simonet. 2007. Similarities of Kawasaki disease and *Yersinia pseudotuberculosis* infection epidemiology. Pediatr. Infect. Dis. J. 26: 629–631.

Virtanen, S.E., L. K. Salonen, R. Laukkanen, M. Hakkinen and H. Korkeala. 2011. Factors related to the prevalence of pathogenic *Yersinia enterocolitica* on pig farms. Epidemiol. Infect. 139: 1919–1927.

Wang, X., Z. Cui, D. Jin, L. Tang, S. Xia, H. Wang, Y. Xiao, H. Qiu, Q. Hao, B. Kan, J. Xu and H. Jing. 2009. Distribution of pathogenic *Yersinia enterocolitica* in China. Eur. J. Clin. Microbiol. Infect. Dis. 28: 1237–1244.

Wang, X., Z. Cui, H. Wang, L. Tang, J. Yang, L. Gu, D. Jin, L. Luo, H. Qiu, Y. Xiao, H. Xiong, B. Kan, J. Xu and H. Jing. 2010. *Canis familiaris* of farmers shares the same subtype pathogenic *Yersinia enterocolitica* with humans and may be a potential source of human infection in Jiangsu Province, China. J. Clin. Microbiol. 48: 1604–1610.

Wauters, G., K. Kandolo and M. Janssens. 1987. Revised biogrouping scheme of *Yersinia enterocolitica*. Contrib. Microbiol. Immunol. 9: 14–21.

Wauters, G., M. Janssens, G. Steigerwalt and D.J. Brenner. 1988. *Yersinia mollaretii* sp. nov. and *Yersinia bercovieri* sp. nov., formerly called *Yersinia enterocolitica* biogroups 3A and 3B. Int. J. Syst. Bacteriol. 38: 424–429.

Wauters, G., S. Aleksic, J. Charlier and G. Schulze. 1991. Somatic and flagellar antigens of *Yersinia enterocolitica* and related species. Contrib. Microbiol. Immunol. 12: 239–243.

Weynants, V., V. Jadot, P.A. Denoel, A. Tibor and J.J. Letesson. 1996. Detection of *Yersinia enterocolitica* serogroup O:3 by a PCR method. J. Clin. Microbiol. 34: 1224–1227.

Index

Color Plate Section

Chapter 11

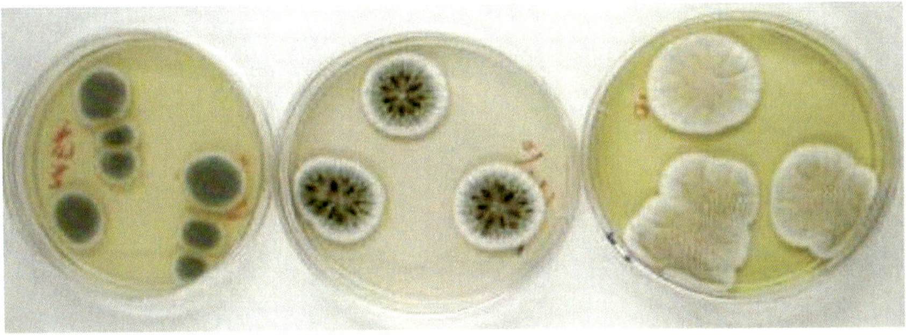

Figure 2. *Penicillium nalgiovense* (from the left side) on Malt Extract Agar (MEA), Czapek Yeast Extract Agar (CYA), and Yeast Extract Sucrose (YES) agar medium (Photo: Dereje T. Asefa2007).

Chapter 18

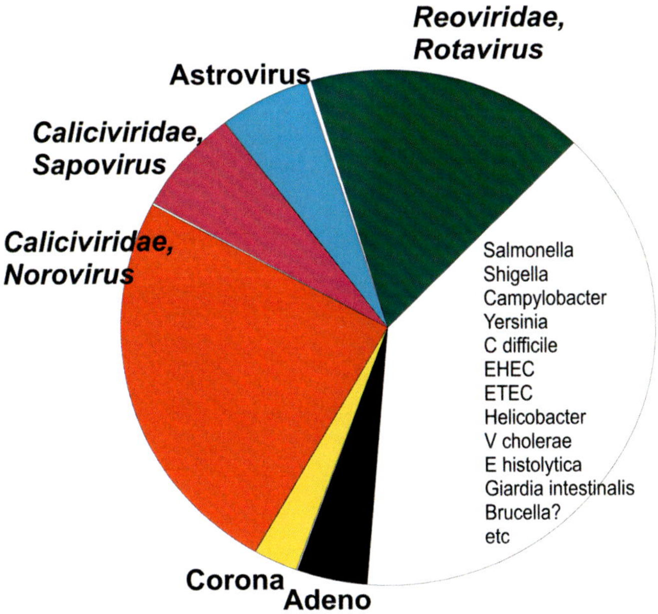

Figure 3. Approximate proportions of viral causes of human gastroenteritis. The figure is based on the Swedish situation, but also has global relevance. Bacteria and protozoa are shown without reference to frequency. Virus names are explained below.

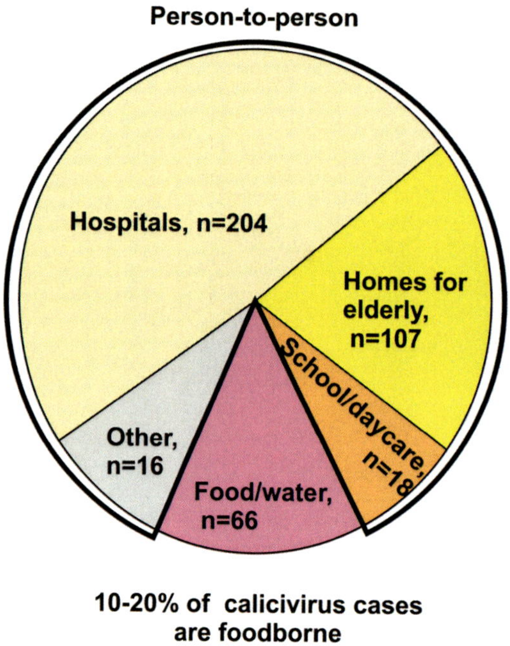

Figure 11. Place and route of infection in 407 outbreaks of gastroenteritis in Sweden 1994–1998 (SICDC). The majority of cases are spread person-to-person.